Igneous and Metamorphic Rocks under the Microscope

Classification, Textures, Microstructures and Mineral Preferred-Orientations

David Shelley

Department of Geology
University of Canterbury
New Zealand

CHAPMAN & HALL
London · Glasgow · New York · Tokyo · Melbourne · Madras

Published by Chapman & Hall, 2–6 Boundary Row, London SE1 8HN

Chapman & Hall India, R. Seshadri, 32 Second Main Road, CIT East,

Blackie Academic & Professional, Wester Cleddens Road, Bishopbriggs, Glasgow G64 2NZ, UK

Chapman & Hall, 29 West 35th Street, New York NY10001, USA

Chapman & Hall Japan, Thomson Publishing Japan, Hirakawacho Nemoto Building, 6F, 1-7-11 Hirakawa-cho, Chiyoda-ku, Tokyo 102, Japan

Chapman & Hall Australia, Thomas Nelson Australia, 102 Dodds Street, South Melbourne, Victoria 3205, Australia

Chapman & Hall India, R. Seshadri, 32 Second Main Road, CIT East, Madras 600 035, India

First edition 1983

Typeset in 10/12 pt Bembo by Excel Typesetters Company, Hong Kong
Printed in Great Britain by the University Press, Cambridge

0 412 44200 0

A catalogue record for this book is available from the British Library

Library of Congress Cataloging-in-Publication data
Shelley, David.
Igneous and metamorphic rocks under the microscope: classfication, textures, microstructures, and mineral preferred orientations/David Shelley. – 1st ed.
p. cm.
Includes bibliographical references and index.
1. Rocks, Igneous – Classification. 2. Rocks, Metamorphic – Classification. 3. Rocks, Igneous – Nomenclature. 4. Rocks, Metamorphic – Nomenclature. 5. Petrofabric analysis. 6. Polarizing microscope. I. Title.
QE641.S4815 1992
552′.1 – dc20 92-4337
CIP

To Iola
and
to the memory of F. Coles Phillips
who introduced me to the subject of structural petrology.

Contents

Preface

This book covers aspects of igneous and metamorphic petrology best studied with the polarizing microscope. It is written in four parts:

Part One: igneous and metamorphic rock classification, terminology and description;
Part Two: textures and microstructures, and their origins;
Part Three: mineral preferred-orientations, and how to measure them;
Part Four: glossary and index.

This text is emphatically a 'petrology', not just a 'petrography'. In other words it deals with the origins of rocks, not just their description.

The book assumes familiarity with the polarizing microscope and an ability to identify minerals in thin section. Otherwise, the book is designed for all levels. The student beginning petrology should first use Parts One and Four as a basic 'petrography' and compendium of necessary information. As familiarity with the subject develops, Part Four should be used as a route into the discussions and explanations of Parts Two and Three. The subjects of textures and fabrics are discussed in terms of recent research, and much of Parts Two and Three will constitute a guide for the advanced student. At the same time, I hope that the advanced student will continue to find Part One a useful reminder of some of the rudiments of petrology.

The 'petrography' of Part One is deliberately concise and to the point. The recommendations of the IUGS subcommission for igneous rock terminology are adopted, and the IUGS hierarchy of classification provides the rationale for the order in which the rocks are dealt with. In fact, the hierarchy constitutes a useful teaching tool because it provides a logical sequence of steps that students can follow when faced with the problem of naming an unknown rock; one consequence, which might seem strange at first, is that the most common rocks are dealt with last. In the absence of corresponding 'international' recommendations for metamorphic rock terminology, but inspired by the IUGS igneous scheme, I have organized metamorphic rocks into a hierarchy different from that of most textbooks.

I hope that the logical sequence of description for both igneous and metamorphic rocks will help the student approach the business of identifying and naming them in an efficient and intelligent way.

Part Two consists of comprehensive explanations and discussions of some common rock textures and microstructures. It is divided into three chapters, the first covering general textures such as twinning, zoning and intergrowths, the second covering igneous textures and microstructures and the third dealing with metamorphic textures and microstructures. The term 'texture' is used in the conventional geological manner for spatial relationships between mineral grains in a rock, and for such features as grain shape and size; combined with 'microstructures' such as layering, xenoliths, vesicles and orbs, the subject matter of Part Two covers many of the most visible aspects of petrology.

Part Three comprises a discussion of mineral preferred-orientations, a subject that is, of course, central to metamorphic petrology. How can one understand what a schist is without understanding what schistosity is? However, the subject is not restricted to metamorphic rocks, and common igneous features such as igneous lamination or foliation are dealt with here. There has been a tendency in some of the modern research literature to use the word 'texture' in its metallurgical sense for preferred orientations. Much as we owe to metallurgy the concepts of mineral preferred orientation development, I regret the borrowing of this usage when 'texture' already has such a long-standing and different usage in geology.

There is a surprising lack of modern texts that integrate the subjects covered in Parts One, Two and Three, subjects which have in common the use of the polarizing microscope. My hope is that this book will fill this gap usefully.

SOME OUTSTANDING PROBLEMS IN NAMING ROCKS AND TEXTURES

We now have a clear guide to igneous rock nomenclature in the latest IUGS recommendations, but the principal outstanding problems lie in two areas. First, despite the recommendation that a mineralogy-based classification be used first and foremost for common volcanic rocks, the reality is that such a classification often does not work. Indeed I suspect that many volcanic petrologists will continue to put geochemical criteria first. In my view, consideration should be given to adopting a simpler mineralogy-based classification (the IUGS field classification, for example) in tandem with TAS. Also, the use of 'charnockitic' is still ill defined, and this issue spills into one of the major problem areas of metamorphic rock nomenclature: when exactly does a gneiss become a granulite, when does a granulite become a charnockite and when is the latter an 'igneous' rather than 'metamorphic' rock? I make some suggestions in this text, but the problem requires further clarification by the geological community at large.

The definition of gneiss is another contentious issue. My own view is that gneiss, by definition, has a gneissosity which, by definition, is a mineral-shape preferred orientation of metamorphic origin which does not allow the rock to split easily (in contrast to schistosity). Schists may develop a pronounced metamorphic layering (spectacularly shown in chlorite-zone schists of New Zealand) and therefore it is not helpful to use the presence of layering to define gneiss.

In the area of textures, an outstanding problem is the question of when to use cumulate terminology. Can it be used for rocks that represent *in situ* crystal growth? Parsons (1987) says yes, MacKenzie *et al.* (1982) say no. What has become clear is that it makes no sense in terms of modern concepts of petrology to use terms like adcumulate and orthocumulate for gabbroic rocks but not for granites. Dr S. D. Weaver recently showed me some A-type granites from Antarctica that display a beautiful array of adcumulate textures involving, in turn, quartz, K-feldspar and plagioclase, and of course the orthocumulate 'ophitic texture' has a parallel in granites in poikilitic K-feldspar megacrysts. In this book I do not attempt to impose a solution. However, my view is that 'adcumulate texture', in particular, does provide a useful general term that should be more widely adopted.

Finally, I note that this book introduces a new textural term: 'poikilomosaic'. I shall await comment from the readers.

ACKNOWLEDGEMENTS

This book is a compilation of other researchers' work, acknowledged in the usual way throughout text, figure captions and the list of references. I am grateful to numerous authors and publishing houses that gave permission to reproduce copyright material, and some specific acknowledgements are given in the figure captions. The University of Canterbury, New Zealand, is thanked for providing a period of study leave during which most of the book was written. I am indebted to the following: K. G. Cox, R. A. Howie, W. S. MacKenzie and R. H. Vernon for being sufficiently interested in my original proposal that the publishers were persuaded to proceed; S. Brown, M. W. Rennison, R. T. Smith, D. Smale, W. A. Watters and S. D. Weaver for supplying specimens for photography; and S. P. Halsor, J. Keller, R. J. Norris and A. R. Philpotts, for generously supplying photographs for use in the book. Especial thanks are due to Albert Downing and Lee Leonard for their efforts in producing most of the photographs and line drawings, and to my igneous petrologist colleague, Stephen D. Weaver, for his patient reading of the manuscript, and for his encouraging comments which were greatly appreciated.

David Shelley
Christchurch, 1992

Part One

Classification, Terminology and Description

Introduction to the scope and organization of this book

This book is written in four parts:

Part One: Outlines the way to describe, classify and name an igneous or metamorphic rock.

Part Two: Discusses and explains common textural and structural features, especially grain shape, crystal-size distributions, the spatial arrangement and distribution of minerals and layering.

Part Three: Discusses the preferred orientation of minerals, including such features as schistosity in metamorphic rocks and the foliation of igneous rocks.

Glossary/Index: Defines and explains briefly all technical terms, and provides a page reference to fuller explanations.

It is important to note that textural and structural features are noted rather than explained in Part One. The reader should use the Glossary/Index to find explanations in Parts Two and Three.

Readers approaching the subject of igneous and metamorphic rocks for the first time require patience and perseverance to become familiar with all concepts and terms. The beginning student may at first use Part One and the Glossary as an introduction to petrography, but the intention of this book is more than that, and the reader should follow up glossary references to Parts Two and Three, and ultimately start to dip into the research literature. The discussions in Parts Two and Three hopefully will persuade students that petrology with the microscope involves a lot more than purely descriptive petrography, and is a subject well worth pursuing for its own sake. It will also become clear that no petrological study can be complete without an understanding of textural and structural development.

Finally, it should be remembered this book is mainly about what can be done with the polarizing microscope. For discussions of field relations and geochemistry, the reader must look elsewhere.

1

Igneous rocks – an introduction

1.1 INTRODUCTION

A systematic approach to classifying and naming igneous rocks has been plagued by several long-standing problems. One is the unnecessary proliferation of names, often a response to the fact that the essential minerals of a rock can show wide variation in spatial distribution and orientation, grain size, shape, colour and alteration; in addition, the type of accessory mineral present may also have a significant effect on the physical appearance of a rock. Consequently, geologists have coined innumerable varietal names which serve little useful purpose (yamaskite, for example, is merely a local name for a variety of pyroxenite).

Another problem has been the uncertain and changing definitions for even the most common rock names, some of which have been used for a long time. 'Granite', for example, derived from the Latin *granum* (= grain), was used in print in Italy as long ago as 1596 (Mitchell, 1985), but within the geological community, there has seldom been precise agreement as to what granite is. In terms of modal mineralogy, definitions have varied over the years, but worse still, geologists have used granite in two distinct ways, the first (*sensu stricto*) as defined in terms of modal mineralogy, the second (*sensu lato*) as a loosely defined field term for coarse-grained rocks (most of which contain some visible quartz) within a large batholith or pluton (granitoid or granitic are now accepted terms for granite, *s.l.*). The population at large uses granite in an even wider sense, often for any hard rock.

Yet another problem is that mineral-based classifications cannot be applied to volcanic rocks containing significant amounts of glass.

In 1970, a subcommission of the International Union of Geological Sciences began the task of formulating an internationally acceptable classification and terminology of igneous rocks. Much of its work is associated with the name of Albert Streckeisen, chairman of the subcommission until 1980. Its recommendations have recently been presented in a single comprehensive volume, edited by Le Maitre (1989). Some geological journals now insist that authors use the terminology recommended by the subcommission, and I fully endorse this approach. There must, of course, be room for future modification, but the time for purely individualistic, arbitrary or capricious use of rock names is surely past.

1.1.1 Two principal factors in classification

(a) Modal mineralogy

The mode or actual mineral content in volume % is one of the two main bases of classification. Exceptions include:

1. glassy or very fine-grained rocks for which it is permissible or necessary to use alternative criteria such as chemistry; and 2. pyroclastic rocks which are named according to grain-size characteristics (as in ash), or genesis (as in ignimbrite). In addition, rocks are often qualified according to their geochemical characteristics. These points will be discussed further throughout the text.

Estimates of the volume % of minerals present in a rock are made using the charts provided (Fig. 1.1). More accurate measurements can be made with point-counters (Hutchison, 1974) or computer image analysis (Allard and Sotin, 1988). Estimating modes requires a good deal of practice, and the best control on accuracy is the simplest one, that is the requirement that the total adds up to 100%.

(b) Grain size

Igneous rocks fall naturally into two grain-size categories:

1. **Plutonic rocks** which contain crystals distinguishable with the naked eye, and which crystallized below the Earth's surface.
2. **Volcanic rocks**, which contain glass and/or a high proportion of crystals not distinguishable with the naked eye, and which were emplaced at or near the Earth's surface.

Plutonic rocks judged to be finer than usual are given the prefix **micro-**, as in microgranite. The only exception is the use of dolerite (British) or diabase (US) for fine-grained gabbro.

Problems in categorizing igneous rocks this way are few, but include chilled margins to plutonic rocks (usually called chilled margins rather than the volcanic name), essentially monomineralic volcanic rocks such as trachyte (which cannot always be distinguished from syenite, if completely crystalline) and minor intrusives such as dykes and sills which may contain rock of truly intermediate grain-size character. A few rock names (e.g., carbonatite) are used for both intrusive and extrusive varieties.

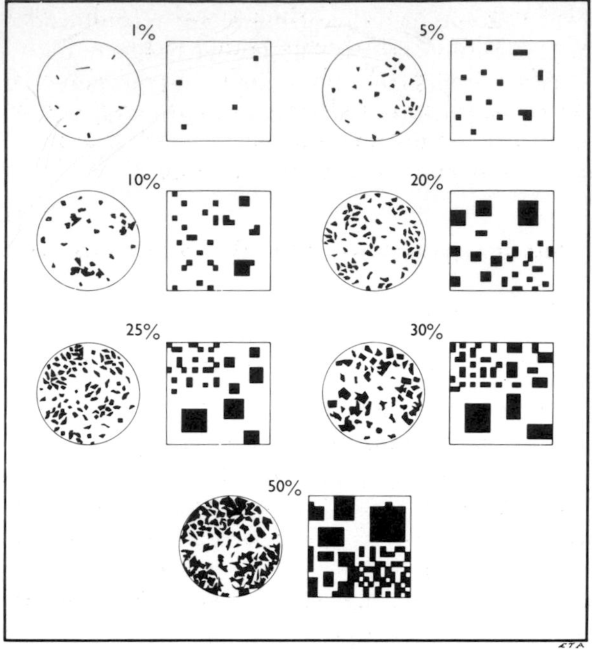

Fig. 1.1 Charts for estimating volume % of minerals in thin section. Reproduced from Folk *et al.* (1970) with permission of DSIR Publishing, New Zealand.

The twofold subdivision may seem rather vague and unsophisticated to students of sedimentology who are accustomed to a more systematic, intensive approach to grain-size studies. The fact is that greater pigeon-holing is inappropriate for igneous rocks, which in sharp contrast with typical sediments are characterized by inequidimensional grains and a highly variable grain-size range. There is, for example, no simple way of describing the grain-size of very thin, yet very elongate, mineral plates, or of categorizing the grain size of a rock that contains a mixture of glass, microlites, microphenocrysts and phenocrysts.

Grain-size descriptions of igneous rocks should include the following information, for each mineral in turn:

1. Range of typical grain lengths (e.g., feldspars range from 5 to 10 mm in length)
2. Typical length/width ratios.

Grain size is measured roughly by comparison with the known diameter of the field of view for each objective lens of the microscope, or more accurately using an eyepiece micrometer or computer image analyser (e.g.,

Starkey and Simigian, 1987). Cashman and Marsh (1988) review the problems of grain-size measurements in thin section.

Inextricably linked to grain size is grain shape and spatial distribution, subjects to be discussed fully in Part Two. Also discussed in section 4.1.2 is the use of crystal-size distribution curves which help in the understanding of processes of nucleation and crystallization.

1.1.2 Recommendations of the IUGS subcommission

(a) Modal parameters

The following five mineral groupings are used to classify rocks:

Q = quartz, tridymite, cristobalite.
A = alkali-feldspar, including albite up to An_5.
P = plagioclase (An_5–An_{100}) and scapolite.
F = feldspathoids ('foids') including nepheline, leucite, kalsilite, the sodalite group of minerals, cancrinite and also the zeolite analcime*.
M = mafic and other minerals such as mica, amphiboles, pyroxenes, olivine, opaque minerals, accessories such as apatite and zircon, melilite and primary carbonates.

The first four are generally called **felsic** mineral groups, the last rather more loosely **mafic** (loosely because not all the minerals are mafic, as for example muscovite, apatite and calcite).

Q and F are mutually exclusive mineral groupings because feldspathoid always reacts with excess silica in a melt to form feldspar. Therefore the maximum number of mineral groupings possible in any one rock is four.

(b) Root-names and qualifiers

Various root-names, such as rhyolite, can be assigned on the basis of modal parameters and grain size, and these can be qualified in any useful way. Typical qualifiers are mineral names (garnet rhyolite), textural terms (spherulitic rhyolite) and general descriptive terms (brown, altered rhyolite). There are no restrictions. If more than one qualifying mineral name is given, the order should be one of increasing abundance: a pyroxene biotite dacite, for example, contains more biotite than pyroxene.

Rocks containing glass are described as: glass-bearing (0–20% glass),

* The F minerals cancrinite and analcime are often secondary, replacing primary magmatic feldspathoids. The presence of seemingly euhedral analcime has often been taken to indicate a primary igneous origin. However, Karlsson and Clayton (1991), using oxygen, hydrogen and nitrogen isotope data, show that such analcime can result from the replacement of leucite in meteoric (not magmatic) water, even in rocks that otherwise show no alteration.

glass-rich (20–50%) or glassy (50–80%). Special names such as obsidian are used if glass >80%. If a chemical classification is used, then the prefix **hyalo-** indicates the presence of glass.

The colour index is defined in terms of the relative percentage of pale minerals (felsic minerals and M-group minerals such as apatite, muscovite, calcite) to dark minerals. Terms used are **leucocratic** (65–100% pale minerals), **mesocratic** (35–65%), **melanocratic** (10–35%) and **ultramafic** (0–10%).

Root-names proposed include the well-established ones such as granite, and the redefinitions of these are made to conform as closely as possible with previous recent usage.

The range of rocks considered to be 'igneous' for the purposes of terminology is broad, and includes pyroclastics (which may also be regarded as sediments), charnockites which are often more-or-less metamorphosed (**meta-**) igneous rocks, and peridotite, which is a name also used for thoroughly metamorphic material (most peridotites and dunites are in fact metamorphic, not igneous).

(c) Seven steps to identifying a rock: the basis of a hierarchical classification

The IUGS classification is hierarchical, and a series of seven steps is to be followed when identifying an igneous rock. Le Maitre (1989) presents the steps in flow-chart form suitable for wall mounting. Rocks with 'special' distinctive characteristics are considered first, leaving the more-common 'ordinary' rocks until the end. The steps provide a sound basis for students learning how to name rocks, and for this reason I use them as the basis for the sequence of rock description in this book. The steps are:

1. If the rock is pyroclastic (fragmental), go to section 1.2.
2. If the modal primary carbonate is >50%, go to section 1.3 (carbonatites).
3. If the rock is mafic or ultramafic, if the rock is a minor intrusive (or a minor leucite-bearing extrusive) and if A, P or F minerals are absent or restricted to the matrix (except leucite which may be a phenocryst), then the rock may be a lamprophyre, lamproite or kimberlite. Go to section 1.4.
4. If the rock has M > 90% and modal melilite >10%, go to section 1.5 (ultramafic melilitic rocks).
5. If the rock is felsic and contains orthopyroxene (or fayalite plus quartz), go to section 1.6 (charnockitic rocks).
6. If none of the above applies, and the rock is plutonic, go to section 1.7.
7. If the rock is volcanic, go instead to section 1.8.

1.2 PYROCLASTIC ROCKS

Magma is commonly fragmented at the Earth's surface during release of volatiles, and the fragmentation can vary from the continuous slow process of strombolian eruptions to the devastatingly rapid process of a major ignimbrite eruption. Strombolian eruptions typically eject large blobs of magma that are still molten or plastic; many ejecta twist when travelling through the air to form spindle-shaped bombs. In contrast, ignimbrite is usually formed from a finer spray of fragmented pumice.

Much of the subject of pyroclastic rocks is beyond the scope of this book. This is obvious when in thin section a spindle-shaped bomb cannot be distinguished from lava flow basalt. A field perspective is essential, and because pyroclastic deposits are as much sediment as igneous rock, many relevant laboratory techniques are those of the sedimentologist and out of place in this book. The reader is referred to Fisher and Schmincke (1984) and Cas and Wright (1987) for detailed information.

Pyroclast terminology

Bombs: >64 mm mean diameter, wholly or partly molten during formation (evidenced by spindle and cow-pat shapes or bread-crust surfaces).

Blocks: >64 mm mean diameter, not molten during formation (as evidenced by angular shapes).

Lapilli: 2–64 mm mean diameter.

Ash grains: <2 mm mean diameter, qualified as **coarse** if >1/16 mm, **fine** if <1/16 mm mean diameter.

Pyroclasts can be fragments of crystal, glass or rock. Fragments of glass are called **shards**, and are usually of ash grain size.

Pyroclastic deposit terminology

Tephra: Any unlithified pyroclastic deposit, often qualified to indicate fragment size (e.g., **block tephra, lapilli tephra**).

Agglomerate: Any deposit made of bombs and/or blocks. Lithified agglomerate can be called **pyroclastic breccia**. (Excluded from agglomerate are the blocky parts of *aa* lava flows, and the screes and blocky deposits that form a carapace around viscous flows. These are considered to be an integral part of the flow.)

Ash: Any unlithified deposit made of ash grains.

Tuff: Any lithified deposit made of ash grains.

Common qualifiers: Ash and tuff may be described as **coarse** or **fine**, depending on ash grain size. In addition they may be qualified as **vitric,**

crystal or **lithic**, according to the proportion of ash that is made of crystal, glass or rock fragments (Fig. 1.2). Any other useful qualifier can be added. Hence **basaltic vitric tuff, air-fall ash, vent pyroclastic breccia**, etc.

Mixed pyroclastic and epiclastic deposits

Reworked pyroclastic material is termed **epiclastic**, and is no longer in the igneous arena (hence a reworked tuff becomes a sandstone). In practice it may be extremely difficult to draw a sharp line of distinction between what is reworked or not. The following mixtures of pyroclastic and epiclastic material are possible (epiclastic refers to other clastic material as well as reworked pyroclasts):

Pyroclastic deposit names, as given above, are to be used if pyroclasts are >75%.
Tuffite is the general term if pyroclasts are 25–75%. Other rock names are sedimentary ones qualified by **tuffaceous**; hence **tuffaceous conglomerate, breccia, sandstone, mudstone**, etc.
Sedimentary rock names are used if pyroclasts are <25%.

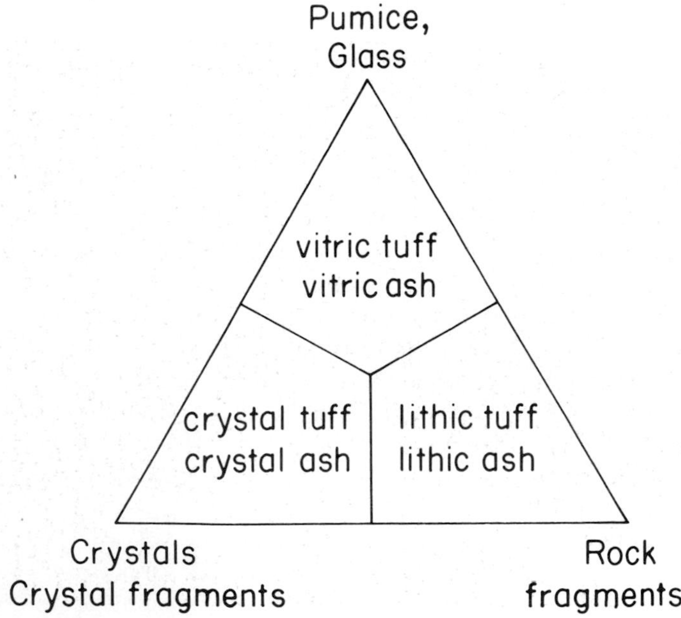

Fig. 1.2 Terminology for tuff and ash (IUGS recommendations from Schmid, 1981; Le Maitre, 1989).

Distinctive textures

Accretionary lapilli are spheroidal bodies made up of concentric layers of fine ash material (Fig. 1.3) accreted during the flight of a pyroclast in wet conditions.

Shard shapes: Explosive magmatic or phreatomagmatic eruption produces cuspate forms (Fig. 1.4A) from broken pumiceous or scoriaceous material. In contrast, cracking of glass by quenching in water tends to produce blocky shards with more-or-less planar surfaces.

Eutaxitic texture and structure: A rapidly accumulated pile of hot plastic glass fragments may deform plastically under its own weight or in response to gravitational forces down a slope, and the resulting textures and structures are called eutaxitic. Rheomorphic flows of basaltic spatter, for example, have a layered appearance due to the flattening and flowage of spatter fragments. In thick rhyolitic or dacitic pyroclastic deposits, pumice fragments may be flattened to discoidal shapes; together with flattened cuspate shards they form the eutaxitic texture characteristic of many welded ignimbrites (Fig. 1.4B).

Glass alteration: Glass is inherently unstable. Basaltic glass (also called sideromelane) hydrates to **palagonite**, a waxy clay-like mixture, and

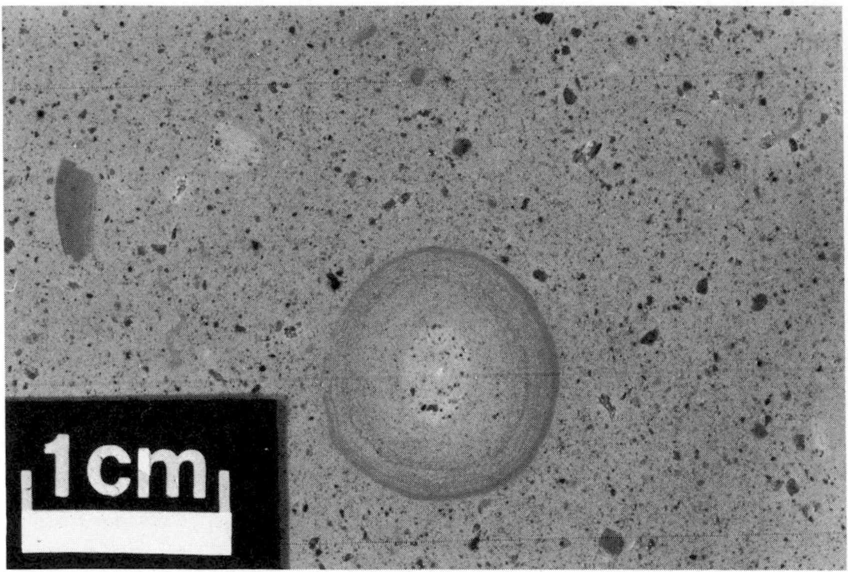

Fig. 1.3 Cut surface of a Plinian air-fall deposit to illustrate the typical concentric layers of accretionary lapilli. Rotorua, NZ.

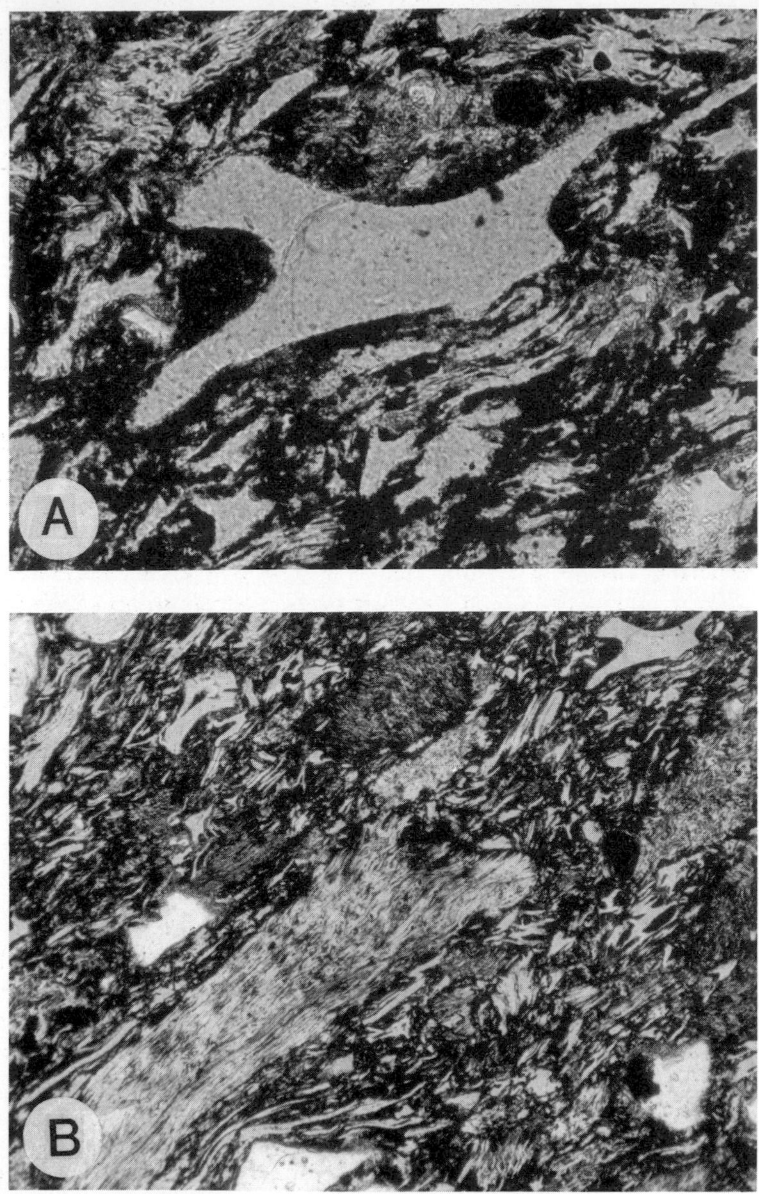

Fig. 1.4 (A): Cuspate shards of glass in ignimbrite, Taupo, NZ. (B): More general view of specimen shown in (A) to illustrate eutaxitic texture formed by flattened pumice fragments and glass shards. (Lengths of view measure 0.71 mm (A) and 2.75 mm (B).)

this alteration may take place very quickly during phreatomagmatic or subaqueous eruptions. Alteration of rhyolitic glass promotes the lithification and cementation of pyroclastic deposits by redistribution of silica. Glassy deposits are prone to zeolitization during very low-grade metamorphism (sections 2.3 and 5.1.1): in some cases individual shards are replaced by a mosaic of zeolite grains, but sometimes a single crystal replaces several shards (Fig. 2.3).

Genetic rock names

Root-names in the IUGS classification are descriptive, but some genetic rock names, such as ignimbrite, are extant too. Both types can be used together, so that it is also possible to describe some parts of an ignimbrite as lapilli tuff, for example.

Ignimbrite is a deposit formed from a hot pumiceous pyroclastic flow ('flow' here means a high particle concentration gas–solid mixture). Central zones of thick ignimbrite sheets are commonly welded with the distinctive eutaxitic texture, but many ignimbrites are not welded and superficially resemble air-fall deposits. The diagnostic feature of an ignimbrite is not the eutaxitic texture (which occurs in non-ignimbritic deposits and does not occur in all ignimbrites) but evidence of strongly directed flow away from an eruptive site, combined with evidence of high temperature such as carbonization of entrained wood: this must be judged in the field.

Hyaloclastite is a shattered glassy deposit formed during subaqueous quenching. Glass shards tend to be bound by flat surfaces, and are cuspate only if the magma was already vesicular.

1.3 CARBONATITES

The possibility of carbonate-rich magmas was not universally accepted until the 1960s when the eruption of carbonatite lava was witnessed in Tanzania (Oldoinyo Lengai volcano). In fact, carbonatites display most typical igneous features including euhedral phenocrysts, chilled margins, vesicles, and *aa* and *pahoehoe* lava types. Most carbonatites are found in intra-plate regions, often in areas of high relief and associated with major fault zones, and most known examples are from Africa, particularly the East African Rift system where they form intrusive and extrusive bodies. Associated rocks are feldspathoidal and melilite-bearing rocks, kimberlites and fenites which are metasomatically depleted in Si and enriched in Fe and alkalis. Carbonatites form hydrothermal vein deposits and replacement bodies due to the interaction of groundwater with primary carbonatite; it is not always easy to distinguish primary dykes from secondary veins. An up-to-date review is given by Bell (1989).

Carbonatite mineralogy and terminology

Carbonatite is either plutonic or volcanic. It contains >50% carbonate.
Calcite carbonatite: Most carbonate is calcite. **Sövite** is coarse grained, **alvikite** is fine grained.
Dolomite carbonatite: Most carbonate is dolomite. Also called **beforsite**.
Ferrocarbonatite: Most carbonate is iron rich.
Natrocarbonatite: Mostly Na–K–Ca-carbonates (these are very rare, but make up some of the lavas in Tanzania).

The usual order of crystallization is calcite, dolomite, ankerite.

The presence of a minor phase (<10%) can be specified as '**ankerite-bearing**', for example. If carbonates are 10–50% the rock is **carbonatitic** (e.g., carbonatitic ijolite), not carbonatite.

Optically it may be difficult to specify the type of carbonate, especially if fine-grained intergrowths are present; a chemical classification can then be used (Fig. 1.5).

Other minerals present: Prominent among the non-carbonate minerals are the Nb-rich oxide pyrochlore, F-rich apatite and amphiboles varying from the common calcic types to sodic–calcic and alkali-amphiboles

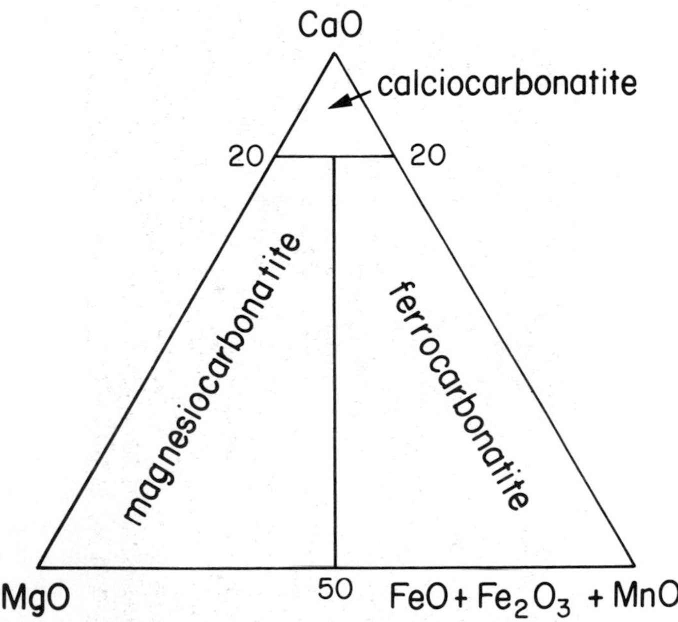

Fig. 1.5 Chemical classification for carbonatites (IUGS recommendations from Le Maitre, 1989, based on Woolley and Kempe, 1989).

such as riebeckite and arfvedsonite. The presence of pyrochlore confirms the carbonatite as igneous (as distinct from marble). Other common primary minerals include diopside, acmite and aegirine–augite, albite, biotite and phlogopite, olivine, fluorite and various opaque minerals, especially magnetite.

Carbonatites are economically very important sources of Nb and Ta (pyrochlore), phosphate (apatite) and rare-earth elements (bastnaesite, parisite and monazite).

Distinctive textures

Volcanic carbonatites may contain tabular, oscillatory-zoned calcite phenocrysts (Fig. 1.6), as described by Keller (1989). Phenocrysts replaced by mosaics of calcite were probably Na–Ca-carbonate (nyerereite) that reacted with groundwater. If apatite is abundant nyerereite was not (Gittins, 1989). Dolomite phenocrysts are rhomb shaped (some workers believe these to be secondary). Elongate, dendritic forms of calcite are found in some intrusives. Evidence for immiscibility in some carbonatites and

Fig. 1.6 Tabular calcite phenocrysts in carbonatite lava bomb, Kirchberg, Germany. View length is 2.1 mm. Photograph courtesy of J. Keller.

associated rocks are carbonate globules or blobs (some diapiric) in a silicate matrix, and flow layering of carbonatite and silicate-rich material. All carbonatites are susceptible to two forms of alteration: 1. reaction with groundwater, and 2. plastic deformation and recrystallization. The first is likely to remove phases such as nyerereite, and also ankeritic carbonate which alters to iron oxides (sometimes in economically commercial quantities), calcite and dolomite. The second is likely to produce rocks that mimic sedimentary-derived marbles. In either case, igneous textures may be extensively modified or destroyed.

1.4 LAMPROPHYRE, LAMPROITE AND KIMBERLITE

These mafic or ultramafic rocks occur mainly as phenocryst-rich minor intrusives (lamproites also as minor extrusive bodies), and they form a distinctive mineralogical and compositional group. Feldspars and/or foids are absent (in kimberlite) or restricted to the groundmass except in the case of leucite phenocrysts in some lamproites. Most are K- and Ba-rich, and contain phlogopite, biotite, olivine or amphibole phenocrysts, and many contain the unusual combination of Mg-rich olivine or pyroxene with alkali feldspar. Lamprophyres and lamproites are often Ti rich, and lamproites and kimberlites are Mg rich.

Lamprophyres occur as dykes and sills in a wide variety of settings: feldspar-bearing, foid-free varieties are commonly associated with granites, feldspar–foid-bearing varieties with alkaline igneous provinces and feldspar-free varieties with carbonatites. Kimberlites occur as explosive vent fillings (diatremes), dykes and sills in intra-plate settings; they often contain a high proportion of mantle or deep-crustal xenolithic material, are associated with carbonatites and are economically important as a source of diamonds (as in Africa). Lamproites form explosive vent fillings, dykes, sills and minor extrusives at the margins of cratonic regions above fossil subduction zones; they may contain diamonds (as in Australia).

Recent reviews are given by Dawson (1987), Bergman (1987) and Rock (1987).

Lamprophyres

Mineralogy and terminology: Lamprophyres are rich in mafic minerals, often Ti-rich or alkali varieties of biotite, phlogopite, amphibole or pyroxene, and are mesocratic to melanocratic. The IUGS recommendations on names are summarized in Table 1.1. The classification is principally based on the type of felsic minerals present or absent.

Distinctive textures: Euhedral mafic phenocrysts (Fig. 1.7A) are often set in a groundmass of zeolites (analcime), feldspars, feldspathoids or

Table 1.1 Lamprophyre terminology based on mineral content (after Le Maitre, 1989)

feldspar	foid	biotite, Mg-rich augite, (± olivine)	hornblende, Mg-rich augite, (± olivine)	amphibole (especially kaersutite or oxyhornblende), Ti-rich augite, olivine, biotite	melilite, biotite, (± Ti-rich augite), (± olivine), (± calcite)
A > P		minette	vogesite		
P > A		kersantite	spessartite		
A > P	A + P > foid			sannaite	
P > A	A + P > foid			camptonite	
	glass or foid			monchiquite	polzenite
					alnöite

carbonates. Hydrothermal alteration of phenocryst phases, especially olivine, is very common. Dark-brown Ti-rich melanite garnet, if present, is often zoned to pale-green or colourless rims.

Lamproites

Mineralogy: Mafic-rich, similar to lamprophyre, but the dominant minerals include: Ti-rich phlogopite, Al-poor diopside, leucite, Al-poor amphiboles, especially K-richterite or alkali amphiboles, olivine and Fe-rich sanidine. Accessory minerals include unusual minerals such as priderite, jeppeite, other K–Ba–Ti–Fe-oxides and wadeite (a K–Zr-silicate), as well as perovskite, apatite and Cr-spinel. Plagioclase, melilite, monticellite, melanite-garnet or feldspathoids other than leucite are absent in lamproites.

Distinctive textures: Possible phenocryst phases include phlogopite (often resorbed), pyroxene, olivine and leucite (Fig. 1.7B). Phlogopite also occurs in the matrix. Amphiboles tend to be late crystallized, and are restricted to the groundmass or as a lining to vesicles. Mantle xenolithic material is not common, but cognate olivine, biotite, pyroxene-rich xenoliths are found, and pyroxene is often xenocrystic and rimmed by phlogopite.

Kimberlites

Mineralogy: Primary igneous minerals include olivine, usually phlogopite, carbonates (usually calcite), and any of diopside, spinel, ilmenite, perovskite and monticellite. Due to the common presence of xenolithic material, many crystals are likely to be xenocrystic, especially olivine

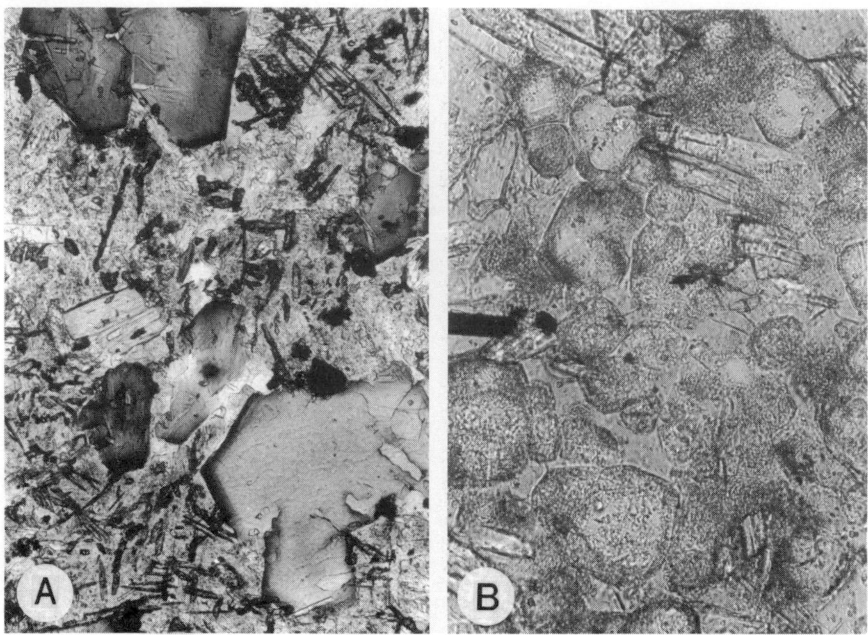

Fig. 1.7 (A): Lamprophyre (minette) rich in euhedral phlogopite set in a matrix of alkali feldspar. (B): Leucite microphenocrysts set in glass in wyomingite, a variety of lamproite from Leucite Hills, Wyoming. (Lengths of view measure 0.82 mm (A) and 0.34 mm (B).)

that is more Mg-rich than Fo_{90}, but also Mg-rich garnet, phlogopite, Cr-diopside, Cr-spinel, orthopyroxene, ilmenite and diamond.

Distinctive textures and structures: The contrast between large, rounded xenocrysts (or megacrysts) of olivine and smaller euhedral (micro)phenocrysts is distinctive of kimberlite (Fig. 1.8A). Other xenocrystic and xenolithic material of mantle origin is common. The matrix is usually carbonate rich, and serpentine and carbonate alteration of silicate phases is extremely common. Primary carbonate material is indicated by small, immiscible diapiric bodies in some kimberlites. The possible abundance of xenolithic material combined with the brecciated nature of many kimberlite deposits leads to an extremely variable petrography.

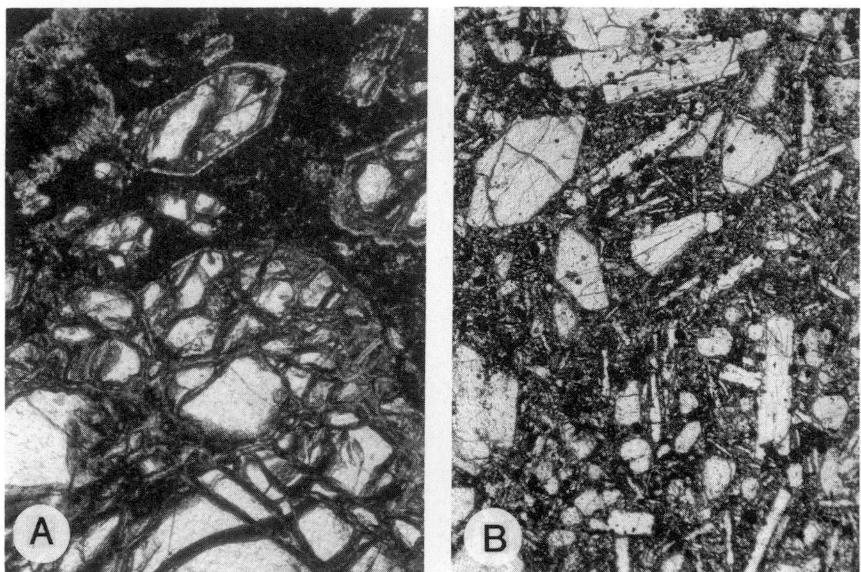

Fig. 1.8 (A): Kimberlite (from Kimberley) showing the typical juxtaposition of large rounded megacrysts and smaller euhedral phenocrysts of olivine. All olivine is altered to some extent. (B): Melilite (tabular) and olivine phenocrysts in olivine melilitite, from Germany. (View lengths measure 1.85 mm (A) and 2.9 mm (B).)

1.5 ULTRAMAFIC MELILITIC ROCKS

Ultramafic melilitic rocks contain >10% melilite and >90% M-group minerals. They are the most ultrabasic of silicate rocks with silica sometimes as low as 30%. These rocks occur in both continental and oceanic intra-plate regions: they are best known from the East African Rift system, but also occur as a late phase on Hawaii, for example. Associated rocks are nephelinites, kimberlites, carbonatites, the lamprophyre alnöite and various nepheline-, kalsilite- or leucite-bearing alkaline igneous rocks.

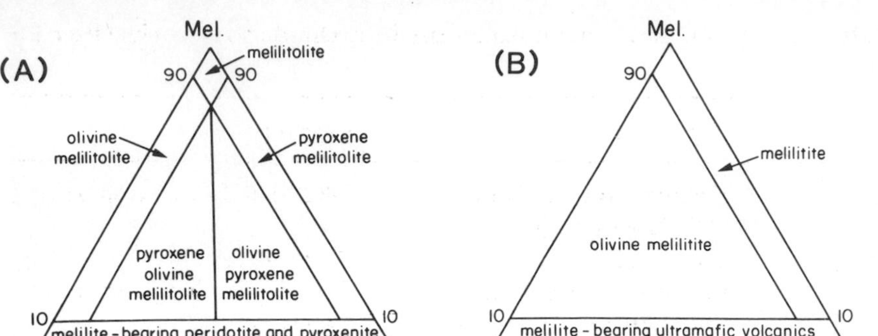

Fig. 1.9 Terminology for melilitic igneous rocks based on content of melilite (Mel), olivine (Ol), and clinopyroxene (Cpx) (IUGS recommendations from Le Maitre, 1989 and Streckeisen, 1979).

Mineralogy and terminology: Volcanic varieties have the general name **melilitite**, plutonic varieties the name **melilitolite**. Apart from melilite, the principal minerals present are olivine and/or clinopyroxene, and varietal names based on the proportion of melilite, olivine and pyroxene are given in Fig. 1.9. Pyroxene melilitolite is also called **uncompahgrite**, olivine melilitolite **kugdite**, and olivine–pyroxene melilitolite is called **olivine uncompahgrite**. Common accessory minerals include biotite or phlogopite, apatite, perovskite, calcite, Fe–Ti-oxides. The Na-rich varieties may have small amounts of nepheline, K-rich varieties, leucite and kalsilite. Ultramafic melilitic rocks never contain feldspar. Non-ultramafic melilitic rocks are classified elsewhere, but are usually qualified by melilite if melilite is >10%: for example, **melilite nephelinite**.

Distinctive textures: Euhedral tablets of melilite are naturally enough the most distinctive feature of melilitic rocks (Fig. 1.8B), and these generally form part of a matrix to larger olivine and clinopyroxene phenocrysts, and often display a preferred orientation similar to the trachytic texture of feldspars. Melilite also commonly exhibits the characteristic 'peg-structure' which consists of oriented rod-like inclusions at right angles to {001}. Melilite plates may be pseudomorphed by carbonates, but it has also been suggested that some are actually pseudomorphs of the unstable Na–Ca-carbonate nyerereite.

1.6 CHARNOCKITIC ROCKS

Charnockitic rocks are characterized by a combination of orthopyroxene (or fayalite plus quartz) and feldspar and possibly quartz. Charnockitic rocks may be igneous, meta-igneous or thoroughly metamorphic, and it

Table 1.2 Special and general names used for charnockitic rocks (after Le Maitre, 1989)

QAPF field	General name	Special name
2	Orthopyroxene alkali-feldspar granite	Alkali-feldspar charnockite
3	Orthopyroxene granite	Charnockite
4	Orthopyroxene granodiorite	Opdalite or charno-enderbite
5	Orthopyroxene tonalite	Enderbite
6	Orthopyroxene alkali-feldspar syenite	
7	Orthopyroxene syenite	
8	Orthopyroxene monzonite	Mangerite
9	Monzonorite	Jotunite
10	Norite (anorthosite if $M < 10$)	

may be difficult to decide which. Igneous charnockitic rocks can also be given more standard names, qualified as in orthopyroxene granite, and metamorphic charnockitic rocks are also properly described as granulites. Despite this confusion, or perhaps because of it, 'charnockitic' and the special names given in Table 1.2 are in very common usage.

Charnockitic rocks are widespread in Precambrian cratonic regions of the world, and often the entire range of compositions from anorthositic to granitic varieties occur in close proximity, sometimes with the more basic ones surrounded by the more acidic. The type locality of charnockite itself is unusual, being the tombstone of Job Charnock in Calcutta, the rock having been quarried in Madras.

Geobarometry on most charnockitic rocks indicates high pressures consistent with formation in the lower parts of the crust, but it is worth remembering that orthopyroxene itself indicates high temperatures and dry conditions, not necessarily high pressures.

Mineralogy and terminology: The mineralogy of charnockitic rocks corresponds with that of the common plutonic rocks of QAPF fields 2–10 (Fig. 1.11), with the additional proviso that orthopyroxene† or fayalite plus quartz should be present. Special charnockitic names exist for six of these fields (Table 1.2). Perthitic intergrowths characterize charnockitic rocks, and this is allowed for in two ways: first, perthite (*sensu stricto*) is counted as A (alkali feldspar), antiperthite as P, and

† Hypersthene, as specified in Le Maitre (1989), is no longer an approved mineral name.

mesoperthite as 50/50 A/P; secondly, the presence of mesoperthite can be noted by the prefix 'm', as in **m-charnockite**.

Distinctive textures: Given that charnockitic rocks can be igneous, meta-igneous or metamorphic, the range of textures is correspondingly wide. Nevertheless, the fact that most charnockites formed under high pressures means that cooling times were long and igneous textures are inevitably modified in some way (section 4.2.2 includes discussion of textural maturity). Charnockites tend, therefore, towards being even grained, lacking strong obvious preferred orientations. However, many anorthosites in particular display igneous feldspar fabrics (grains elongate parallel to {010} with a preferred orientation). Strain episodes during cooling may cause plastic deformation and recrystallization of quartz, and orthopyroxene may be cracked and rimmed with hornblende, biotite or garnet, products of retrogressive metamorphism. Feldspars are typically perthitic, plagioclase is often antiperthitic and mesoperthite is common in alkali-feldspar bearing varieties. Mesoperthite requires an initial hypersolvus crystallization of an intermediate composition alkali feldspar, possible at high pressures only because of the lack of water (most granites are subsolvus). Despite long cooling times, perthite coarsening or complete separation of the intergrown feldspars is not achieved because relatively high An levels in charnockitic perthite inhibit diffusion, and lack of water inhibits late coarsening. Many charnockitic rocks contain a 'greasy-green' plagioclase due to the presence of very fine-grained chlorite and calcite in brittle fractures, and quartz may be bluish due to the presence of numerous tiny exsolved rutile needles.

Distinguishing igneous from metamorphic varieties: In thin section the most distinctive feature of magmatic crystallization is euhedral or subhedral feldspar (shapes dominated by {010} as in Fig. 1.10A), and possibly a preferred orientation of such crystals. Carlsbad twinning is also distinctively igneous, as is ophitic or subophitic texture. Igneous feldspar fabrics may be completely modified by a thorough strain- or stress-induced metamorphic recrystallization, but are likely to survive simple (although prolonged) cooling. It may be almost impossible to distinguish texturally a mildly deformed but dehydrated 'ordinary' igneous rock from a primary charnockitic rock. The presence of thoroughly metamorphic textures (Fig. 1.10B) does not eliminate the possibility of some earlier magmatic stage, and evidence of magmatic crystallization may be related to local melting within a gneissic complex rather than widespread igneous activity. Evidence for a magmatic vs. metamorphic origin may need to be found in the field. For example, Duchesne *et al.* (1989) report charnockitic dykes cutting anorthosite as well as chilled margins to mangerite and jotunite. Rocks that are simply

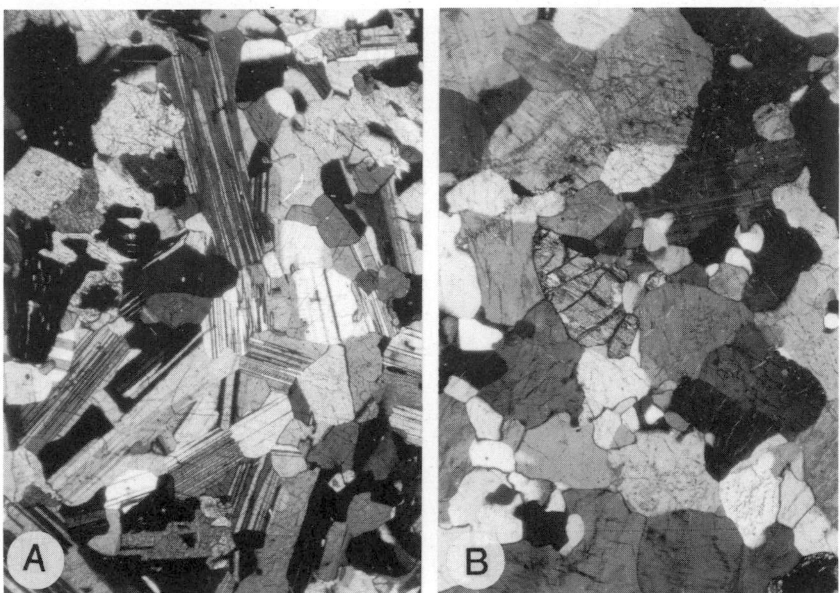

Fig. 1.10 (A): Subhedral plagioclase in norite, Bluff, NZ. (B): Charnockite with metamorphic textures, Kerala, India. (View lengths measure 3.0 mm (A) and (B).)

massive on a large scale are most likely to be fundamentally of igneous origin; other charnockitic rocks form part of obviously metasedimentary successions as described by Treloar and Kramers (1989), for example, although, in this Zimbabwean occurrence, the rocks underwent partial melting generating charnockitic dykes as well. It may be necessary to resort to geochemistry to find further evidence as to origin. Nedelec *et al.* (1990), for example, show that heavy rare-earth element impoverishment is incompatible with crystallization of magmatic orthopyroxene: the charnockites in question are primary igneous rocks, but the charnockitic character is secondary.

1.7 PLUTONIC ROCKS

'Plutonic rocks' refers to those plutonic igneous rocks not already dealt with as special cases (charnockite, melilitolite, etc.), and they fall into two main groups: those with M < 90% and classified according to their position in the double QAPF triangle (Fig. 1.11), and ultramafic rocks with M > 90%.

1.7.1 The QAPF double triangle

Q, A, P and F are defined in section 1.1.2, and we have seen for charnockitic rocks that perthite is counted as A, antiperthite as P, and mesoperthite as A/P = 50/50. To plot a rock within the double triangle the mode % of minerals belonging to the Q, A, P and F groups must be recalculated to total 100%. For example, a rock with modal quartz = 10%, K-rich feldspar = 10%, plagioclase = 30%, biotite = 30%, hornblende = 20% plots at Q = 20%, A = 20% and P = 60%. The double triangle is used in two ways:

Field classification: There is a need to name rocks provisionally in the field, and geologists prefer to use the more common names inexactly rather than coin a whole set of different names. The IUGS subcommission recommends that these most-common names be used with the suffix -oid to give nouns such as granitoid (Fig. 1.12) or -ic to give adjectives as in granitic rocks. The choice is a matter of personal taste.

Laboratory classification: The QAPF double triangle is subdivided into 15 fields for the purposes of thin section work (Fig. 1.11). The principal rock names assigned to each field are given in Fig. 1.13; numerous additional varietal names are well established and approved by the subcommission. Field 1 is not represented by any common igneous rock. The other fields are discussed below.

1.7.2 Fields 2–5 (granitic rocks)

Granitic rocks are the most common plutonic rocks in continental areas, and they dominate the great batholiths of the world. Most are predominantly felsic, although M-rich varieties also exist. Leaving aside the question of the ultimate origin of continental crust, most granitoids originate as crustal melts; indeed, granitoids make up the 'igneous' portion of most migmatite complexes (sections 2.5.3 and 5.3.6).

Mineralogy and terminology: At least 20% of the felsic mineral content of all granitic rocks is quartz (note that this is not the same as >20% modal quartz). Most granitic rocks are subsolvus (section 4.2.3) so

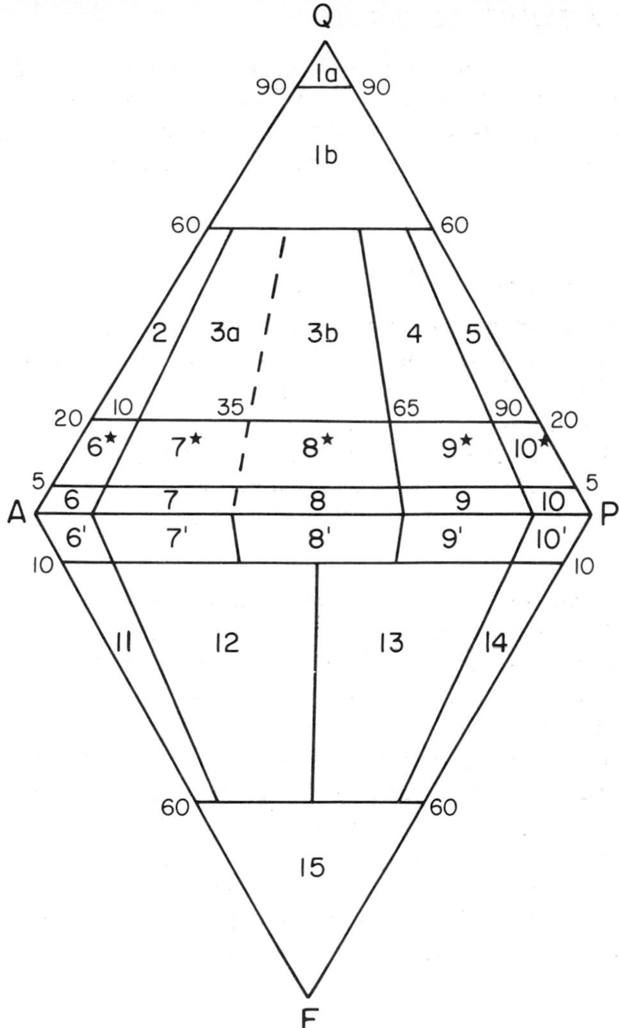

Fig. 1.11 The double QAPF triangle used for plotting the more common plutonic and volcanic rocks with M < 90% according to the IUGS recommendations (Le Maitre, 1989; Streckeisen, 1976). The numbered fields are referred to throughout the text. Fields 6*–10* are slightly oversaturated variants of fields 6–10, and 6′–10′ are slightly undersaturated variants.

K-rich feldspar (orthoclase or microcline) and plagioclase form independent crystals. The specific names **alkali feldspar granite, granite, granodiorite** and **tonalite** depend on the relative proportions of A and P (Fig. 1.13). Granite can be subdivided into **syenogranite** (field 3a) and **monzogranite** (field 3b). **Alaskite** is a varietal name for leucocratic

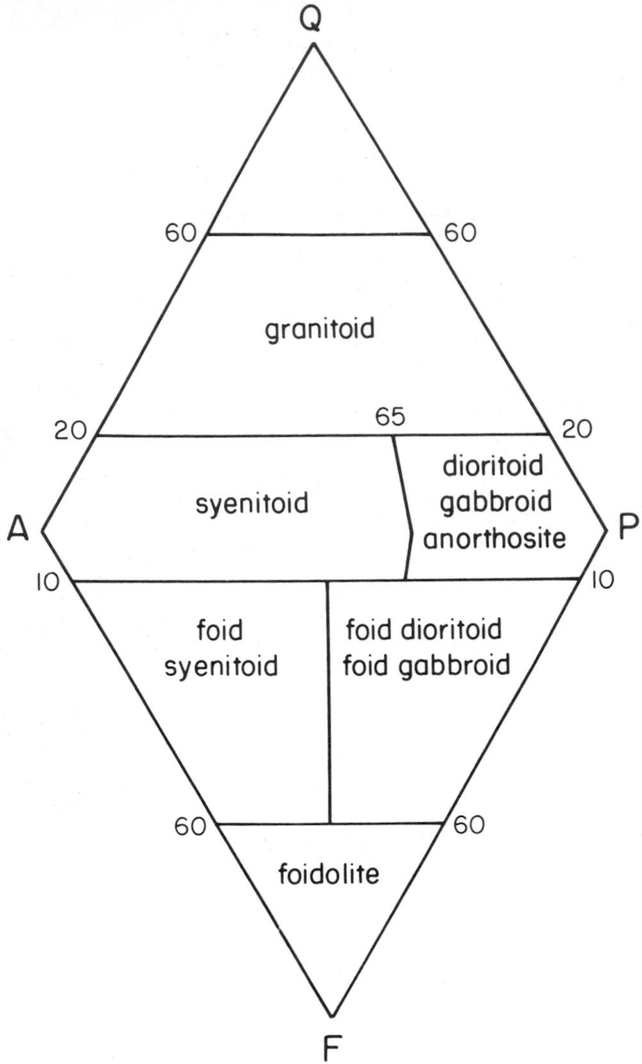

Fig. 1.12 IUGS recommendations (Le Maitre, 1989) for a provisional (field) classification of plutonic rocks that plot in the double QAPF triangle.

alkali-feldspar granite, and **trondhjemite** or **plagiogranite** are alternative names for leucocratic tonalite. Hypersolvus granitoids form only at low pressures or at higher pressures under very dry conditions, and they often contain mesoperthite rather than independent K-rich and plagioclase feldspars: in the low-pressure case, granophyric intergrowths usually develop and the rock is called a **granophyre** (section 3.5.1); in

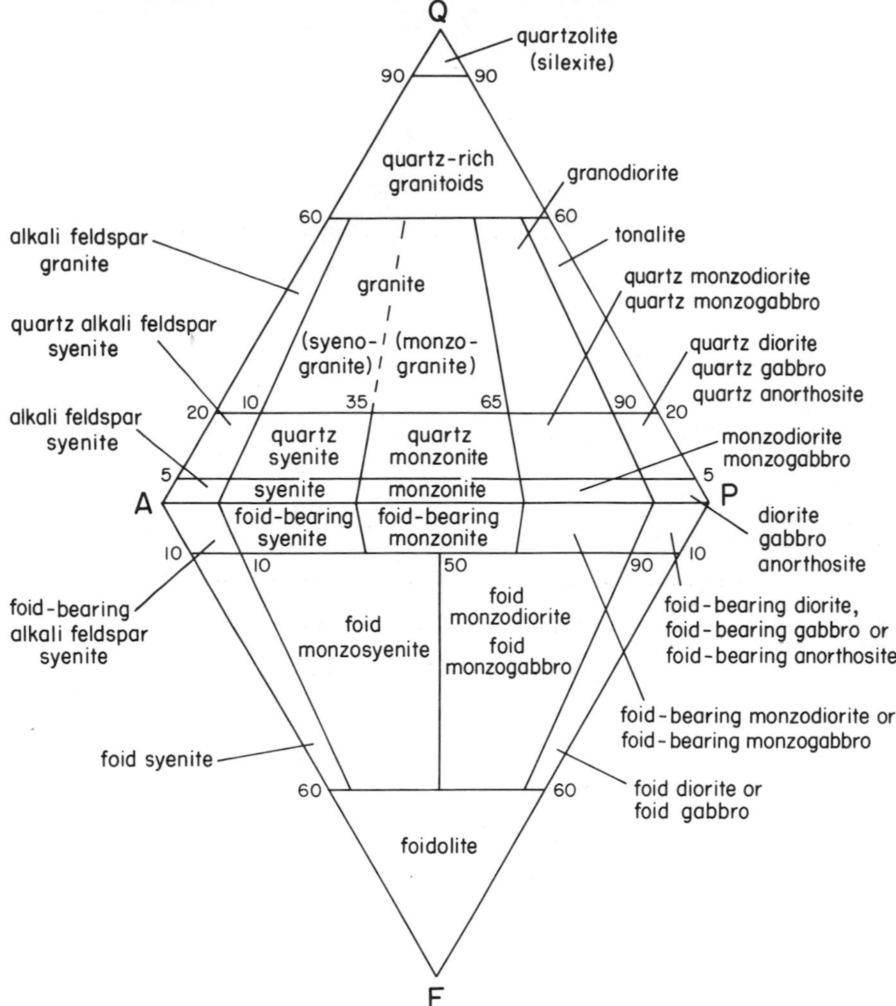

Fig. 1.13 Terminology of plutonic rocks with M < 90% and using the QAPF double triangle (IUGS recommendations from Le Maitre, 1989).

the second case the granitoid contains orthopyroxene and is classified as an m-charnockite (section 1.6). Hornblende, biotite and muscovite commonly accompany the felsic minerals, less commonly pyroxene or fayalite, and cordierite and garnet may also be present. Accessory minerals include apatite, zircon, titanite, allanite, monazite, tourmaline, topaz and opaque iron–titanium oxides.

Fig. 1.14 (A): I-type granite (Separation Point Batholith, NZ) showing hornblende (left), biotite (top), titanite (lower right) and plagioclase set in microcline. (B): S-type granodiorite (Kosclusko Batholith, Australia) showing biotite, muscovite, plagioclase and quartz. (View lengths measure 3.3 mm (A) and (B).)

Common qualifiers: Any mineral name may be used as a qualifier, such as in **hornblende granite**. Structural terms are also used: hence **foliated tonalite**, **gneissic granodiorite**. Chappell and White (1974) proposed that granitoids should be categorized according to the crustal material that they are derived from:

I-type granitoids (e.g., Chappell and Stephens, 1988) derive from melting of igneous rocks, often those underplating the crust, or immature sediments derived from igneous rocks, and have low K/Na, high Ca, and a high state of oxidation. Typical mineralogy (Fig. 1.14A) includes hornblende, green, brown or straw-coloured biotite (biotite and hornblende usually contain apatite inclusions), titanite, possibly allanite and magnetite, sometimes with ilmenite.

S-type granitoids (e.g., White and Chappell, 1988) derive from weathered sedimentary material, have high K/Na, low Ca and Na (removed by weathering), high Al, and a low oxidation state (due to C and S in the weathered sediments). In consequence they contain 'foxy-red–brown' biotite, no hornblende, monazite instead of allanite and titanite, muscovite (Fig. 1.14B), possibly cordierite and garnet, and ilmenite not magnetite. Apatite tends to occur as discrete crystals. The peraluminous character also means that sillimanite or andalusite may be present, perhaps associated with pelitic xenoliths, as well as topaz (indicating F-rich fluid activity) and tourmaline.

Two other granitoid types have now been proposed:

A-type granitoids (e.g., Whalen *et al.*, 1987; Eby, 1990) are K plus Na, F, Zr-rich varieties, poor in Al and Ca, and with high Fe/Mg. They were first named after 'anorogenic' and their occurrence in rift zones and stable continental areas, but they are now known to be more widespread but not as common as I- and S-types. Mineralogically they are characterized by Fe-rich micas, amphiboles and pyroxenes, and in peralkaline varieties by alkali amphiboles (arfvedsonite, riebeckite, etc.). Granophyric intergrowth, fluorite and an abundance of Zr-bearing minerals are common characteristics. They are often ascribed to melting of granulitic terranes already depleted by earlier partial melting, or to primary differentiation of alkali basalt magma; the source of F is an important consideration in determining A-type granitoid origins.

M-type granitoids (e.g., Whalen, 1985) are a subcategory of I-types found only in oceanic island arcs. They are derived from the mantle, possibly from subducted oceanic crust, and are characterized by hornblende, biotite and pyroxene, basic igneous xenoliths, and K-feldspar as a late-stage interstitial granophyric intergrowth.

Distinctive textures and structures: These are discussed in detail elsewhere (e.g., sections 3.5, 4.2.3–4.2.5 and 5.3.6). Granitic texture (Figs 1.14 and 4.42A) is typical, with euhedral–subhedral mafics, subhedral feldspar and anhedral quartz. The K-feldspar, if present, can be either orthoclase or microcline (with cross-hatch twinning), often forms large phenocrysts which may poikilitically enclose plagioclase (Fig. 4.42B), and is likely to be perthitic. Quartz tends to be anhedral, filling spaces between subhedral feldspar, but euhedral bipyramidal quartz may be present, either as inclusions in feldspar, or as clusters with an adcumulate-type texture. Myrmekite (Fig. 3.21) commonly accompanies K-feldspar in subsolvus granitoids, especially deformed ones. Low-pressure hypersolvus granitoids usually contain granophyric intergrowth (Fig. 3.20) and are typically mesoperthitic. Granitoids are susceptible to deformation and gneissic granitoids often display ribbons of quartz bent around broken feldspar phenocrysts (Figs 2.6 and 7) with micas smeared out along shear planes. Feldspars may be aligned by primary flow processes or secondary strain. Igneous phase layering is relatively rare, but granitoids form the 'igneous' layers in migmatite complexes (section 5.3.6). Some granitoids are orbicular (Fig. 4.47). The S-type granitoids are characterized by metapelitic xenoliths, and rounded zircons (Fig. 4.50) may possibly represent refractory relics of the parent sediment; I- and M-types are characterized by igneous xenolithic material. The S-types commonly contain greisen and pegmatites rich in topaz, tourmaline, Li-rich mica and W and Sn minerals. Granitoids alter by chloritization of biotite and hornblende, and sericitization and clay mineral alteration of feldspars. Cordierite in S-type granitoids is often extensively altered to fine-grained pinite or replaced by coarse clots of muscovite.

1.7.3 Fields 6–9 (alkali feldspar syenite, syenite, monzonite, monzodiorite and monzogabbro)

Rocks of fields 6–9 are much less abundant than either granitoids or the gabbroic rocks of field 10. They form plutons in close association with their volcanic equivalents in a wide variety of geological settings wherever alkali basaltic volcanoes or alkaline volcanics occur. In addition, the metasomatic 'fenites', formed by desilication of crustal rocks adjacent to carbonatites, are often syenitic in composition. Syenitic bodies are often phase layered (e.g., Parsons, 1987).

Mineralogy and terminology: These rocks consist essentially of alkali feldspar and variable amounts of plagioclase. The ratios A/P that characterize **alkali feldspar syenite**, **syenite**, **monzonite**, **monzodiorite** and **monzogabbro** are indicated in Fig. 1.13. Rocks of field 9 are called monzodiorite if plagioclase has An $< 50\%$, monzogabbro if An $> 50\%$. Rocks of fields 8 and 9 can be given the general names **syenodiorite** or **syenogabbro**. If Q $> 5\%$ and $<20\%$ the names are qualified as in **quartz syenite**. If foids are present (usually nepheline or sodalite) and F $< 10\%$ the rocks are qualified as foid bearing (e.g., **nepheline-bearing syenite**). Mafic minerals generally increase with increase in P and include hornblende and biotite, titanian augite or diopside, especially in the M- and P-rich varieties, and sodium-rich pyroxenes (aegirine, aegirine augite) and alkali amphiboles such as arfvedsonite, especially in the more alkaline syenites. Melanite garnet is found in some syenites. Accessory minerals include apatite, zircon, titanite, opaque iron oxides, and fluorite, and they tend to be more diverse and abundant in the more alkaline syenites. These rocks are particularly diverse in detail, and numerous varietal names have been coined. Perhaps the best known are **larvikite**, the building stone characterized by rhomb-shaped feldspars with complex intergrowths that give a characteristic iridescence, **pulaskite**, a leucocratic nepheline-bearing alkali feldspar syenite, and **albitite**, a syenite made up almost entirely of albite.

Distinctive textures and structures: Most of these rocks have a granitic-type texture with euhedral–subhedral mafic minerals set among euhedral or subhedral feldspars (Fig. 1.15A). Quite commonly, K-feldspar poikilitically encloses plagioclase (Fig. 1.15B), a texture sometimes termed monzonitic. Alkali feldspars are usually perthitic. Augitic pyroxenes may be encased in hornblende and biotite, and replaced by secondary actinolitic hornblende. Feldspars and nepheline may be replaced by micas and clay minerals. Lacking abundant quartz, these rocks do not deform plastically in the same way as gneissic granitoids. Primary preferred alignments of tabular feldspar are common, especially parallel to intrusive contacts or igneous layering. Syenitic rocks are relatively common in layered intrusives (e.g., Parsons, 1987).

1.7.4 Field 10 (diorite, gabbro, anorthosite)

Rocks of field 10 are the most common plutonic rocks in the oceanic crust and, next to granitoids, the most common in continental areas. Large intrusive bodies of gabbroic rock underlie most basaltic volcanoes, and many are spectacularly layered as in the Bushveld and Skaergaard complexes (section 4.2.2). Separation of gabbroic minerals into layers means that the very plagioclase-rich layers are best termed anorthosite, and the very M-rich layers peridotite or pyroxenite. Apart from the layers in gabbroic intrusives, large anorthosite masses occur, particularly in Precambrian terranes (e.g., Miller and Weiblen, 1990), often emplaced at lower crustal levels. Diorite is the plutonic equivalent of andesite; many diorites show field evidence of a hybrid origin by mixing of basic and acid magmas.

Mineralogy and terminology: The essential mineralogy is plagioclase (but not albite) plus varying amounts of M-group minerals. **Anorthosite** (also called **plagioclasite**) has $M < 10\%$, and plagioclase is usually in the range andesine to bytownite. Typical mafic minerals are hornblende, pyroxene and olivine. In **diorite** the plagioclase is $An < 50\%$, and $M > 10\%$, usually hornblende or biotite, sometimes uralitized augite. In **gabbro**, plagioclase is $An > 50\%$, $M > 10\%$, the main minerals being olivine, augitic pyroxene, orthopyroxene and hornblende, and the IUGS subcommission recommends the following varietal names:

Fig. 1.15 (A): Syenite showing subhedral mesoperthitic feldspar and hornblende. (B): Monzonitic texture in monzonite porphyry, Montana. (Lengths of view measure 3.3 mm (A) and (B).)

gabbro (*sensu stricto*) is plagioclase and augitic clinopyroxene; **norite** (Fig. 1.10A) is plagioclase plus orthopyroxene; **troctolite** is plagioclase and olivine; **hornblende gabbro** is hornblende and plagioclase (<5% pyroxene). Other approved names are given in Fig. 1.16. Biotite is a common minor component in gabbroic rocks. Common accessory minerals include apatite, rutile, ilmenite, magnetite and spinel.

Distinctive textures and structures: These are discussed in detail elsewhere (e.g., sections 3.5, 4.2.2 and 4.2.5). Granitic-type textures are common, but gabbroic rocks are also typified by ophitic texture (Fig. 1.17): in cases where pyroxene is dominant, euhedral grains of

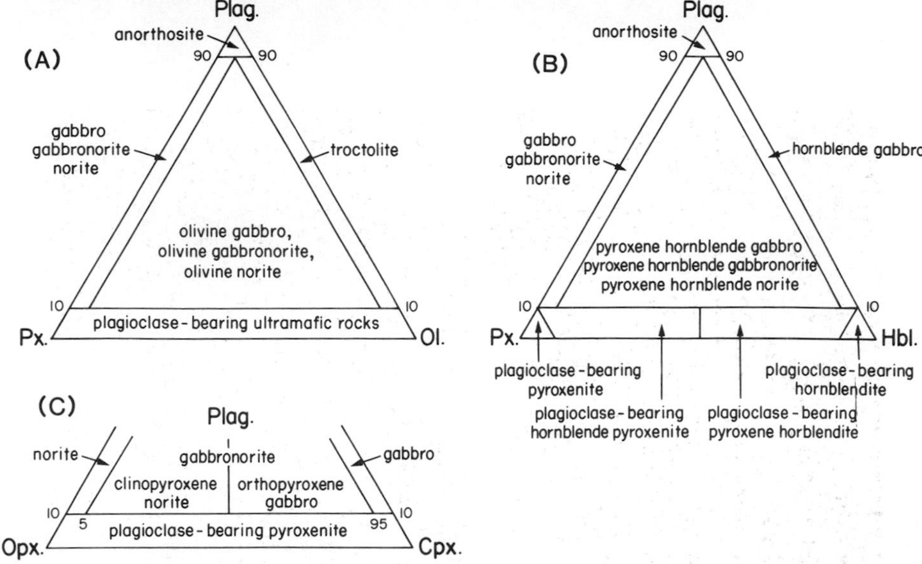

Fig. 1.16 Terminology of gabbroic rocks (IUGS recommendations from Le Maitre, (1989) based on content of plagioclase (Plag), pyroxene (Px), olivine (Ol) and hornblende (Hbl). The options of gabbro, gabbronorite and norite in (A) and (B) are based on orthopyroxene (Opx) and clinopyroxene (Cpx) content, as shown in (C).

plagioclase are set within large augites, but when plagioclase dominates, the feldspar may be joined as an anhedral/subhedral mosaic with pyroxene confined to the corners of feldspars (Fig. 1.17C). Olivine is often enclosed within pyroxene which in turn is often rimmed by hornblende; biotites often grow as a reaction product around opaque iron oxides. Pyroxenes may display prominent exsolution lamellae of another pyroxene (section 3.5.3). Igneous foliation, if present, is usually defined by an alignment of initially tabular plagioclase, often visible even when grains are joined in a mosaic as in anorthosite. Cumulate textural terms such as adcumulate, orthocumulate and crescumulate are fully discussed along with grain-size sorting and density grading under 'layered intrusives' (section 4.2.2). Gabbros do not metamorphose easily, and the typical response at high temperatures is the development of corona textures along grain boundaries (text and figures in sections 2.3.1 and 3.5.4). At lower temperatures, common alteration products include serpentine after olivine (the resulting expansion causes cracking of adjoining plagioclase – Fig. 1.18), uralite and chlorite after pyroxene, and saussurite after plagioclase.

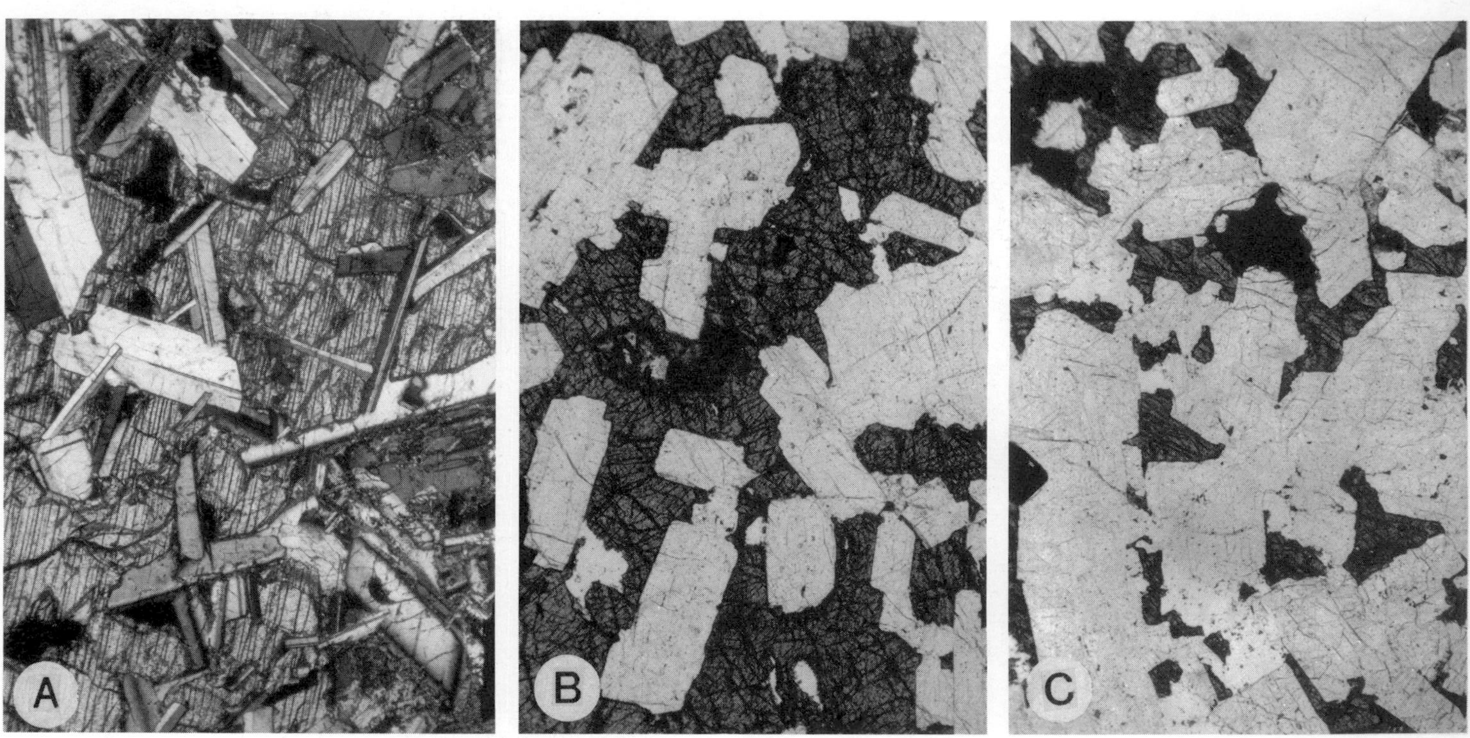

Fig. 1.17 Ophitic texture. (A): Dolerite, Tholei, Germany. (B) and (C): Gabbro, Banks Peninsula, NZ. (View lengths measure 3.2 mm (A), (B) and (C).)

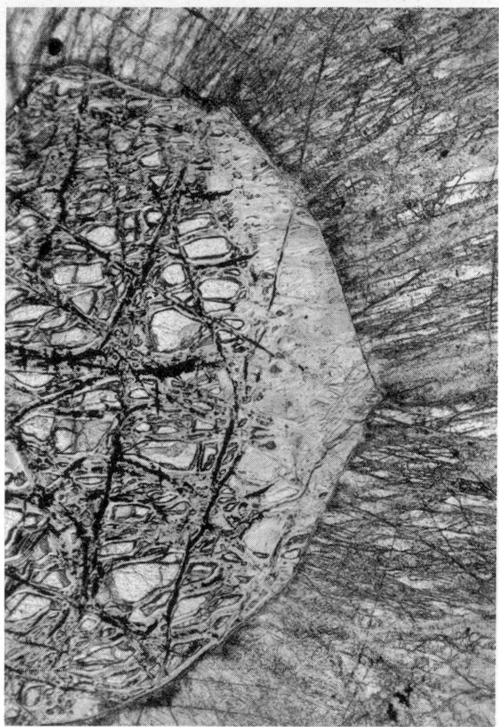

Fig. 1.18 Radiating expansion cracks in plagioclase surrounding serpentinized olivine in troctolite. (Length of view measures 3.3 mm.)

1.7.5 Fields 11–14 (foid syenite to foid gabbro or foid diorite)

Rocks of fields 11–14 plot between 10% and 60% F in the APF triangle. They occur in association with alkaline volcanics in a wide variety of geological settings, and are commonly phase layered in large intrusive complexes (e.g., Parsons, 1987). Some metasomatic 'fenites', formed by desilication of crustal rocks around carbonatite complexes, also fall into these compositional fields.

Mineralogy and terminology: Root-names are **foid syenite**, **foid monzosyenite**, **foid monzodiorite** (or **foid monzogabbro**), and **foid diorite** (or **foid gabbro**), and their positions within the APF triangle are given on Fig. 1.13. Use of diorite versus gabbro is on the basis of plagioclase An % being <50 or >50. **Foid plagisyenite** is a synonym for foid monzodiorite. In practice the word 'foid' is replaced by whichever feldspathoid is dominant: hence **nepheline syenite**, **can-**

crinite monzosyenite. As well as feldspars and the feldspathoids, they may contain any of a wide variety of mafic minerals, varying from the more common types of clinopyroxene (augite) and amphibole to the strongly alkaline species. In addition, melanite garnet is frequently present; relatively unusual minerals such as eudialyte (or eucolite) and aenigmatite may be abundant; peraluminous varieties may contain muscovite and corundum; varieties associated with carbonatite often contain carbonate. Accessory minerals may include a wide variety of rare Zr- and Ti-bearing minerals as well as the more usual apatite, titanite, zircon and opaque Fe–Ti oxides. Not surprisingly, such a wide possible variety of mineral compositions has attracted numerous varietal names. While not wishing to encourage the pursuit of name collecting, mention must be made of the following best-known varieties: **foyaite**, a nepheline syenite with a conspicuous alignment of platy alkali feldspars; **agpaite**, a peralkaline nepheline syenite with Zr–Ti-rich minerals such as eudialyte; **malignite**, a mesocratic aegirine-augite-rich nepheline syenite; **shonkinite**, a melanocratic augite-rich nepheline syenite; **miaskite**, a leucocratic biotite oligoclase nepheline monzosyenite; **essexite**, a nepheline monzodiorite or nepheline monzogabbro with titanian augite and Ti-rich hornblende or biotite; **teschenite**, an analcime gabbro; **theralite**, a nepheline gabbro.

Distinctive textures and structures: The enormous variety of mineralogical combinations possible in rocks of fields 11–14 results in a similarly wide variety of textural types. In general, the granitic type of texture naturally dominates, as in most plutonic rocks, that is, a mixture of euhedral, subhedral and anhedral grains joined together in an arrangement which is even grained when compared with volcanic equivalents. Euhedral tabular or prismatic phenocrysts are often aligned more-or-less parallel to intrusive contacts or phase layering. In the more-common nepheline syenites, euhedral nepheline may be poikilitically enclosed within alkali feldspar, and alkali feldspars are generally perthitic. Of the 'foid' minerals, nepheline, sodalite and leucite are primary and commonly euhedral; in contrast, cancrinite and analcime may be anhedral, late-stage igneous or secondary, replacing feldspars or even other 'foids'. Leucite (or primary?‡ analcime) may be replaced by complex intergrowths of nepheline, feldspar and analcime (called pseudoleucite).

1.7.6 Field 15 (foidolites)

Rocks of field 15 plot at >60% F in the APF triangle. Foidolites most commonly occur in association with carbonatites, kimberlites and melilitic

‡ See footnote to section 1.1.2

rocks in intra-plate oceanic and continental settings, and they are particularly well known from the East African Rift system (Le Bas, 1977). Metasomatic fenites may include rocks of field 15 composition.

Mineralogy and terminology: 'Foid' in foidolite can be replaced by the name of any dominant feldspathoid: hence **nephelinolite** or **leucitolite**. Depending on the relative proportions of F and M-group minerals, the following names can be used. If the dominant foid is nepheline: **urtite** (M < 30%), **ijolite** (M = 30–70%), and **melteigite** (M > 70%). If leucite is the dominant foid: **italite** (M < 30%), **fergusite** (M = 30–70%), and **missourite** (M > 70%). Ijolite is sometimes loosely used for all those types with nepheline dominant. The most common M-group mineral is aegirine or aegirine augite, but Na-rich hornblendes, alkali amphiboles, biotite, melanite garnet and melilite are also found. A diverse range of accessory minerals (including apatite, titanite, eudialyte, carbonates, perovskite and various opaque minerals) are found in these alkaline rocks, and sometimes they are in sufficient quantity to be classed as major constituents of the rock. For example, almost pure apatite layers alternate with nephelinolite in the Kola Peninsula alkaline rocks (Kogarko, 1987). Secondary replacement minerals include wollastonite, cancrinite, andradite, pectolite and zeolites; some wollastonite seems to be primary.

Distinctive textures: Given the wide range of possible mineral components, textures vary considerably in detail. In general, it is still the granitic-type texture – the mixture of euhedral to anhedral grains of similar size – that characterizes most of these rocks. Leucite (or primary?[§] analcime) may be replaced by complex intergrowths of nepheline, feldspar and analcime (called pseudoleucite).

1.7.7 Ultramafic plutonic rocks

These rocks contain >90% of various mixtures of olivine, pyroxene, and hornblende, possibly biotite or phlogopite, and often garnet or spinel. They dominate the upper mantle of the Earth, and are found most commonly at the Earth's surface in ophiolite complexes. In addition, ultramafic plutonic rocks occur in association with gabbroic and anorthositic rocks, particularly in phase-layered intrusive complexes (e.g., Parsons, 1987). The latter are clearly igneous, but mantle rocks are mainly metamorphic in character (e.g., Nicolas and Violette, 1982). Regardless of their ultimate origin, mantle peridotites undergo cycles of deformation and recrystallization, sometimes accompanied by partial melting, related to plate movements. Exotic (mantle) and cognate (cumulate) ultramafic xenoliths

[§] See footnote to section 1.1.2.

are found in volcanic rocks, particularly in association with alkali basalt, and in diatremes, especially those associated with kimberlite. In kimberlite, most xenoliths are garnet lherzolite, in alkali basalt most are spinel lherzolite.

Mineralogy and terminology: The IUGS subcommission recommends a terminology based on the relative proportions of olivine, orthopyroxene, clinopyroxene and hornblende, as given in the two triangles of Fig. 1.19. General names are **peridotite** for rocks with >40% olivine, and **pyroxenite** or **hornblendite** for rocks with <40% olivine and mainly pyroxene or hornblende respectively. Peridotites are subdivided into **dunite** (pure olivine peridotite), **harzburgite**, **lherzolite** and **wehrlite**, pyroxenites into **orthopyroxenite**, **websterite**, and **clinopyroxenite**. If garnet or spinel are <5% the rock is qualified as in **garnet-bearing peridotite**; if >5% the name becomes instead **garnet peridotite**. Olivine is usually Fo-rich, especially in those peridotites depleted in Fe and Ti by partial melting (where Fo > 90%). The spinel in peridotite is typically chromite, sometimes concentrated in economically important layers, and pyroxenes may be significantly enriched in chromium, as in chrome diopside.

Distinctive textures and structures: Some general features include the exsolution lamellae developed in pyroxene (section 3.5.3), and the propensity of these rocks to alter partially or completely to serpentine. Distinction between igneous and metamorphic ultramafic plutonic rocks is not always easy, in part because olivine is susceptible to plastic deformation and recrystallization which means essentially igneous rocks can easily develop secondary minor metamorphic features. In addition,

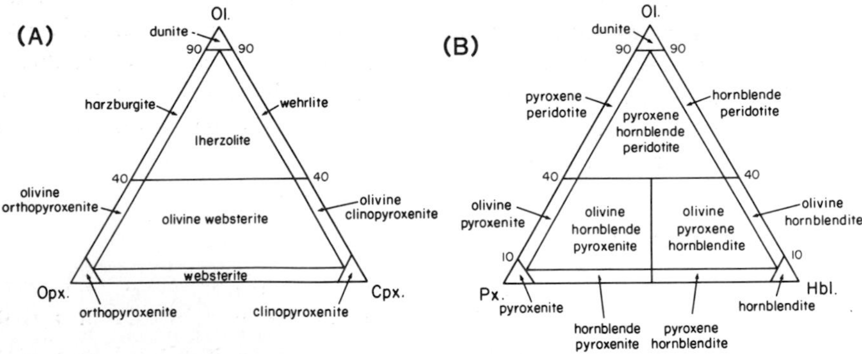

Fig. 1.19 Terminology for rocks with M > 90% (IUGS recommendations from Le Maitre, 1989 and Streckeisen, 1976) based on contents of olivine (Ol), orthopyroxene (Opx), clinopyroxene (Cpx), pyroxene (Px), and hornblende (Hbl).

the deep-seated and high-temperature origin of many ultramafic rocks makes it possible for significant annealing and textural maturation to take place (section 4.2.2 – Hunter, 1987), thus blurring distinctions between what is igneous and metamorphic.

Igneous features: A clear indication of an igneous origin is association with gabbroic rocks in a layered complex (section 4.2.2). Texturally such rocks tend to be coarse grained, and characterized by at least some development of olivine or pyroxene crystal faces (Fig. 1.20A). In pyroxenites, euhedral olivine may be poikilitically enclosed by pyroxene in an orthocumulate-type texture (Fig. 1.20B). Adcumulate-type textures may mimic a metamorphic mosaic, and a study of lattice preferred-orientations (Part Three) then may be necessary to determine origin.

Metamorphic features: Strain features such as the common {100} kink-band boundaries or subgrain boundary development are so easily produced that they do not in themselves signify the rock is 'metamorphic'. Mosaics due to dynamic recrystallization may be confused with adcumulate-type textures, but in general the metamorphic mosaics are finer grained, totally lack crystal shapes and faces, have distinctive lattice preferred-orientations (Part Three), and are associated with porphyroclasts, possibly of olivine (Fig. 1.21) but more commonly of pyroxene (Fig. 1.22) and garnet. Fine-grained dunite mylonites (Fig. 1.21) are particularly distinctive. Pyroxenes are significantly less ductile than olivine, and hornblende is less ductile again, and they typically form relatively large undeformed porphyroclasts set in a finer recrystallized mass of olivine. In some cases, pyroxene tails of ultra-fine-grained material behaved superplastically (Fig. 1.22), as discussed in section 5.3.3, indicating high-temperature deformation; the lack of subsequent grain coarsening suggests deformation in association with magma formation in the mantle, and rapid rise to the surface with that magma.

Textures related to partial melting: Superplastic textures may relate to deformation connected with partial melting of the mantle. More direct evidence can be found in mantle xenoliths with films of glass along grain boundaries or other clear evidence of melt generation such as spongy textures (section 4.1.7e).

1.8 VOLCANIC ROCKS

'Volcanic rocks' refers to those not already dealt with as special cases (carbonatite, melilitite, etc.). Volcanic rocks pose a special problem in that many contain glass or are so fine grained that a mode cannot be determined, and the only proper way to proceed further is with a chemical analysis. This is so frequently a problem that chemical classifications of volcanic rocks have become more important, certainly more useful, than

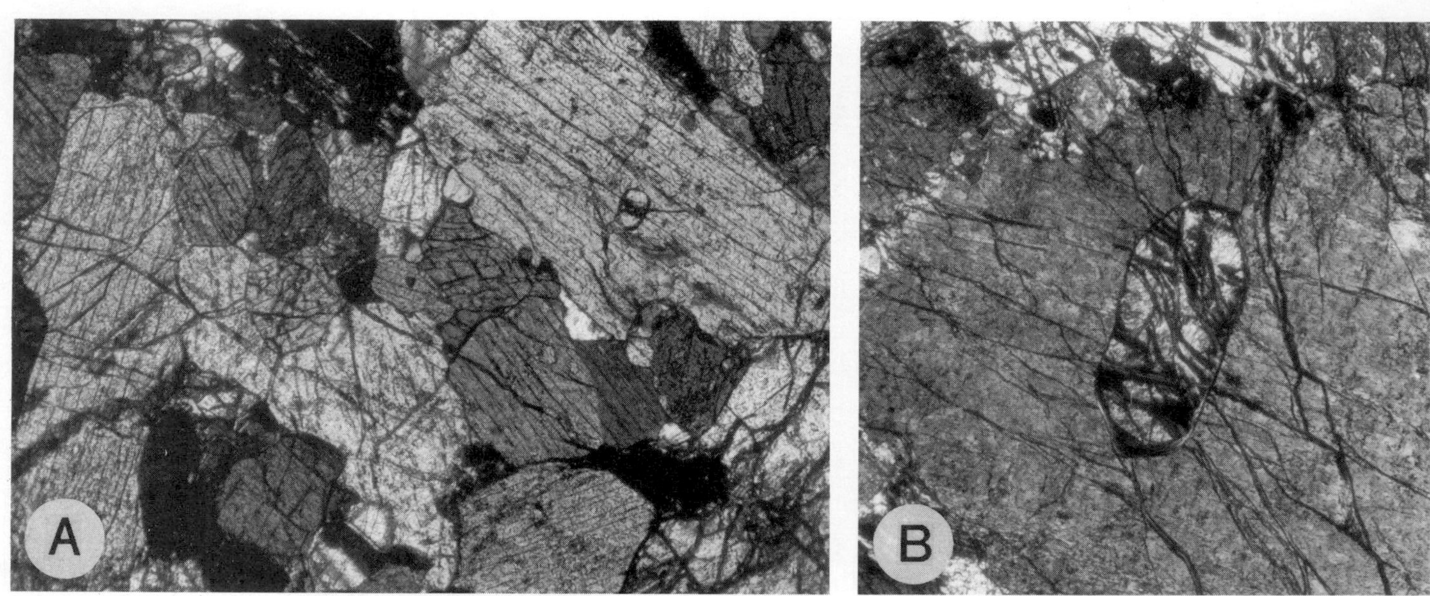

Fig. 1.20 (A): Subhedral pyroxene in pyroxenite indicates igneous crystallization. Note also the small euhedral olivine enclosed in the largest pyroxene. (B): Subhedral olivine enclosed in pyroxene in pyroxenite. View lengths measure 3.0 mm (A) and 1.7 mm (B). Both specimens from Bluff, NZ.

mineralogical ones. Some names such as basalt and rhyolite (or the field equivalents basaltic, rhyolitic) are used regardless of whether the classification is mineralogical, chemical, or field-based, and the student must learn to cope with the subtle differences in meaning in each case. Basalt, for example, as defined mineralogically encompasses both basalt and hawaiite as defined chemically.

The presence or absence of glass, and the general grain size can have such a marked effect on colour that the idea of using colour index is not a useful one for volcanic rocks. Thus a crystalline rhyolite or trachyte may be white or pale in colour, and glassy varieties may be black; fine-grained trachybasalt is usually very dark, but well-crystallized dyke samples can be pale grey.

There is always a need for preliminary field classifications, and indeed, for volcanic rocks, I would suggest that it is preferable to use only the field classification if microscopic examination does not yield sufficient modal

Fig. 1.21 Dunite mylonite (Anita Bay, NZ) with deformed olivine porphyroclast surrounded by recrystallized fine-grained olivine. (View length measures 3.3 mm.)

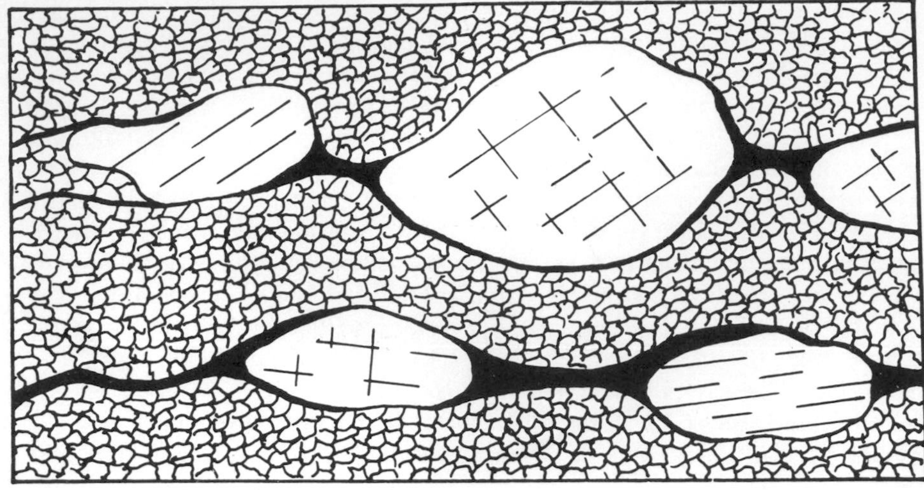

Fig. 1.22 Schematic illustration of orthopyroxene porphyroclasts (*c.* 1–2 mm long) set in a recrystallized mosaic of olivine. Ultrafine pyroxene, less than 10 microns in grain size, forms tails and rims to the pyroxene. As described by Boullier and Gueguen (1975).

data and a chemical analysis is not available. Even if a chemical analysis is available, one may need to fall back on the field classification if the rock is significantly altered. I therefore present field classifications first, go on to outline the approved chemical classification, and finally present the full mineral-based classification, commenting throughout on the typical petrography of chemically defined names.

1.8.1 Field classification, and classification for microscope work with glassy and otherwise difficult specimens

The classification is based on the QAPF double triangle (Fig. 1.23), and uses the most familiar rock names but with the suffix -oid for nouns, or -ic for adjectives; hence **foiditoid** or **foiditic rock**. As explained above, it may not be feasible to use anything more sophisticated than the 'field' classification. The precision with which one can classify difficult materials is clearly in inverse relationship to the amount of difficult material present. For example, a thoroughly glassy rock with the occasional phenocryst of alkali feldspar would probably be classified as a 'trachytoid', but this would certainly leave a great deal of uncertainty. When dealing with altered material it is essential to have a good knowledge of the typical alteration products and shapes of important igneous minerals.

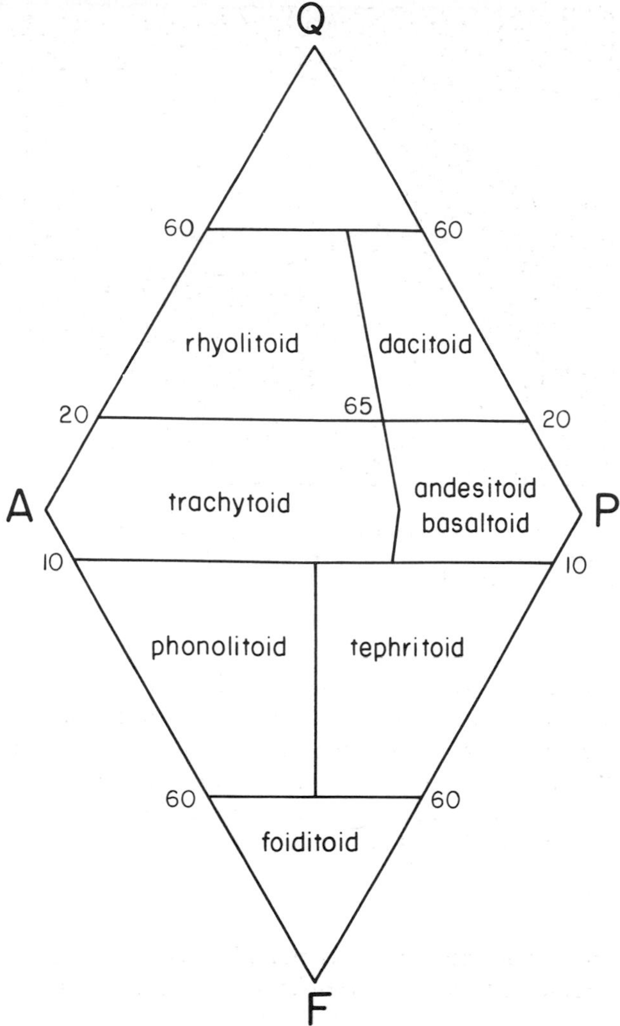

Fig. 1.23 IUGS recommendations (Le Maitre, 1989) for a provisional (field) classification of volcanic rocks that plot in the double QAPF triangle.

Mineralogy and terminology: Rhyolitoids and **dacitoids** may have quartz phenocrysts plus alkali feldspar and/or plagioclase in the appropriate proportions. Either may have siliceous vesicle infillings, and mafic minerals are likely to include orthopyroxene, hornblende, biotite, or garnet (varieties that are colourless in thin section). **Trachytoids**, **andesitoids** and **basaltoids** do not have phenocrysts of quartz or 'foids', and the choice of name is on the basis of the relative proportions

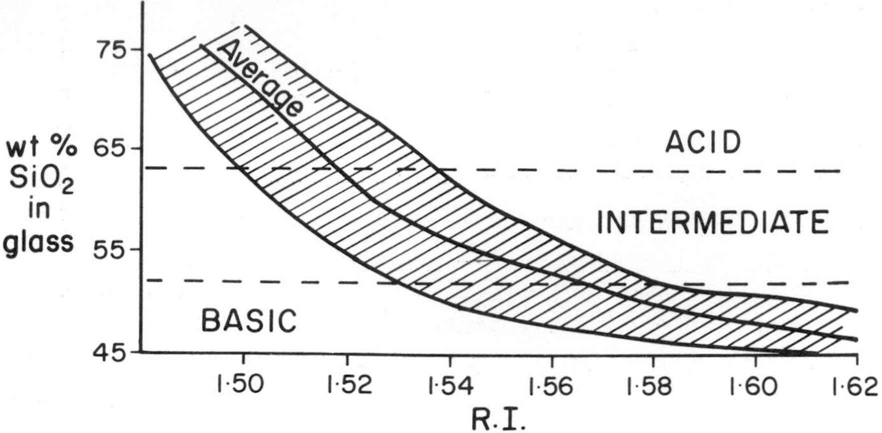

Fig. 1.24 Refractive indices of natural volcanic glasses (based on George, 1924). The broad range for any one silica content reflects mainly the effect of water. More accurate determinations can be made using calibrated curves for artificially fused glasses for specific volcanic suites (Mathews, 1951).

of determinable A and P. Mafic phenocrysts most likely to be present are: (altered) olivine and/or augite in basaltoids; augite and orthopyroxene, or hornblende in andesitoids; aegirine augite, hornblende or biotite in trachytoids. Basaltic glass alters readily to green–brown clayey palagonite. **Phonolitoids**, **tephritoids** and **foiditoids** are named on the basis of the visible proportions of A, P and F, and mafic phenocrysts possibly include Ti and/or alkali-rich varieties of pyroxene, amphibole, and biotite or phlogopite.

Volcanic glass–refractive index (RI) as a measure of SiO_2: Determination of glass RI is a useful adjunct to the field or provisional classification. An inverse relationship exists between RI and SiO_2 (Fig. 1.24), and although a precise correlation does not exist, it is sufficiently precise for glasses to be approximately categorized as **acid** (SiO_2 > 63%), **intermediate** (SiO_2 = 52–63%), or **basic** (SiO_2 = 45–52%). Rock made up of >80% acid glass is known as **obsidian**, or **pitchstone** if the water content is >4%. Basic glass is known as **tachylyte**. More generally, glass can be qualified by the name of its crystalline equivalent, as in **trachytic glass**, for example. **Pumice** is the term for highly vesicular and usually acid glass that will float in water. **Scoria** is a less vesicular and usually more basic glass.

1.8.2 The TAS chemical classification

TAS refers to the Total Alkali Silica diagram which is the basis of the classification, although in addition to potash, soda and silica, one also

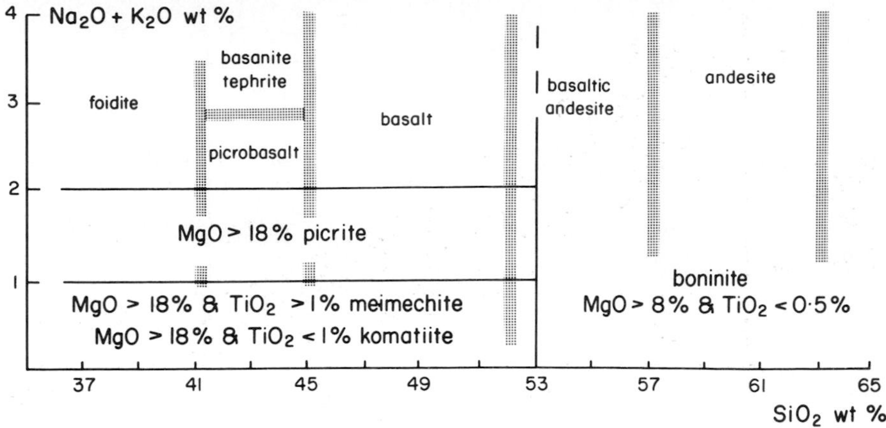

Fig. 1.25 Terminology of 'high-Mg' volcanic rocks (picrite, komatiite, meimechite and boninite) using the TAS chemical classification plus wt. % MgO and TiO_2 (IUGS recommendations from Le Maitre, 1989). The thick stippled lines indicates TAS fields (Fig. 1.26).

requires a CIPW norm, as well as MgO, total iron and alumina contents. Analyses are recalculated to 100% on a volatile-free basis. Details are given in Le Maitre (1989), from which the following is a summary.

TAS step 1: If the rock is unusually rich in Mg, it is either a **picritic rock** (SiO_2 < 53%) or a **boninite** (SiO_2 > 53%). Picritic rocks are further subdivided into **picrite**, **komatiite** or **meimechite**, as shown in Fig. 1.25.

TAS step 2: Other rocks are classified in the TAS diagram (Fig. 1.26). Further subdivisions include: **basalt** into **alkali basalt** that contains normative nepheline and **subalkali basalt** that does not; the group **trachybasalt** to **trachyandesite** is subdivided further according to the Na/K content, as shown in Table 1.3; the group **basalt** through **andesite** to **rhyolite** can be described as **low-**, **medium-** or **high-K**, as shown in Fig. 1.27; **rhyolite** and **trachyte** are qualified as **peralkaline** if $(Na_2O + K_2O)/Al_2O_3 > 1$; **peralkaline rhyolites** and **trachytes** can be subdivided into **comendite**, **pantellerite** or **comenditic or pantelleritic trachyte** as shown in Fig. 1.28; **trachyte** and **trachydacite** are distinguished on the basis of normative quartz being <20% or >20%. Le Maitre (1989) notes that the boundaries between foidite and basanite–tephrite are unsatisfactory, and the subcommission will be making further recommendations on this.

1.8.3 The QAPF classification for volcanic rocks (Fig. 1.29)

This classification is to be used if M < 90%, and if a complete or nearly complete modal determination is available. As discussed above, the reality

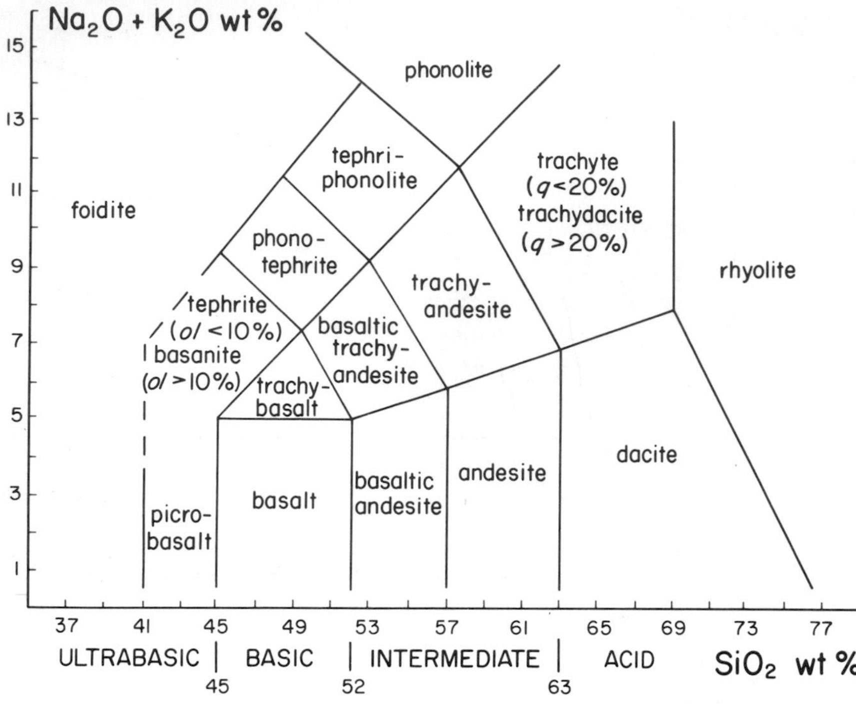

Fig. 1.26 TAS (total alkali vs. silica) diagram for the chemical classification of volcanic rocks (IUGS recommendations from Le Maitre, 1989, based on Le Bas *et al.*, 1986, and by permission of Oxford University Press). The dashed line separating foidite from tephrite and basanite indicates an unsatisfactory boundary requiring further definition. Key: ol = normative olivine, q = normative quartz.

Table 1.3 Further subdivision of trachybasalt and trachyandesite

	Trachybasalt	Basaltic trachyandesite	Trachyandesite
$Na_2O - 2.0 \geqslant K_2O$	Hawaiite	Mugearite	Benmoreite
$Na_2O - 2.0 \leqslant K_2O$	Potassic trachybasalt	Shoshonite	Latite

is that many volcanic rocks do not lend themselves easily to accurate modal analysis, and many researchers exclusively use the TAS classification instead.

The QAPF 'fields' are essentially the same as for plutonic rocks (Figs 1.11 and 1.13), but fields 4 and 5 and 9 and 10 are not separated, whereas field 15 is further subdivided for volcanic rocks.

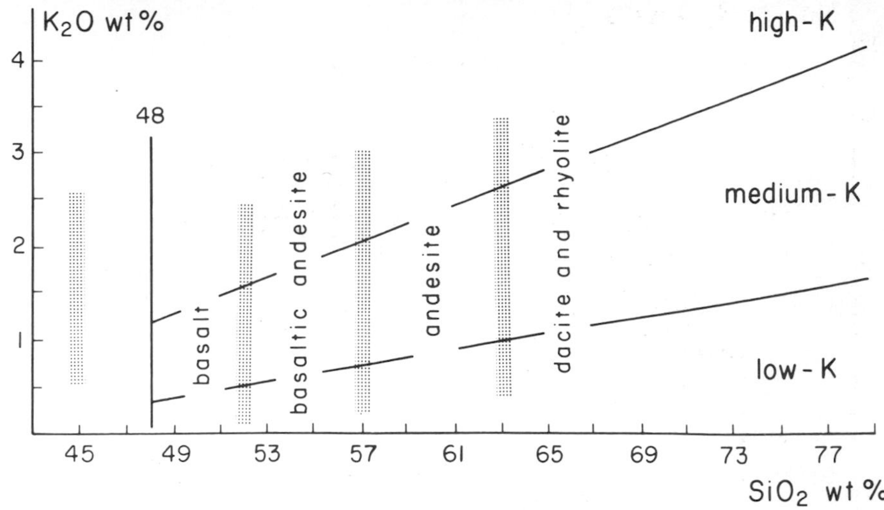

Fig. 1.27 Definition of low-, medium-, and high-K types of basalt through andesite to rhyolite (IUGS recommendations from Le Maitre, 1989). Thick stippled lines indicate TAS fields (Fig. 1.26).

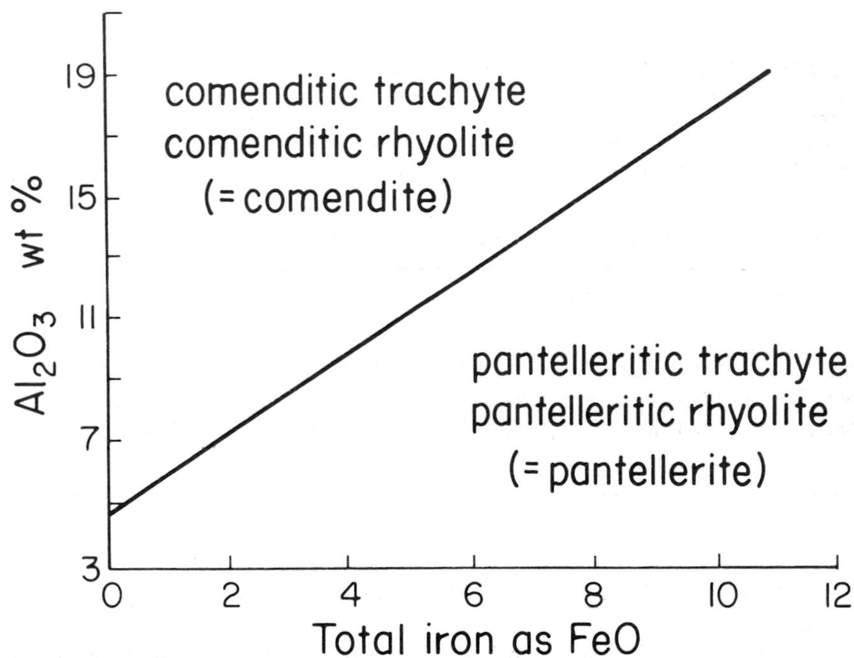

Fig. 1.28 Definition of comenditic and pantelleritic types of trachyte and rhyolite (IUGS recommendations from Le Maitre, 1989).

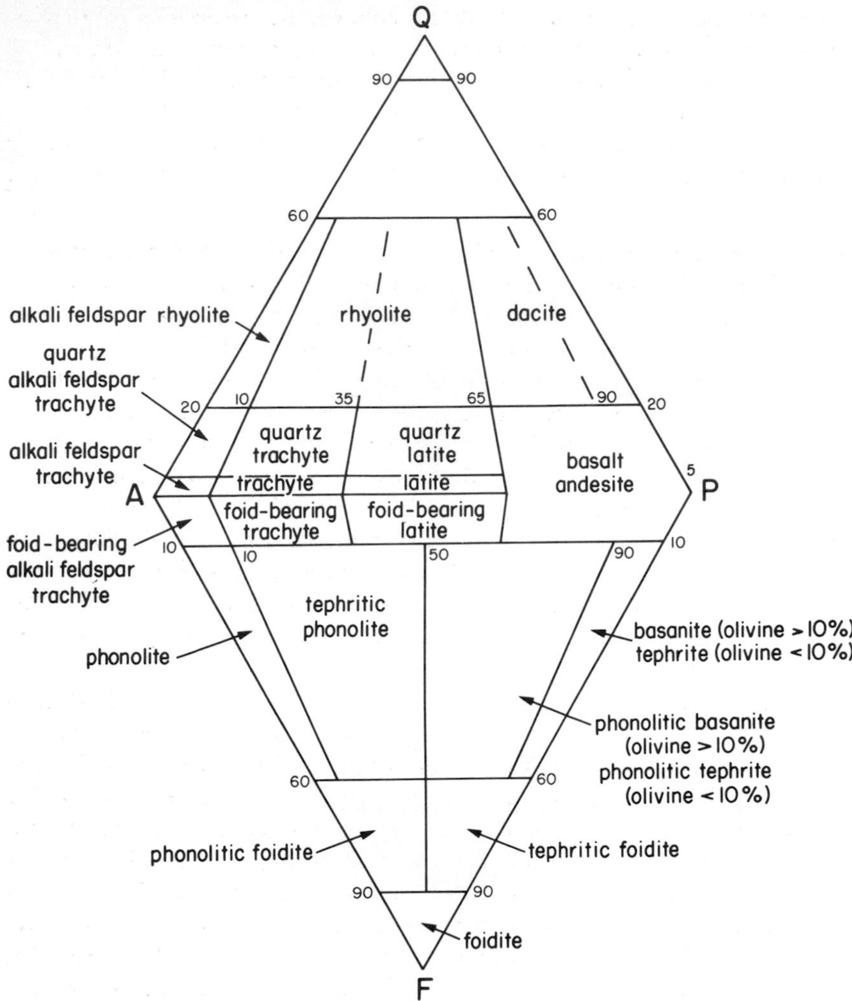

Fig. 1.29 Terminology of volcanic rocks with M < 90% and using the QAPF double triangle (IUGS recommendations from Le Maitre, 1989).

1.8.4 Fields 2–5 (rhyolitic and dacitic rocks)

Rhyolites and dacites are the volcanic equivalents of granitic rocks, and like granitic rocks they occur principally in continental areas above active subduction zones where they are the result of crustal melting. The main exceptions to this are the peralkaline rhyolites (comendite, pantellerite), found in continental rift-valley systems and as differentiates of alkali basalt in oceanic island volcanic settings. In the continental setting, rhyolitic/

dacitic magma is erupted in batholithic volumes but mainly as pyroclastic material, either as extensive ignimbrite sheets or widely scattered air-fall tephra. Smaller volumes are erupted or intruded near the surface in the form of rhyolite or dacite domes. Compared with subjacent granitoids, a disproportionately small volume of rhyolitic/dacitic material is preserved in the geological record because much of it is easily eroded away in active orogenic belts.

Mineralogy and terminology: Root-names are **alkali feldspar rhyolite**, **rhyolite** and **dacite**, as defined in Fig. 1.29 on the basis of >20% Q and varying proportions of A and P. Possible phenocrysts include quartz (typically in the shape of the bipyramidal high-temperature form), alkali feldspar and/or sodic plagioclase (usually An < 50%), orthopyroxene (especially in dacite), hornblende or biotite. Occasionally, plates of tridymite occur instead of quartz, but these are often replaced by mosaic quartz on cooling. Most of these volcanics are glass bearing, or have a groundmass that crystallized at high degrees of undercooling. If glass is >80% the rock is called **obsidian**, or if the glass is water rich (>4%), **pitchstone**. The peralkaline rhyolite **comendite** typically contains alkali amphiboles, aenigmatite, aegirine and biotite, and **pantellerite** usually contains aegirine augite and aenigmatite; fayalite may also be present.

Distinctive textures: These are discussed in detail in section 4.1. Phenocrysts represent pre-eruptive crystallization at a low degree of undercooling. Quartz phenocrysts are commonly embayed (Fig. 1.30A), and this may be a disequilibrium growth form or a solution effect due to decrease in pressure as magma rose to the surface (section 4.1.6). Phenocrysts may be set in glass or a fine-grained crystalline matrix containing spherulitic (Figs 1.30B and 4.9), snowflake (Fig. 4.8) or felsitic textures (Fig. 4.10) that represent high degrees of undercooling (section 4.1.4). Glass may display perlitic cracking (Fig. 1.30C), and this spheroidal cracking may be preserved even after crystallization of the glass. Vesicles and lithophysae are typically filled with fine-grained silica or celadonite. Flow layering is common (section 4.1.7).

1.8.5 Fields 6–10 (trachytes, latites, andesites and basalts)

Volcanic rocks that plot near the A–P join of the QAPF double triangle are more abundant and conspicuous in the geological record than all other volcanic types. To some extent this reflects the fact that rocks of fields 6–10 dominantly form massive flows, contrasting with the often unlithified pyroclastics and widely scattered tephra typical of voluminous but easily eroded rhyolitic products. Basalts are formed in a very wide

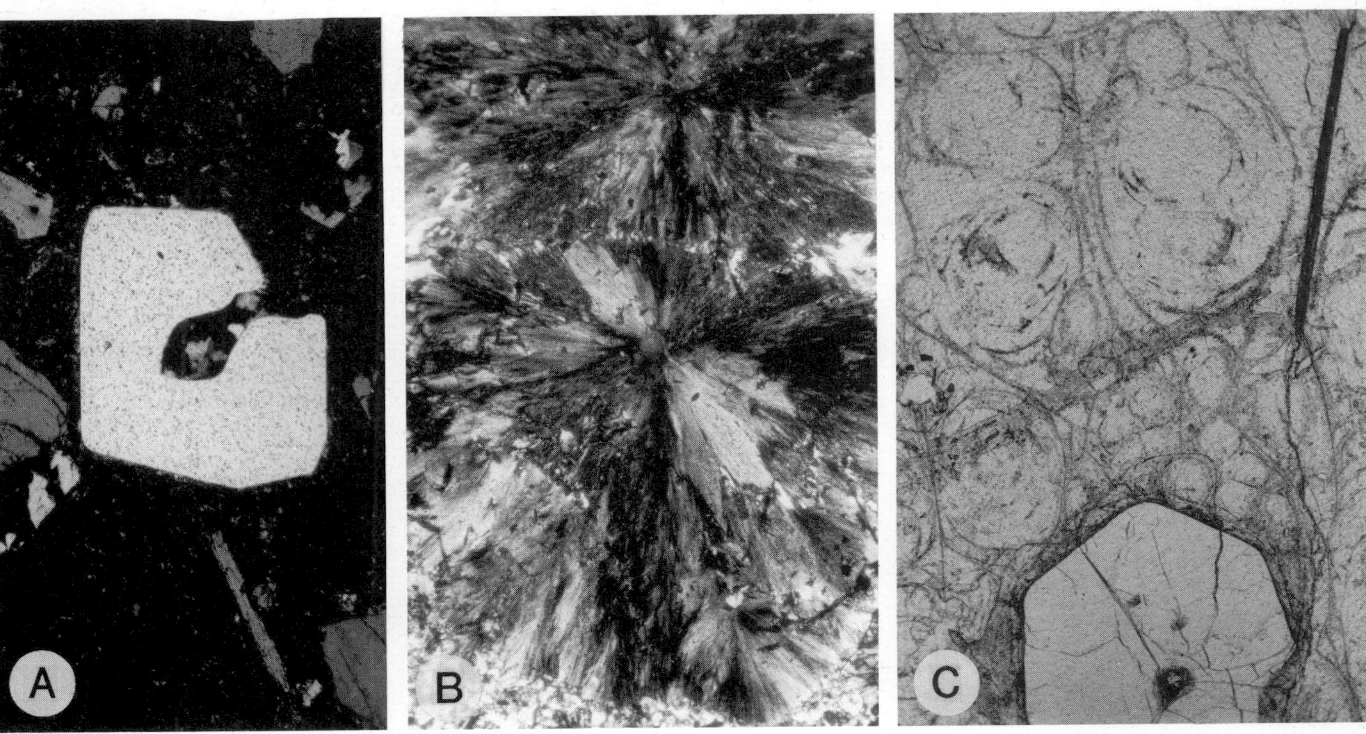

Fig. 1.30 Textures in rhyolite. (A): Embayed quartz phenocryst. (B): Spherulites. (C): Perlitic cracking of glass (note also the quartz phenocryst). (View lengths measure 3.2 mm (A), (B) and (C).)

range of environments including mid-oceanic ridges, above subduction zones, and intra-plate continental and oceanic volcanic settings. Andesites are typically found in volcanic arcs erupted above subduction zones in both oceanic and continental environments. Latites (perhaps the most unsatisfactory term in the IUGS recommendations – shown below) and trachytes are typically found in association with and in the same wide range of settings as basalt.

Mineralogy and terminology: **Basalt** and **andesite** both plot at the P corner of the QAPF diagram, and the difference between them has never been satisfactorily defined petrographically. Pointers to andesite include a paler colour than basalt, the presence of orthopyroxene (often coexisting with augite), and the lack of abundant olivine, but these pointers are far from infallible. Some andesites contain significant amounts of hornblende, possibly accompanied by augite or biotite, as, for example, in the Taranaki andesites of New Zealand. These same andesites also contain plagioclase that overall has An $<50\%$ (phenocrysts zoned from labradorite to andesine plus groundmass andesine), similar to the plutonic equivalent diorite, but more commonly the only determinable plagioclase in an andesite is more calcic than An_{50} (often bytownite), and it is not always feasible to use An % as a determinative guide. Andesites often contain mixed phenocryst assemblages that display complex zoning patterns suggesting magma mixing (section 4.1.6). **Boninites** (TAS) are essentially Mg-rich andesites, and are usually glassy (SiO_2-rich), with abundant phenocrysts of Mg-pyroxene and olivine. **Subalkali basalt** (TAS) is called **tholeiitic basalt** or **tholeiite** petrographically, contains a Ca-poor pyroxene such as pigeonite or orthopyroxene and if completely crystalline usually contains interstitial granophyric intergrowth; olivine is absent from the groundmass and any olivine present as phenocrysts may show evidence of a reaction relationship with Ca-poor pyroxene. **Alkali basalt** (TAS) is petrographically **olivine basalt** with abundant olivine and augite (often purple–brown and Ti rich), as phenocrysts and in the groundmass. **Hawaiites** (TAS) are feldspar-rich basalts usually containing phenocrysts of labradorite, groundmass laths of andesine and interstitial anorthoclase. Vesicles in basalt are most often filled with zeolites. **Picrite** (TAS) plots in the basalt field petrographically, but contains more abundant mafic material than common basalt: it can be subdivided into **ankaramite**, rich in titanian augite, and **oceanite**, rich in olivine. **Spilite** is basalt altered by low-grade metamorphism or interaction with sea-water to a greenschist or sub-greenschist-facies assemblage of albite (replacing calcic plagioclase) and chlorite and epidote replacing the mafic material and groundmass (section 2.3.1). Calcite is common in spilite, especially filling vesicles. **Latite** is perhaps the most inappropriate term in the QAPF diagram. It is usually considered to contain K-rich feldspar, as indicated in the TAS use of the name (Table 1.3), and this means no provision is made for the Na-rich alkali-feldspar-bearing mugearites and benmoreites that plot in field 8, but which few petrographers would call latite. It seems to me that the best solution is to use the general TAS term **trachyandesite** rather than latite for rocks that plot in field 8: K-feldspar-bearing varieties should be ascribed provisionally to the **shoshonite–latite** series, anorthoclase-bearing varieties to the **mugearite-benmoreite** series. **Trachytes** may contain any kind of alkali-feldspar from Na-rich anorthoclase to K-rich sanidine. Usually feldspar is completely dominant; subsidiary mafic material may include aegirine augite, hornblende or biotite. **Alkali feldspar trachyte** is the root-name for trachytic rocks that contain virtually no anorthite component in the feldspars. The petrographic indicators of **peralkaline trachyte** (TAS) are alkali pyroxenes and/or amphiboles, and possibly aenigmatite and fayalite. Rocks of fields 6–8 may be qualified by 'quartz' or 'foid-bearing' (e.g., **quartz trachyte** or **nepheline-bearing trachyte**) as indicated in Fig. 1.29.

Distinctive textures: Rocks of fields 6–10 are dominated by feldspar, except for the most ultramafic picritic varieties. Textures are therefore dominated by the way that inequidimensional feldspar crystals are arranged. If they are aligned due to flow, the texture is called **trachytic** (Fig. 1.31A). Rocks that display a marked trachytic texture in sections parallel to the flow direction may show an apparent random texture in sections perpendicular to flow (Fig. 6.2). The arrangement whereby relatively small equidimensional pyroxene and opaque minerals are set between the dominant feldspar laths is called **intergranular texture** (Fig. 1.31B), and this is particularly characteristic of basalt. If pyroxenes are larger and envelop feldspar laths, intergranular texture merges into **subophitic** and **ophitic texture** (Fig. 1.17). The presence of glass between the feldspar laths leads to **intersertal texture**, and if glass dominates over feldspar, this merges into **hyalopilitic texture**. Rapidly quenched varieties may contain swallow-tail plagioclase and pyroxene (section 4.1.3). **Spherulitic** and **felsitic textures** are not common in basaltic rocks, but do occur in some trachytes. Of the mafic minerals, olivine is most commonly altered, typically to iddingsite and/or bowlingite (section 3.5.4) in the primary magmatic setting (or possibly during later alteration), and to serpentine as a result of later metamorphism. Olivine, hornblende and biotite often display evidence of reaction with magma: biotite and hornblende may be lined or crowded with reaction products, often opaque, as shown in Figs 1.31C and D, and reaction may produce 'corroded' shapes (Fig. 1.31D). However,

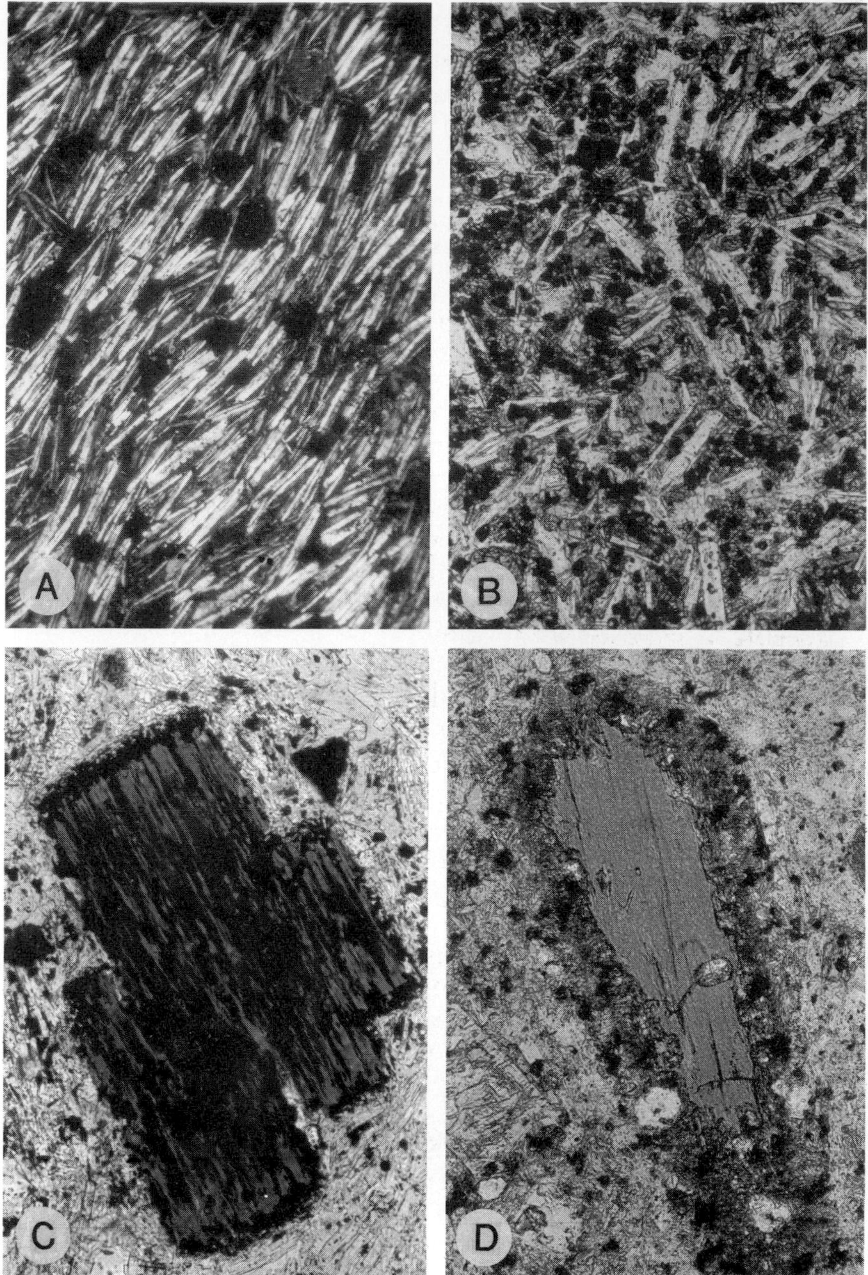

Fig. 1.31 (A): Trachytic texture in trachyte. (B): Intergranular texture in basalt. (C): Biotite crystal in trachyte corroded and crowded with opaque reaction products, especially near the margin of crystal. (D): Hornblende phenocryst in dacite with corroded shape and lining of reaction products. (View lengths measure 3.3 mm (A) and 0.85 mm (B), (C) and (D).)

'corroded' shapes may sometimes be due to disequilibrium growth rather than solution (section 4.1.6). Corroded olivine may be encased in orthopyroxene which precipitated as olivine was dissolved, but more commonly any such orthopyroxene is dispersed.

1.8.6 Fields 11–14 (phonolites, tephrites and basanites)

These feldspathoidal rocks occur in alkaline volcanic provinces in a wide variety of geological settings, but are best known in either the continental or oceanic intra-plate setting.

Mineralogy and terminology: Most **phonolite** is sodic and contains Na-rich feldspathoids and alkali feldspar; use of the name without qualification implies the presence of nepheline ± sodalite. If other feldspathoids or analcime dominate, the name should be qualified, as in **leucite phonolite**. Mafic minerals typically include aegirine augite or aegirine, alkali amphibole, aenigmatite and biotite. Common accessory minerals include titanite, Ti-magnetite, zircon, apatite and eudialyte. **Basanite** and **tephrite** are distinguished on the basis of *normative* olivine being >10% or <10%. Petrographically, basanites should have significant modal olivine. The plagioclase in basanite and tephrite is usually very calcic, and pyroxene phenocrysts are purple–brown titanian augite, often zoned to green sodic varieties at the rim. Amphibole and biotite are less common, but if present tend to be alkali- and Ti-rich varieties. Accessory minerals include titanite, apatite and Ti-magnetite. The dominant F-group mineral may be indicated in the name, as in **nepheline tephrite**. **Limburgite** (determined chemically as basanite) lacks modal feldspar, and contains olivine and pyroxene phenocrysts in a glassy base. Intermediate types between phonolite and basanite or tephrite are called **tephritic phonolite** or **phonolitic basanite** or **tephrite**, as shown in Fig. 1.29.

Distinctive textures: The feldspar-rich varieties may display textures similar to the rocks of fields 6–10, but as the 'F' content increases so textures change, the principal effect being a loss of preferred orientation due to the lack of highly inequidimensional mineral grains (Fig. 1.32). This effect helps to explain the massiveness and sonorous tone of typical phonolite when hit with a hammer – hence its name. Foids and feldspars are often present as phenocrysts, and if groundmass feldspar is important it typically has a trachytic texture wrapped around the various phenocrysts. In the basanites and tephrites, olivine and pyroxene commonly form phenocrysts, but the only common mafic phenocryst in phonolite is aegirine augite.

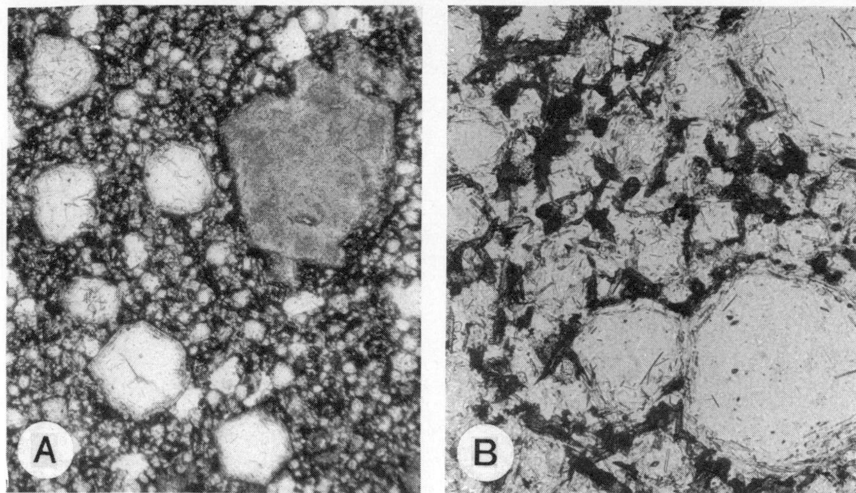

Fig. 1.32 Two examples of foidite that lack anisotropic fabrics due to the dominance of equidimensional feldspathoids (cf. more common volcanic rocks with abundant feldspar laths). (A): One larger sodalite phenocryst (cloudy) and numerous leucite (clear) and nepheline (square) crystals. (B): Mainly leucite phenocrysts with smaller amounts of nepheline (square sections). (View lengths measure 2.6 mm (A) and 0.67 mm (B).)

1.8.7 Field 15 (foidites)

Volcanics dominated by 'foids' are typical of intra-plate continental and oceanic settings. The most common rock type is nephelinite, and according to Le Bas (1987), olivine-poor nephelinites occur in association with carbonatites, as, for example, in the East African Rift system, whereas olivine-rich nephelinites occur in association with basanitic and alkali basalt volcanism.

Mineralogy and terminology: The root-name **foidite** is changed to indicate whichever F-mineral dominates: hence **nephelinite**, **leucitite**, **sodalitite**, etc. **Basanitic**, **tephritic** or **phonolitic foidites** contain >10% but <40% A or P. The most common rock-type nephelinite usually contains olivine and/or clinopyroxene phenocrysts, sometimes in abundance. The olivine-rich ones usually contain purple–brown titanian augite, and the olivine-poor ones a pale-coloured Ti-poor augite, often so abundantly that the rock is called **melanephelinite** because of its dark colour. Phlogopite and melilite are common constituents of nephelinites, and accessory minerals include magnetite, apatite, perovskite and titanite. **Leucitite** is the most common K-rich variety of foidite, and the mafic phases usually include a Ca-rich

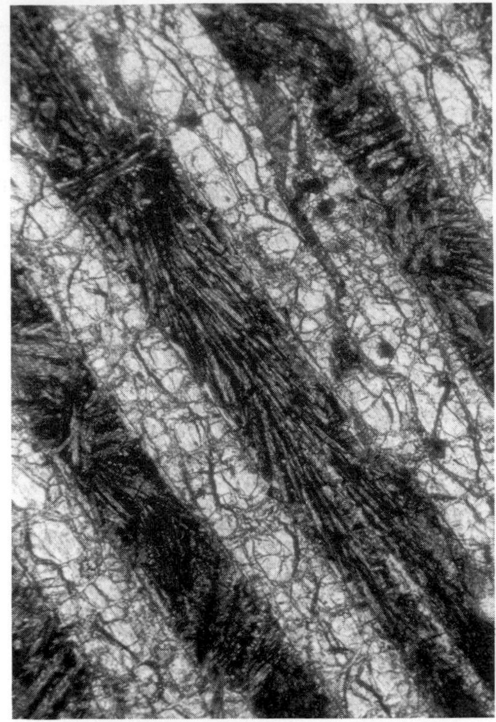

Fig. 1.33 Spinifex texture in komatiite from Ontario, Canada. The lighter areas that extend across the view are elongate olivines; set in between are finer grained and highly elongate clinopyroxenes. (View length measures 3.3 mm.)

pyroxene. Mafic varieties of leucitite can be separately categorized as **lamproite**, and are described as such in section 1.4.

Distinctive textures: Often lacking strongly inequidimensional minerals, the textures of foidites are simply a consequence of the packing of the particular minerals and grain sizes present (Fig. 1.32). Mafic minerals usually form phenocrysts, and nepheline and leucite occur both as phenocrysts and in the groundmass. The primary feldspathoids and melilite, if present, may be altered to cancrinite, zeolites and carbonates.

1.8.8 Ultramafitites (M > 90%)

Ultramafic volcanic rocks are rare except in Archaean greenstone belts, as found in Australia, South Africa and Canada, for example. The varieties **komatiite** and **meimechite** are defined in the TAS system (Fig. 1.25), and while some flows contain ordinary-textured olivine, especially at the base of flows, they are best known for the spectacular spinifex texture, which consists of highly elongate bladed crystals of olivine (Fig. 1.33), sometimes with skeletal forms, that represent very rapid growth during quenching (section 4.1). Clinopyroxenes (Fig. 1.33) and chrome spinels may be set in between the elongate olivines, all of which may be embedded in glass. Olivine-rich rocks associated with komatiites but containing significant feldspar are often loosely referred to as basaltic komatiite, but this additional name seems unnecessary.

2

Metamorphic rocks – an introduction

2.1 INTRODUCTION

Metamorphic rocks, by their very nature, are more diverse and complex in character than igneous rocks. Surprisingly then, metamorphic petrology has not been beset by the same proliferation of names, extant and extinct, as in igneous petrology. In fact, naming a metamorphic rock is a relatively simple and flexible matter, as will be explained below. What isn't so simple is the requirement that one determine the premetamorphic nature of the rock, the facies of metamorphism, and the metamorphic history. But before we deal with any of these matters it is useful to summarize the essential characteristics of metamorphism.

2.1.1 The principal factors in metamorphism

By definition, metamorphism is a structural/textural modification and/or a mineralogical modification of a pre-existing rock, and it involves one or more of the following:

1. Crystallization of new phases (Chapters 3 and 5)
2. Recrystallization of existing phases (section 5.2)
3. Strain of the rock (Chapters 5 and 7) involving principally:
   (a) Removal of material in solution
   (b) Mechanical rotation of mineral grains
   (c) Plastic deformation
   (d) Cataclasis
   (e) Superplastic deformation

All these are discussed in detail in Parts Two and Three.

Normally excluded from the metamorphic arena are such processes as weathering, deuteric alteration, coal formation and diagenesis.

Crystallization, recrystallization and strain all cause the redistribution of chemical elements on one scale or another. It is obvious that a physical deformation produces a new distribution of matter in the rock, as does the growth of a new phase (formation of a garnet porphyroblast, for example, requires certain elements to be moved to and from the growth site). Even recrystallization and polymorphic transformations involve a redistribution of matter on the very small scale. The loss or gain of volatile components may represent a significant change in bulk chemistry, and as volatiles escape or enter a rock they are likely to transport in solution whatever other elements are available. It is clear therefore that the concept of strictly isochemical metamorphism cannot be supported. Even so, the bulk compositions of most metamorphic rocks (on the hand-specimen scale) are still recognizably those of likely precursors (excepting volatile content), and metamorphism to this extent is roughly isochemical.

If some obvious major chemical change does occur (other than volatiles') then the process is called metasomatism.

2.1.2 The main categories of metamorphism

The three principal categories of metamorphism depend on the relative importance of strain versus crystallization and/or recrystallization.

1. **Dynamic metamorphism** produces conspicuous strain effects but no major mineralogical change.
2. **Thermal metamorphism** is a thorough-going crystallization and/or recrystallization not involving major strain. This is also called **contact metamorphism** because it typically occurs close to intrusive igneous contacts.
3. **Dynamothermal metamorphism** involves thorough-going crystallization and/or recrystallization as well as strain, but the direct effects of strain are not conspicuous on the grain scale because of the modifying effects of crystallization and recrystallization. This is also called **regional metamorphism** because it usually occurs on the regional scale.

However, these familiar categories do not adequately cover all kinds of metamorphism, and most importantly there are several distinctive kinds of 'regional' metamorphism that do not fall within the normal use of that term. For example, the incipient but widespread effects of very low-grade metamorphism are not always accompanied by conspicuous strain on the grain or hand-specimen scale, and the term **burial metamorphism** is sometimes used (Coombs, 1961). The changes in illite crystallinity that characterize progressive very low-grade metamorphism of pelitic rocks, again on a regional scale, are often described as **anchimetamorphism** (Frey and Kisch, 1987). The term **incipient metamorphism** usefully describes the incomplete effects of metamorphism, and it is particularly common in the very low-grade regional metamorphism of sediments and at all grades affecting massive igneous rocks. Igneous rocks gain water during progressive metamorphism (contrasting with the general loss of water from metasediments), and if such metamorphism is more-or-less strain free, it is usefully termed **hydrothermal metamorphism. Ocean-floor metamorphism** is one variety of hydrothermal metamorphism that affects oceanic crust immediately following its formation at a spreading ridge; sea-floor spreading ensures that the effects are dispersed on a regional scale. Hydrothermal metamorphism is also found in volcanic arcs, as in the Taupo Volcanic Zone of New Zealand, where again it is essentially regional in character.

Whatever categories and terms are found to be useful, it is not always possible to draw sharp boundaries between them. The blastomylonite of one worker may well be another's schist, and so on. The phenomenon of polymetamorphism may also make strict categorization for any one rock a difficult affair. What category, for example, does a partly mylonitized hornfels belong to?

2.1.3 The boundaries of metamorphism

Sediments are often subject to mineralogical modification and lithification at or very close to the sediment/water interface, and this process is called diagenesis. As sediments are more deeply buried and/or subject to strain or significant thermal perturbations, diagenesis merges into what we call metamorphism. Defining the position of that boundary is a difficult if not impossible task. Normally one considers a change to be metamorphic if it involves modification that could not have occurred in the superficial sedimentary layers. Thus one can argue that laumontite, not heulandite, marks the onset of zeolite-facies metamorphism because heulandite can and does form in the most superficial of sedimentary layers. What can be called the onset of metamorphism depends to a large extent on lithology. Thus at the one extreme, metamorphism of glassy tuffs at very low grades may be quite obvious and readily studied with the microscope, but an orthoquartzite from the same zone could not realistically be described as metamorphic in any sense. Within such a spectrum lie the anchi-metamorphic changes involving illite crystallinity, which must be studied with X-rays rather than the microscope.

The other boundary of metamorphism is with igneous processes, and is just as difficult to define. At high temperatures metamorphic rocks start to melt, but the temperature required and the amount of melting depends on bulk composition, volatile content and pressure. In a complex metamorphic sequence, the distribution of melt-derived rock may be similarly complex. Migmatite is the name used for rocks that appear to be partly 'igneous' partly 'metamorphic', but confirming the magmatic origin of the igneous-like part has proved to be a contentious matter (section 5.3.6). Some high-grade paragneisses ('para' = sedimentary origin) are difficult to distinguish from orthogneisses ('ortho' = igneous origin), and with the latter it is often a moot point as to whether the rock should be given a metamorphic or igneous name (for example, granitic gneiss or gneissic granite).

2.2 DESCRIBING AND NAMING METAMORPHIC ROCKS

The goal of microscope work with metamorphic rocks should be more than simply providing a rock name. Indeed, all of the following should be aimed at:

1. Deciding what the rock was before metamorphism.
2. Determining the metamorphic sequence of events that together produced the minerals, structures and textures observed.
3. Determining the P/T conditions (in essence the metamorphic facies) prevailing at the thermal climax of metamorphism, but also as far as possible for the complete sequence of metamorphic events.
4. Giving the rock a name.

Point 1. requires a knowledge of the chemical composition of common rocks and minerals, and is discussed throughout sections 2.3–2.6. Point 2. is discussed further in section 5.5. With regard to point 3., the positions of metamorphic facies in P–T space are given on Fig. 2.1, and typical mineralogical assemblages for the facies are listed throughout sections 2.3–2.6. The approach to point 4., taken in this book, is introduced here.

A hierarchical system for naming metamorphic rocks

No recommendations on metamorphic rock nomenclature equivalent to those from the IUGS subcommittee on igneous rocks exist. The traditional way of setting out a metamorphic petrography text is to take rocks of a certain bulk composition, and gradually deal with the effects of dynamic vs. thermal vs. dynamothermal metamorphism, as well as the changes of mineralogy according to the facies of metamorphism. Here I organize things differently and, inspired by the hierarchical flow-chart system of identification for igneous rocks (Le Maitre, 1989), I propose the following series of steps for metamorphic rock identification, steps designed to be most helpful for the beginning student:

1. If the rock retains an essentially igneous aspect, lacks any major strain effects on the grain scale, and contains a mineral assemblage due to a gain in water, go to section 2.3.1 (hydrothermal, often incipient metamorphism). The general name is meta-(igneous rock name), as in metagabbro. Mineral-derived qualifiers such as serpentinized

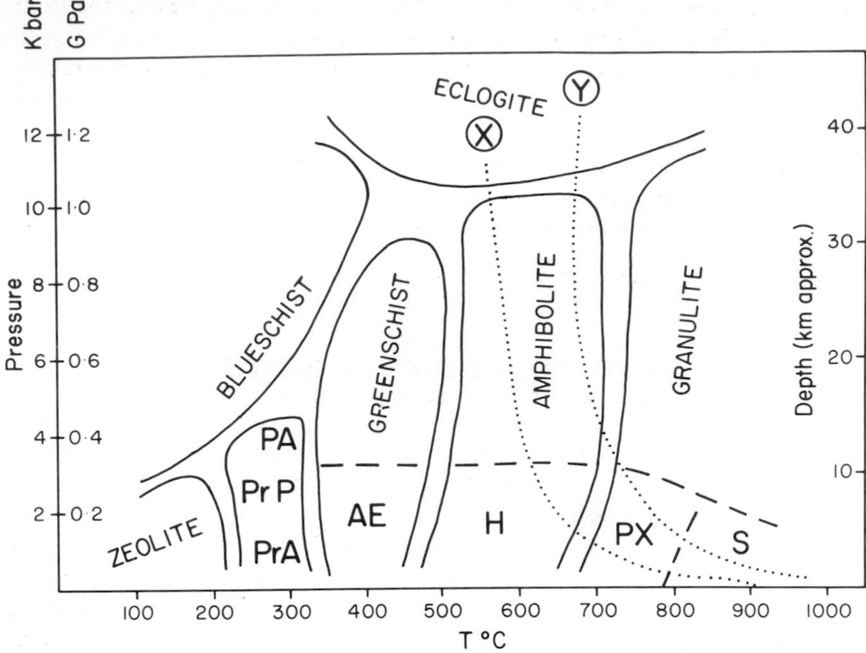

Fig. 2.1 Approximate positions of metamorphic facies on a pressure–temperature diagram. PA, PrP and PrA = pumpellyite–actinolite, prehnite–pumpellyite and prehnite–actinolite facies; AE = albite–epidote hornfels facies; H = hornblende-hornfels facies; PX = pyroxene-hornfels facies; S = sanidinite facies. Gaps are left to indicate transitional facies and the high degree of uncertainty in positioning some boundaries. Dashed lines reflect the often arbitrary decision as to whether a 'contact metamorphic' or 'regional metamorphic' facies name is applied. Curves X and Y show the beginning of water-saturated melting for granitic and basaltic compositions.

may replace 'meta' if they best describe the dominant effect. Special names for igneous rocks with conspicuous effects of hydrothermal metamorphism include greenstone, spilite, propylite, epidiorite and coronite.

2. If the rock is sedimentary (including pyroclastic rocks), and mainly displays premetamorphic textures, go to section 2.3.2. The general name is meta-(sedimentary rock name), as in metagreywacke.
3. If the rock exhibits profound strain effects on the grain scale, go to section 2.4 (dynamically metamorphosed rocks: mylonite, cataclasite, phyllonite).
4. If the rock is well crystallized or recrystallized, and:
   (a) If the rock is composed mainly of quartz, or carbonates, or a mixture of olivine and pyroxene, or serpentine, or a mixture of

pyroxene and garnet, go to section 2.5.1 (special cases: quartzite, marble, peridotite, serpentinite and eclogite).

(b) If the rock lacks any strong obvious mineral preferred-orientation, and if the rock is not one of the special cases in (a), go to section 2.5.2 (hornfels, and some granulites and charnockites).

(c) If the rock displays an obvious mineral preferred-orientation, go to section 2.5.3 (slate, schist, greenschist, blueschist, amphibolite, gneiss, some granulite and charnockite and migmatite).

Some common metasomatic effects are listed in section 2.6.

2.3 HYDROTHERMALLY AND/OR INCIPIENTLY METAMORPHOSED ROCKS

We can define an incipiently metamorphosed rock as one that displays some significant mineralogical change, yet retains enough of the pre-metamorphic characteristics to warrant a non-metamorphic rock name. This type of metamorphism is appropriately indicated by the use of the prefix 'meta-' before the rock name: hence metagreywacke and metagabbro. Sedimentary rocks generally have high porosity and/or permeability, lose water during metamorphism and are susceptible to strain. With few exceptions, the sedimentary aspect on the grain scale is retained only if the metamorphism is incipient and very low or low grade. In contrast, igneous rocks generally lack high porosity and permeability, require water during metamorphism and often resist strain and mineralogical change. It is not unusual for igneous rocks to survive metamorphism, or to be incipiently altered, even at high grades, and an igneous aspect may be retained even where water is available and able to effect a thorough hydrothermal metamorphism.

Terminology of hydrothermally and/or incipiently metamorphosed rocks is beset by boundary problems: given that most rocks are 'altered' to some degree, when does one use 'basalt' versus 'altered basalt' versus 'metabasalt'? There is no simple answer to this, but one essential factor in any solution is the spatial perspective: is the 'alteration' localized (as in a deuteric alteration, for example) or on a regional scale indicating a significant metamorphism? In this respect, the pioneering works of Coombs (1954) on zeolite zones in undeformed sediments and Walker (1951, 1960) on zeolite zones in basalts of Northern Ireland and Iceland showed clearly that vesicle fillings, 'alteration' of glass, and various incipient effects, may represent a significant regional metamorphism. Another essential factor is the individual petrologist's attitude to 'alteration', and this depends to a large extent on what he or she is aiming to do. An igneous petrologist, for example, may actively seek to avoid 'altered' material in the pursuit of igneous geochemistry, but the consequence may be a loss of metamorphic perspective.

2.3.1 Igneous rocks

(a) Metabasaltic rocks

The combination of high geothermal gradients (100s of °C km^{-1}), abundantly available sea-water, and an active fracture zone, means that ocean floor basalts in the vicinity of a spreading ridge are likely to be hydrothermally metamorphosed soon after formation. Evidence for such metamorphism comes mainly from dredged and drilled ocean-floor samples, and less directly from ophiolite complexes, especially those that provide evidence of a very high geothermal gradient. For example, changes culminating in greenschist-facies mineral assemblages, have been described from very condensed sections (Spooner and Fyfe, 1973) due to geothermal gradients as high as 1300°C km^{-1} (400°C at 300 m depth). Metasomatism during ocean-floor metamorphism of basalt may include significant loss of Ca and Si, and gains in Mg and water. In other ophiolites the metamorphism may be better ascribed to that within an accretionary prism. Less extensive hydrothermal changes occur out of the oceanic environment, as, for example, in the zeolitization of Icelandic basalts, documented by Walker (1960).

Mineralogy and terminology: The general name is metabasalt; green hydrothermally metamorphosed basalt can be called **greenstone; spilite** is used for red or green pillow lavas in ophiolite sequences where plagioclase is albitized, and matrix metamorphosed. In ocean-floor metamorphism, glass alters to palagonite, a smectite-bearing waxy material, then chlorite. Albite replaces calcic plagioclase at an early stage, and primary twins are often destroyed. The Ca from the plagioclase finds its way into Ca zeolites, then pumpellyite, and with increasing temperatures, successively epidote and amphibole. Clinopyroxene is relatively resistant to change, and may survive. Olivine is replaced first by calcite and hematite, later by chlorite and possibly pumpellyite and amphibole. Dispersed hematite in the early stages of metamorphism causes the red colour in pillow lavas. At greater depth, the hematite content drops, and pillow lavas take on a green colour. The Ti-bearing opaque oxides are replaced by titanite.

Textures: Microporphyritic textures usually survive metamorphism, but vesicles are filled, the nature of the matrix is thoroughly changed, and so may be the mineralogy of the phenocrysts, as described above. At high levels, veins and vesicles are filled with zeolites, smectites, celadonite and calcite; at lower levels, chlorite and a variety of Ca–Al-silicate dominate. It is characteristic of the Ca-bearing silicates to occur in veins and vesicles rather than the body of the rock.

Metamorphic facies: Some petrologists might regard the earliest stages of hydrothermal metamorphism as simply 'alteration', but that denies the fact that palagonitization near a spreading ridge is an integral part of the hydrothermal process, not some unrelated alteration. Usually zeolites start to form at the highest levels, and metabasalts with vesicle- and vein-fillings of zeolite typify the zeolite facies. The zeolites may be in sequence so that in ocean-floor metamorphics, analcime, natrolite, chabazite, heulandite and stilbite form at higher levels, and laumontite followed by wairakite form at lower levels. In Iceland, the sequence is from chabazite–thomsonite to analcime to mesolite–scolecite with increasing depth of burial. The incoming of pumpellyite indicates the prehnite–pumpellyite (Pr–P) facies (prehnite is rare), and epidote, chlorite and albite may be present. Some zeolites, notably wairakite, develop or survive higher than the zeolite facies. At low pressures (ocean-floor metamorphism), a prehnite–actinolite (Pr–A) facies is recognized between Pr–P and greenschist facies, and is marked by the incoming of actinolite, and disappearance of pumpellyite. In regional metamorphic terranes, Pr–P is succeeded by the pumpellyite–actinolite (P–A) facies (assemblage = pumpellyite + actinolite + epidote + chlorite + quartz + albite). In moderate to high geothermal gradients, the greenschist facies succeeds Pr–A or P–A, and is typified by actinolite + epidote + albite + chlorite + quartz. In low geothermal gradients, the blueschist facies succeeds the zeolite or Pr–P or P–A facies, and is marked by the incoming of lawsonite.

(b) Metagabbroic rocks

Gabbroic rocks at lower levels of oceanic crust may be subject to amphibolite-facies hydrothermal metamorphism. Gabbro in which most or all of the pyroxene is replaced by amphibole (uralitization – section 3.5.4) resembles a diorite, and is called **epidiorite**. At higher temperatures, particularly at great depth in continental crust, and subject to prolonged penetration by sparse grain-boundary fluids, gabbros develop corona textures (see section 3.5.4 and Fig. 2.2), and if these become an important or dominant feature of the rock it is called a **coronite**.

(c) Serpentinization

Serpentinization of olivine and pyroxene is common (e.g., Fig. 1.18), both as a local alteration product and as the product of regional-scale hydrothermal metamorphism. The process is discussed in section 3.5.4, and it

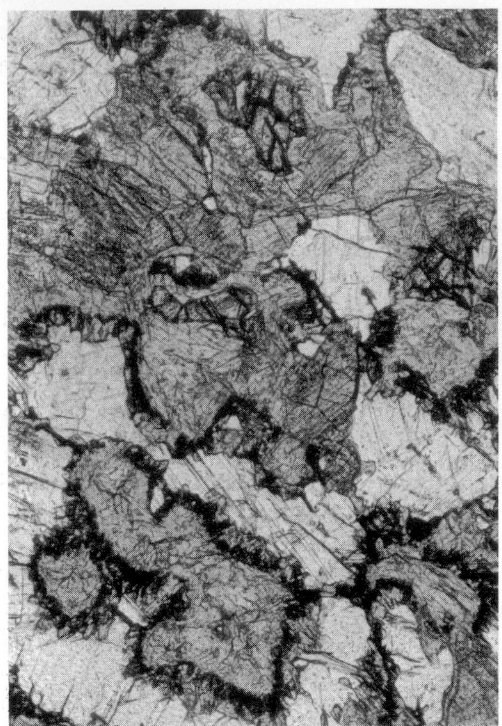

Fig. 2.2 Corona texture in gabbro, Southland, NZ. The dark rims between clinopyroxene and plagioclase consist of amphibole and spinel, some of it symplectitic. Three relic patches of corroded olivine within the pyroxene are visible in the left half of the photo. (View length measures 2.1 mm.)

clearly involves the gain of copious amounts of water (and usually a loss of Mg and a gain in Si) at temperatures within the range <100°C to 500°C. The partial but widespread serpentinization of peridotites may fit the category here of rocks that retain an igneous aspect, but complete metamorphism produces serpentinites which are easily deformed, lose any remaining igneous character, and are better treated as a special case in section 2.5.1.

(d) Propylites, kaolinitization and hydrothermally metamorphosed acid rocks

In continental margin volcanic arcs, circulating groundwater heated by subjacent intrusives commonly effects a regional hydrothermal metamorphism on andesitic, dacitic and rhyolitic material. The intrusive granitoids themselves may become extensively altered during a long sequence of igneous and geothermal activity.

Terminology: Apart from the name **propylite** for thoroughly altered andesite, there are no special names. Rocks are usually referred to in the form **kaolinitized granite**, **hydrothermally metamorphosed rhyolite**, etc.

Mineralogy: Acid igneous rocks are dominated by feldspar and quartz, and at low temperatures (<150°C) the feldspathic material may be changed to kaolinite-group clays: in thoroughly metamorphosed granitoids this forms a soft (in water) matrix to the gritty relic quartz. Montmorillonite and interlayered illite are also common and survive at a higher temperature than kaolin. At $T > 220°C$ illite–chlorite mixtures become dominant. Fractures, cavities and vesicles are filled with silica, zeolites (ranging from mordenite at 50°C to laumontite and ultimately wairakite, especially at $T > 215°C$), adularia and calcite. Impermeable rocks, not affected by fractures, may remain unaffected by the metamorphism, but permeable rocks may be thoroughly metasomatized and replaced above 280°C by an assemblage of albite–adularia–chlorite–epidote–calcite–quartz–illite–pyrite. The mineral alunite becomes important if fluids are very acid, and calcite forms instead of Ca–Al-silicates if fluids are rich in CO_2. According to Browne (1978), in the rhyolitic volcanic arc of North Island, New Zealand, the same mineral assemblage replaces all rocks at $T > 280°C$, be they originally rhyolite, basalt or sandstone. Intermediate andesitic rocks are less silica rich and more mafic than rhyolite; consequently the early alteration products are likely to include greater amounts of smectite, chlorite and serpentine, and the low-temperature zeolites include chabazite, thomsonite and scolecite rather than mordenite.

2.3.2 Sedimentary rocks

Sediments are permeable, usually have abundant fluids circulating within them, and are easily deformed; yet many show little obvious change other than diagenetic lithification below the greenschist facies. This is because many sediments are already made of weathered materials, stable at low temperatures in the presence of abundant water. This contrasts sharply with igneous rocks which are made of unstable minerals which survive only because of a lack of permeability and fluid supply. The sediments that display the most conspicuous mineralogical changes at $T < 300–400°C$ are those that contain unstable igneous-derived materials. Pyroclastic rocks and immature sediments like ‘greywacke’ are particularly vulnerable.

A word of caution on terminology. The prefix ‘meta’ is appropriately used for metasediments with original textures preserved, but it is also used more widely. Thus metapelite and metasediment are used for rocks of any grade and textural state, including schists.

(a) Metatephra

Tephra is highly permeable and contains large quantities of unstable volcanic material, especially glass. In addition to low-pressure hydrothermal metamorphism, tephra shows the most marked metamorphic changes of any sedimentary rock at very low grades of regional burial metamorphism.

Mineralogy and metamorphic facies: The changes during geothermal activity are essentially the same as those outlined above for igneous rocks of the same composition. The changes during burial metamorphism, as exemplified by the type locality in southern New Zealand (Coombs, 1954), start with the complete replacement of intermediate to acid glass by zeolites such as analcime, heulandite and laumontite. One can argue that analcime and heulandite are diagenetic minerals, viewed in a sedimentary perspective, but one can equally well argue that they represent the start of a sequence of zones characterizing the zeolite facies of metamorphism. Plagioclase may be albitized or occasionally replaced by zeolites or carbonates. Basaltic glass is prone to palagonitization, and zeolitization of basalt produces a range of zeolites including analcime, heulandite, stilbite, chabazite, thomsonite and laumontite. It is worth noting that analcime reacts with quartz at 180°C to produce albite within the zeolite facies, and that wairakite occurs only in geothermal areas, not in areas of burial metamorphism. Some zeolites, notably wairakite, develop or survive outside the zeolite facies. As metamorphic grade increases, the Ca-zeolites are replaced by a succession of Ca–Al-silicates including prehnite, pumpellyite, epidote, zoisite, actinolite, Ca-rich garnets or, at higher pressures, lawsonite. They accompany albite and chlorite, usually with some quartz and illite, and bulk composition has an important bearing on the proportions of minerals present. Typical assemblages for the low-grade facies are the same as those indicated above for metabasalts, but minerals often appear out of sequence in burial metamorphic terranes, and this indicates either disequilibrium assemblages, variable fluid composition, or differences between fluid and lithostatic pressures (as is likely in veins and porous rocks). The oxygen and sulphur content of fluids determines whether magnetite, hematite or pyrite is present, and the abundance of CO_2 causes calcite to substitute for some of the Ca–Al-silicates.

Textures: In glassy tephra, all vesicles and pore space may be filled by zeolite, and sometimes the zeolite crystals are much larger than the original shards (Fig. 2.3A). Cuspate shard texture may be visible because of original dusty margins. In other examples, vesicle and pore space may be filled with other minerals producing sharp mineralogical and

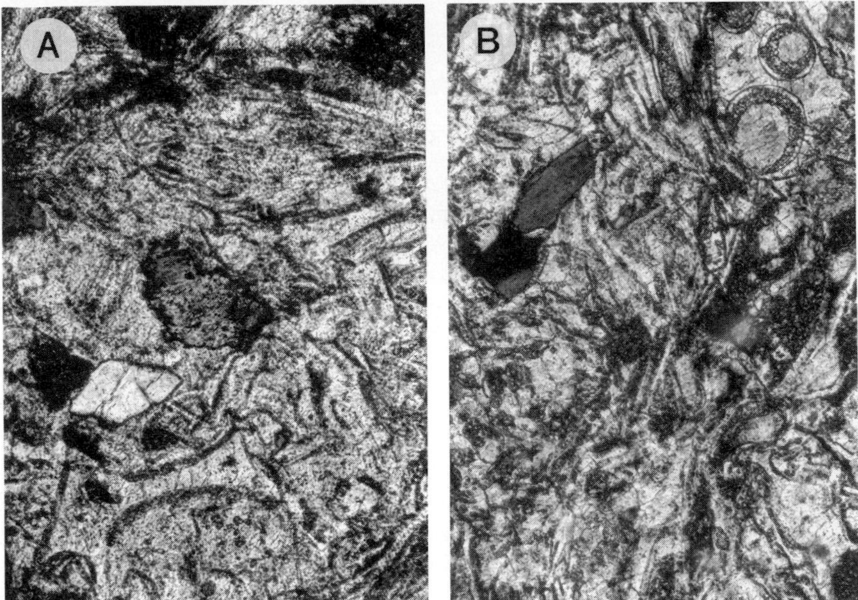

Fig. 2.3 Laumontite replacing shards in Triassic tuffs, Southland, NZ. (A): The cuspate shard shapes (and some unaltered feldspar) are clearly visible within the one large laumontite crystal that covers the entire field of view. (B): An earlier vesicle infilling, and some feldspar, are visible within the large laumontite crystal that has replaced a complex of pumiceous material. (View lengths measure 0.81 mm (A) and (B).)

textural contrasts (Fig. 2.3B). Crystal ejecta or phenocrysts contained within glass fragments often remain more-or-less unaltered (Fig. 2.3). Apart from replacing glass and the earlier-formed metamorphic minerals, new phases grow in veins and vesicles, or in strain-shadow areas around coarser clast fragments. Coarse and/or crystal-rich tephra may retain original pyroclastic textures into the blueschist or greenschist facies, but fine-grained and/or glass-rich tephra are likely to lose most traces of original texture within or soon after the zeolite facies.

(b) Metagreywacke

Metagreywacke often contains feldspar and unstable igneous-derived material such as volcanic rock fragments (possibly including glass), and these are likely to be affected by very low-grade metamorphism. Greywacke can be affected by geothermal activity, but most metagreywacke represents changes that occurred during burial or subduction-zone metamorphism.

Mineralogical changes and metamorphic facies: Zeolites directly replace glass fragments in the lowest-grade rocks. The albitization of plagioclase and consequent release of Ca, the replacement of glass and chloritization of pyroxene, all contribute to the formation of Ca–Al-silicates such as zeolites, prehnite and lawsonite. In the higher-pressure regions of the blueschist facies, albite is replaced by the Na-pyroxene jadeite. The distinctive mineral assemblages for the various low-grade facies are the same as noted above under metabasalts.

Textures: The distinctive textural feature is the general preservation of sand-sized clast shapes. New minerals tend to form at three distinct sites: 1. in veins, 2. in the matrix, 3. replacing unstable sand-sized clasts. New mineral growth affecting clasts tends to be confined within the boundaries of any one clast. If matrix and clast replacement mineralogy is the same, then clast boundaries may become indistinct. Zeolites directly replace glass fragments, but also crystallize in veins. Prehnite (usually accompanied by quartz) is found mainly in veins (Fig. 2.4A), whereas pumpellyite and lawsonite (higher pressures) most commonly crystallize in the matrix (Figs 2.4B and C) or within fine-grained lithic fragments. At very high pressures, sand-sized clasts of albite are replaced by mosaics of prismatic jadeite (Fig. 2.4D), usually accompanied by lawsonite in the rest of the rock. Clast boundaries are sites of solution and reaction: therefore clasts become narrower concomitant with the development of pressure-solution seams (section 5.3.4 and Fig. 2.5A). In semi-pelitic metagreywackes, sheet-silicates (illite + chlorite) crystallize or recrystallize with a preferred orientation to form beards in the strain shadows about sand-sized clasts (Fig. 2.5B). The narrower clasts, the solution seams, and the beards represent the first stages in the formation of schistosity.

2.4 DYNAMICALLY METAMORPHOSED ROCKS

Three conspicuous strain features characterize dynamically metamorphosed rocks:

1. Bending of grains, particularly sheet-silicates, quartz and calcite, either during buckling in compression or in response to differential strain about some other 'hard' grain such as feldspar (Fig. 2.6). Bending may involve plastic deformation, twinning, or cataclastic flow, and may even affect relatively hard minerals such as feldspar (Fig. 2.6A).
2. Breakage of grains (cataclasis), particularly non-ductile minerals such as feldspar, but also relatively ductile sheet-silicates. Typically, smaller fragments are smeared out along shear planes to form tails (Fig. 2.7, also Fig. 5.46), while the parent grain may develop undulatory ex-

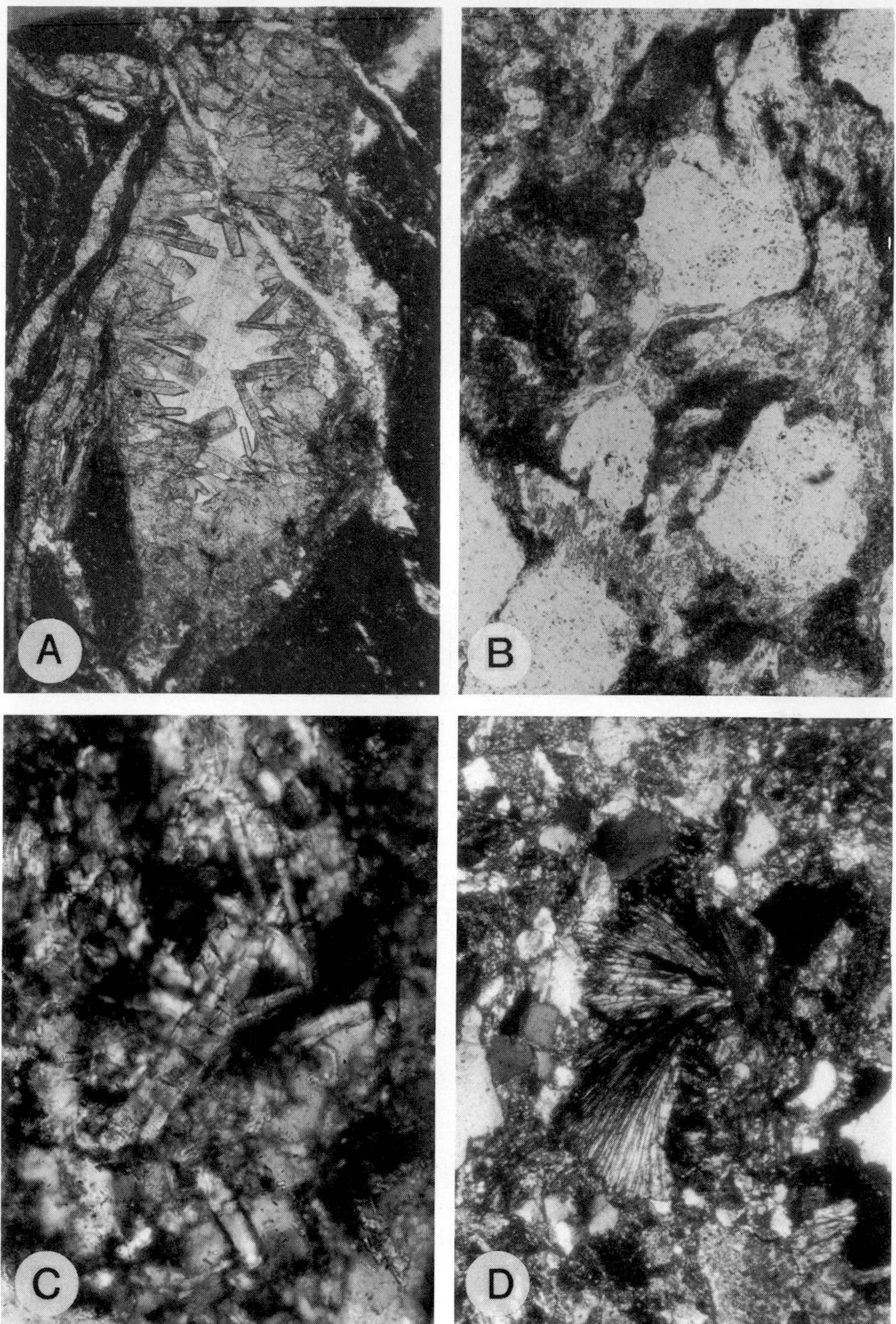

Fig. 2.4 (A): Prehnite in veins. Note the euhedral terminations to prehnite plates set in quartz. Torlesse Group, NZ. (B): Pumpellyite in metasandstone, Torlesse Group, NZ. The pumpellyite forms dense mats of fine grains between the clastic sand grains. (C): Lawsonite plates in metagreywacke, Franciscan, California. (D): Jadeite which has replaced clastic feldspar in metagreywacke, Franciscan, California. (View lengths measure 3.2 mm (A), 0.83 mm (B), 0.3 mm (C), and 1.8 mm (D).)

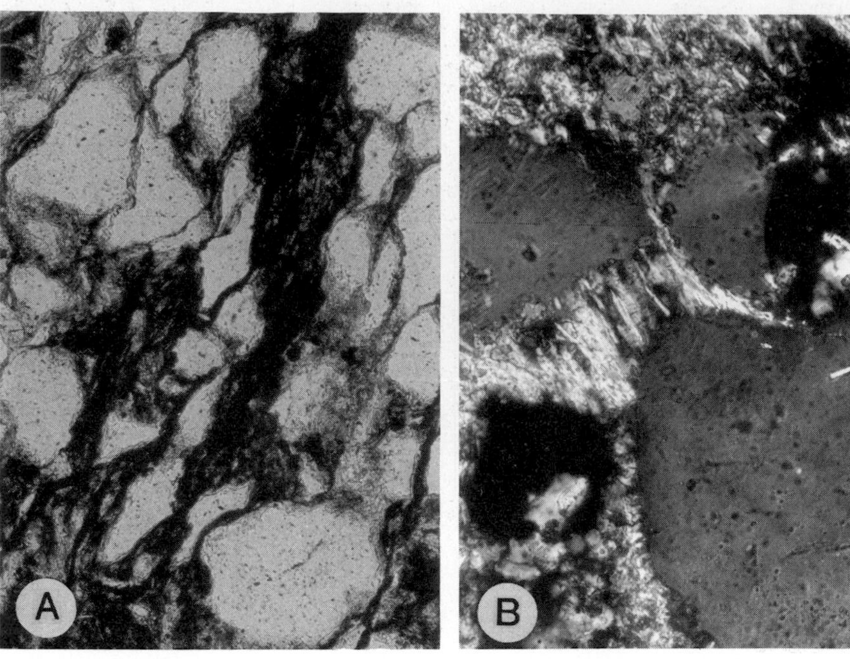

Fig. 2.5 (A): Pressure-solution seams (dark), and characteristic shapes of partially dissolved quartz grains. Torlesse Group, NZ. (B): Sheet-silicate beards growing on clastic grains, Wangapeka Formation, NZ. (Lengths of view measure 0.84 mm (A) and 0.35 mm (B).)

tinction and become elongated due to microfaulting (Tullis and Yund, 1987).

3. Plastic deformation, which increases the aspect ratio of ductile minerals such as quartz and calcite to produce 'ribbons' (Figs 2.6B and 2.7B), and is responsible for some of the bending noted in 1. above.

The process of dynamic recrystallization also plays an important role in producing some of the very fine-grained material that typifies mylonites. Recrystallized grain size decreases with increase in differential stress (section 5.2), and the resultant very fine grain size of highly stressed mylonites may initiate the process of superplasticity (section 5.3.3). Strictly speaking, superplastic textures do not show strain on the grain scale, but the process is so superficially similar to the combination of cataclasis and shearing out of grains, that most writers include superplastically deformed rocks under mylonite.

These processes are discussed further in Parts Two and Three. As usual, the boundaries between categories are not sharp. One could, for example, describe the 'bending' of quartz as inhomogeneous plastic deformation,

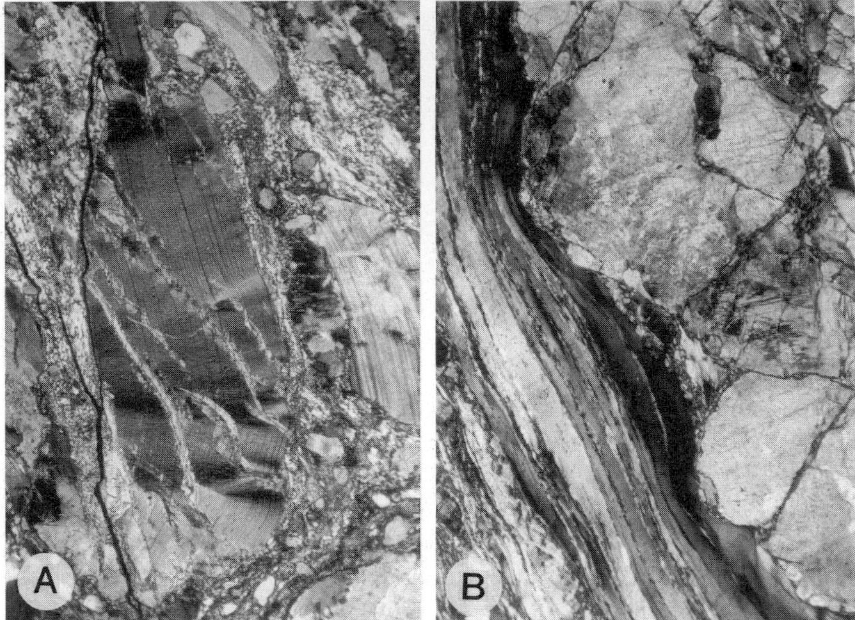

Fig. 2.6 Granite mylonites, Constant Gneiss, NZ. (A): Bent mica (centre) and feldspar (right). Note the extremely fine-grained nature of the quartz (top right and left) due to recrystallization under high differential stress. (B): Ribbons of quartz, produced by plastic deformation, bent around relatively hard feldspar grains. (View lengths measure 3.3 mm (A) and (B).)

and the smearing out of fine fragments of sheet-silicate could be described either as cataclasis or a highly partitioned, inhomogeneous plastic deformation. Certainly the very fine grains resulting from dynamic recrystallization may be difficult to distinguish from those produced by cataclasis.

Dynamically metamorphosed rocks are typically restricted to relatively narrow zones of movement. At the one extreme, these may be on a regional scale, as exemplified by the Alpine Fault mylonites in New Zealand which form a zone some hundreds of metres thick within the fault zone, extending hundreds of kilometres parallel to the fault line (Sibson *et al.*, 1979). At the other extreme, shear zones, perhaps only a few millimetres thick, may cause a dynamic metamorphism in otherwise unaffected rock; these zones usually appear as black, fine-grained, possibly anastomosing seams within the host.

Terminology: Dynamically metamorphosed rocks are subdivided into **mylonite** and **cataclasite**, depending on whether or not the rock has a planar fabric. Mylonites are by far the most common, and many so-called cataclasites, judged as such in the field, prove to be mylonites

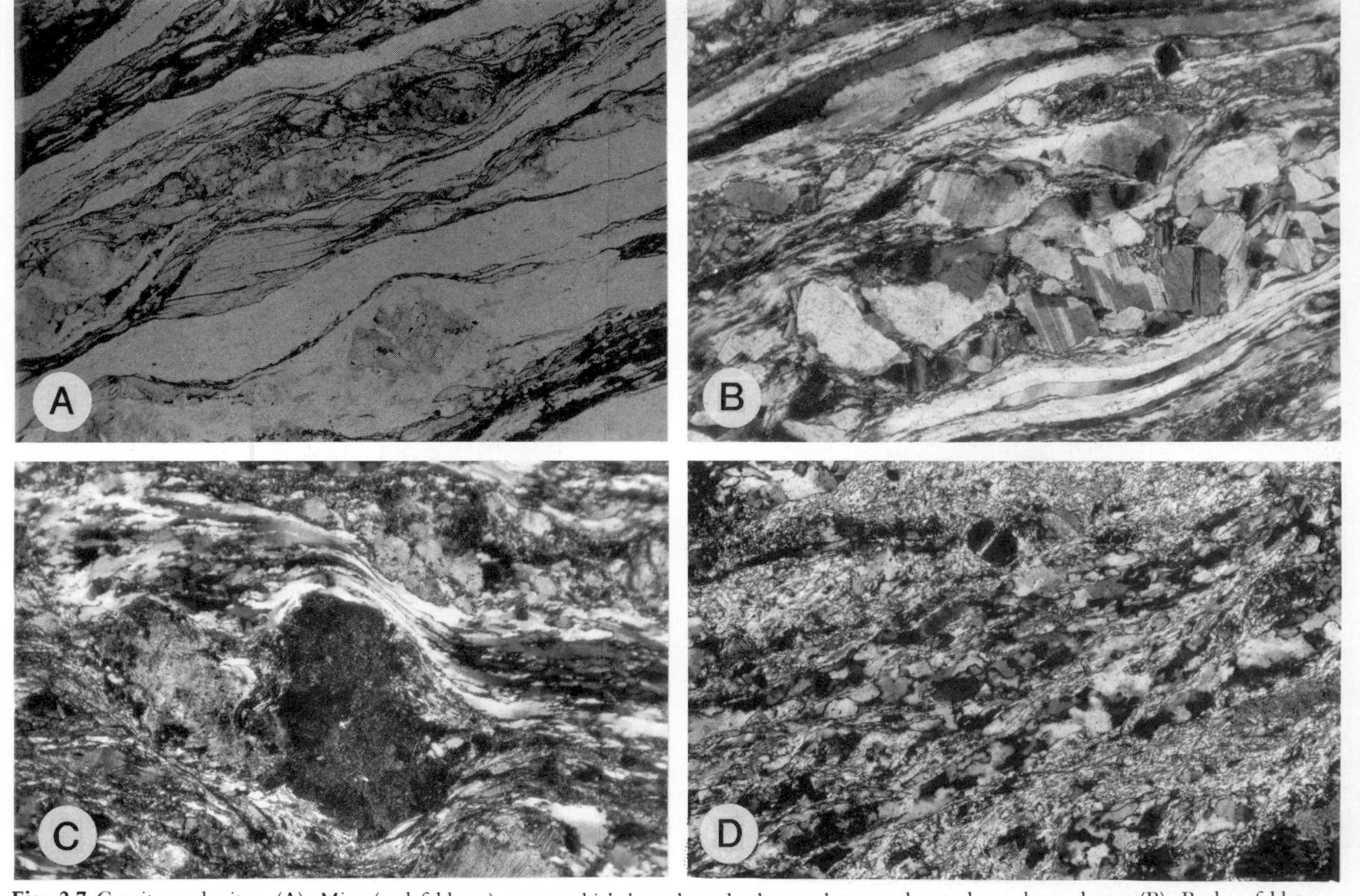

Fig. 2.7 Granite mylonites. (A): Mica (and feldspar) augen which have been broken and smeared out along shear planes. (B): Broken feldspar grains, smeared out along shear planes by cataclastic flow. (C): Feldspar porphyroclast forms an augen surrounded by ribbon quartz. The (A), (B), and (C) are from Constant Gneiss, NZ. (D): Metasomatized granite mylonite from the Moine Thrust, Scotland, consisting of ultra-fine-grained

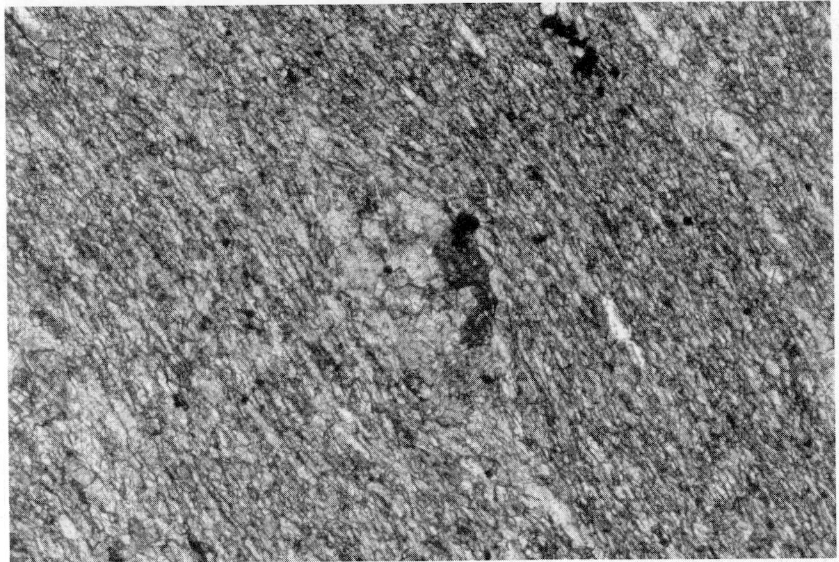

Fig. 2.8 Calcite porphyroclast set in calcite mylonite (a possible alternative interpretation is that this rock represents a strained micrite, the clasts being organic crystal plates!). Arthur Marble, NZ. (View length measures 0.85 mm.)

under the microscope. Both may be qualified in any useful way, but commonly the name of the parent rock is used, as in **granite-derived mylonite** (Figs 2.6 and 2.7), or more simply **granite mylonite**, or the name of the dominant mineral, as in **calcite mylonite** (Fig. 2.8) for mylonite derived from marble. The name mylonite is also qualified according to the proportion of relatively large pre-mylonitic crystals (termed **porphyroclasts**) that have not undergone substantial grain-size reduction. Thus, a **protomylonite** contains >90% pre-mylonitic material, and **ultramylonite** <10% pre-mylonitic material. If there is significant porphyroblastic growth (but not enough to warrant the rock having a non-mylonitic name), the rock is called a **blastomylonite**. The special name **phyllonite** is used for sheet-silicate-rich mylonites that resemble slate or phyllite. Mylonites develop in seismically active regions, and one possible response to rapid fault movement is melting (Sibson, 1975). Typically the melt is generated along foliation surfaces, but also intrudes along adjacent fracture systems as a network of veins. The rapidly undercooled product is a dark volcanic-like glass called **pseudotachylyte** which may contain tiny igneous-like laths of feldspar and other minerals.

Textures: Cataclasis and/or dynamic recrystallization result in a grain size that is finer than in other metamorphic rocks, and the extreme thinning of plastically deformed quartz and calcite adds to the overall effect of grain-size reduction. Indeed many mylonites are so fine grained that they are dark, almost black, in colour, and have a flinty appearance in the field. The word 'mylonite' is derived from the idea that the rock has been 'milled' to a rock flour, and the term 'mortar texture' is often applied to the very fine-grained material in a mylonite, for the same reason. However, much mortar texture is due to dynamic recrystallization, not cataclasis, so the term will not be used further in this book. Of course, one could advocate abandoning the name mylonite for the same sort of reason, but that would be unrealistic in view of its universal usage for dynamically metamorphosed rocks, regardless of the importance of cataclasis. Crystals that resist grain-size reduction (porphyroclasts) form eye-shaped bodies (augen) around which the plastically deformed, recrystallized, or cataclased material is sheared and/or bent (Fig. 2.7C). Of considerable significance is the fact that mylonitic textures provide evidence for the actual sense of shear during deformation. The criteria are discussed in section 5.4. The asymmetry of mineral preferred-orientations can also be interpreted in terms of shear sense (Chapter 7). Mylonites may develop plastically or superplastically, and it is worth noting that strong lattice preferred-orientation characterizes plastic but not superplastic deformation (section 5.3.3 and Chapter 7).

Mylonites are illustrated below by a few selected examples:

Granite-derived mylonite: The most commonly recognized mylonites are derived from crustal granitoids. They provide excellent examples of brittle failure and the cataclastic flow of feldspar (Fig. 2.7B), the cataclastic flow of mica (Fig. 2.7A), the plastic deformation of quartz (Figs 2.6 and 2.7), and grain-size reduction due to dynamic recrystallization and high differential stresses (Fig. 2.6A). In many examples, two foliation planes develop simultaneously within the shear zone (Fig. 5.44). These represent a partitioning of the strain, are known as C and S planes (Berthé *et al.*, 1979; Lister and Snoke, 1984), and are discussed further in section 5.4.

Metasomatic granite ultramylonites: Mylonites may hydrate during deformation, feldspars become 'sericitized', and the rock loses Na and Ca in the process (Dixon and Williams, 1983). The end product is a feldspar-free, finely laminated, quartz–muscovite ultramylonite, in which both mica and quartz are sheared out into ultra-fine-grained layers (Fig. 2.7D).

Quartz or **calcite** or **olivine mylonites:** Calcite, quartz and olivine all deform plastically, and may form ribbons (Fig. 2.12A) in severely strained quartzite, marble or dunite; a few grains of resistant orientation may be left as augen among the ribbons. Dynamic recrystallization under high differential stress may convert the entire aggregate of grains to an ultra-fine grain size; in other instances, residuals of parent grains are mantled by the fine-grained material (e.g., Heitzmann, 1987, and Fig. 2.8). In general, dynamically recrystallized grains do not depart greatly in orientation from the parent, and the parent ribbons already possess a strong lattice preferred-orientation. However, if the grain size is so fine that superplasticity is initiated (for quartz the critical grain size is approximately 0.01 mm), the inherited lattice preferred-orientation will be lost (e.g., Behrmann, 1985).

Schist-derived mylonites: Mica schists normally absorb much of the strain during metamorphism, and they do so easily because of their fine grain size and the ability of sheet-silicates to deform and recrystallize readily. Schist-derived mylonite (called **phyllonite**) may therefore be difficult to identify because it resembles fine slaty or phyllitic material; the main distinguishing features are the occasional relics of former porphyroblasts of garnet or kyanite, for example. Mylonitized quartz-rich schists that contain relatively large mica grains develop characteristic 'mica-fish' (Fig. 5.46) which essentially outline what Lister and Snoke (1984) call Type II S–C mylonitic planes (section 5.4).

Superplastic mantle-peridotite mylonites: In high-temperature shear zones in the mantle, possibly associated with partial melting, dynamically recrystallized pyroxene aggregates deform superplastically. The ultra-fine-grained tails which extend in both directions from the parent grain (Fig. 1.22) are characterized by a lack of lattice preferred-orientation. The lack of grain coarsening probably reflects rapid cooling as mantle blocks were brought to the surface by the associated magma. The phenomenon is discussed in sections 1.7.7 and 5.3.3

2.5 WELL-CRYSTALLIZED OR RECRYSTALLIZED ROCKS

2.5.1 Special cases (quartzite, marble, peridotite, serpentinite and eclogite)

(a) Quartzite

A quartz-cemented quartz sandstone is properly termed an orthoquartzite, as distinct from a well-recrystallized quartz-dominated rock which is called a quartzite. Orthoquartzite may be distinguished from quartzite by:

1. The presence of dusty margins separating overgrowth cement from the original clastic grains
2. The lack of a strong lattice preferred-orientation.

Unfortunately, not all orthoquartzites display dusty grain margins, and not all quartzites have a strong preferred orientation.

Terminology, mineralogy and metamorphic facies: The name **quartzite** is used for metamorphic rocks made mainly of quartz. Commonly used qualifiers include accessory mineral names, as in **micaceous quartzite**, and sedimentary features as in **bedded** or **cross-bedded quartzite**. It may be difficult to judge the metamorphic grade of a quartzite, impossible in the case of an absolutely pure quartz variety. The best way is to look for adjacent lithologies with distinctive metamorphic assemblages, although impure quartzites, or thin layers within the quartzite, may contain sufficient mineralogical indicators. The accessory minerals may give some indication as to the original rock type. Thus Mn-rich minerals such as spessartine and piemontite suggest chert, abundant feldspar suggests an original feldsarenite, and so on. Calcite derived from a calcareous cement or fossil fragments does not react with the quartz except at very high temperatures in thermal aureoles where wollastonite may form; on the other hand, dolomite reacts during metamorphism to form minerals such as talc, tremolite, diopside and forsterite (dolomitic marbles – below).

Grain size and shape: Grain size in many quartzites is much the same as that of an orthoquartzite, but one must never assume that this is any more than a coincidental similarity. Plastic deformation and recrystallization usually effect a complete modification of the original sand-grain shape and size, and one must remember that some quartzites were originally chert, flint or quartz-vein material. The spessartine-bearing, chert-derived quartzite illustrated in Fig. 2.9 has numerous small, oriented sheet-silicate grains enclosed by individual quartz grains: at some earlier stage the quartz was no coarser than the size of spaces between the sheet-silicates, and a substantial coarsening is indicated. Quartz is susceptible to plastic deformation and dynamic recrystallization within the greenschist and higher temperature facies, so regional metamorphic quartzites are likely to have been through protracted cycles of deformation and recrystallization. Deformation creates high dislocation densities and grains with a high aspect ratio, but recrystallization has the opposite effect, and constant interplay between the two may maintain a more-or-less equidimensional and strain-free aggregate of grains (Fig. 2.10). Grain size depends on the strain rate, the differential stress and the temperature, not the grain size of the original sediment. If moderate to high metamorphic temperatures are maintained after strain ceases, static recrystallization further reduces dislocation density and induces grain coarsening.

Sedimentary structures: Primary structures such as cross-bedding are frequently preserved in quartzite, but do not assume that this indicates a lack of metamorphism. The relative competence and massiveness of

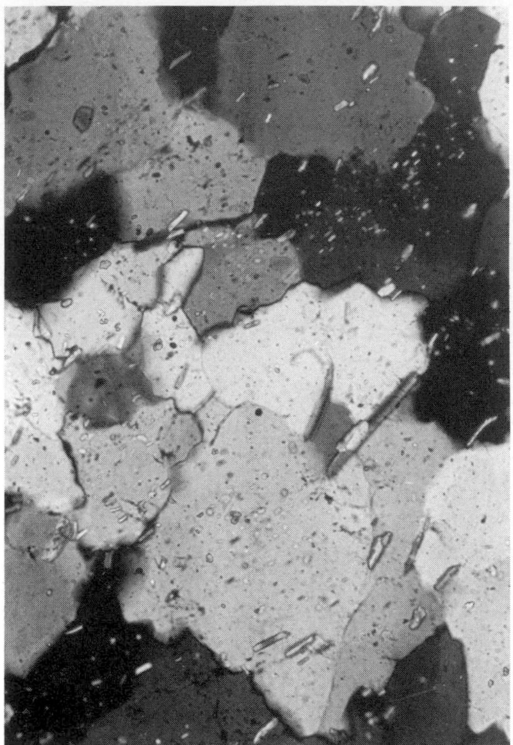

Fig. 2.9 Micas enclosed in quartz coarsened after recrystallization from chert. Pikikiruna Schist, NZ. (View length measures 0.85 mm.)

many quartzite horizons ensures a homogeneous and often less-intense strain than in surrounding metapelites; sedimentary structures often survive even the highest grades of metamorphism.

(b) Marble

Marble is the metamorphic equivalent of limestone, and there is little likelihood of confusing the two. Limestones are commonly very fine grained, or made up of obvious shell fragments, perhaps with a coarse sparry cement. Marbles are seldom as fine grained as micritic limestone, and coarse-grained varieties lack the organic detail of fossils. The only possible area of confusion is between marble and echinoidal or crinoidal limestone which is made of large single-crystal plates, spines or ossicles. Close examination will reveal the distinctive organic structures within those single crystals. Igneous carbonatites may resemble marbles, though the accessory mineral content of carbonatite is distinctive.

Terminology, mineralogy and metamorphic facies: A pure marble reveals little about the metamorphic grade. However, calcite is not

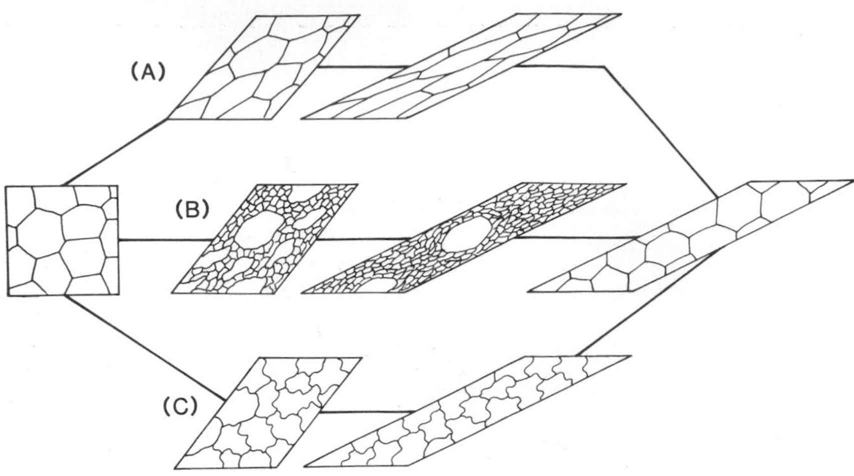

Fig. 2.10 Equidimensional grains in a stable mosaic, modified in various ways. (A): Homogeneous plastic deformation produces ribbons. (B): Heterogeneous plastic deformation accompanied by dynamic rotation recrystallization under high differential stress causes most grains to be replaced by a new fine-grained mosaic; some resistant old grains remain as 'augen'. (C): Migration recrystallization accompanies deformation, serrated grain-boundaries develop, and more or less equidimensional grain shapes are maintained. Subsequent annealing or static recrystallization may in all cases restore a stable equidimensional mosaic with smooth grain boundaries, often of coarser grain size. In (A) and (B) this probably requires an increase in temperature.

stable throughout the entire possible range of metamorphism. At very high pressures the polymorph aragonite (section 5.1.3) becomes stable (Fig. 5.17), and **aragonite marble** is known from high-pressure terranes such as in California. At high temperatures and low pressures, calcite and quartz react to produce wollastonite and CO_2 (Fig. 2.11). The reaction occurs only in high-temperature thermal aureoles, and is inhibited by high fluid pressures of CO_2. **Dolomite marble** is made mainly of dolomite which by itself is stable throughout a wide range of metamorphic conditions. However, at high temperatures and low pressures it loses CO_2 to form periclase which below 900°C reacts with water to form brucite. Therefore the common result of decarbonation of a dolomitic or dolomite marble is a mixture of brucite and calcite. Impurities in marble are most likely to include quartz, sheet-silicates and graphite. In calcite marbles, little reaction with quartz takes place, except at the high temperature necessary to produce wollastonite, but quartzose dolomitic marbles develop a characteristic sequence of Ca- and/or Mg-silicates starting with talc, then tremolite in the greenschist facies, and diopside and/or forsterite in the amphibolite facies. Sheet-silicate impurity in calcite and dolomite marbles adds variety by allowing Al-bearing minerals to feature in the assemblage: typically they include

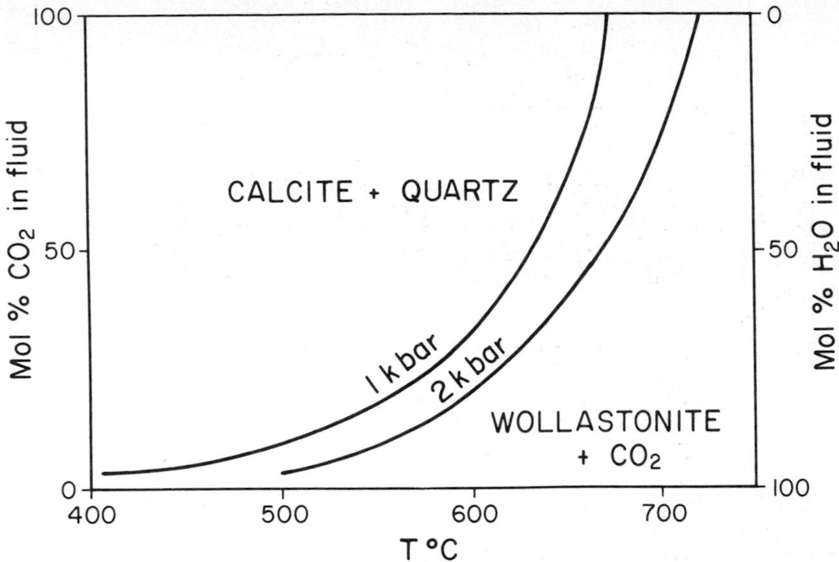

Fig. 2.11 Curves for the reaction calcite + quartz = wollastonite + CO_2. Temperature of reaction varies with pressure and with composition of fluid. Based on the data in Greenwood (1967, Fig. 1) with permission. Copyright: Mineralogical Society of America.

zoisite, epidote and Ca-rich garnet in the greenschist facies, and anorthite in the amphibolite facies. Interpretation of calc-silicate assemblages is hindered by the effects variable fluid compositions (CO_2 vs. water) have on mineral stability.

Textures: Strain-free thermal metamorphism simply produces an equidimensional mosaic of calcite, lacking lattice preferred-orientation. The more common metamorphic situation involves strain, and calcite is particularly prone to plastic deformation (including twinning) throughout the entire range of possible metamorphic temperatures and pressures (Chapter 7). Recrystallization and grain growth are effective except at sub-greenschist-facies temperatures, and grain shapes and sizes can be expected to be utterly different from the original sedimentary grains. As with quartz, shapes vary from ribbons to equidimensional mosaics (Fig. 2.12), but with the additional feature of conspicuous twinning. Grain shape and size depend on strain rates, differential stress, and temperature. Twinning may be profuse (Fig. 2.13), but in almost all cases it proves to be a relatively late-stage feature imposed on the final generation of recrystallized grains (Chapter 7). Dolomite is less susceptible to plastic deformation than calcite, and low-grade dolomite marble is characterized by a lack of twinning, which does not develop until 300–400°C.

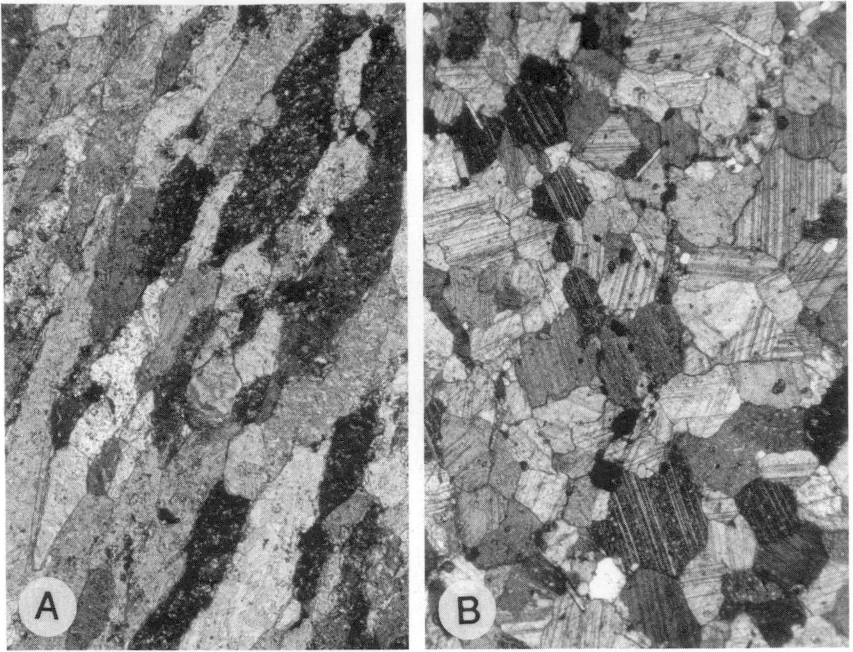

Fig. 2.12 (A): Ribbons of calcite in Arthur Marble, NZ. (B): Mosaic of equidimensional twinned calcite grains in Arthur Marble, NZ. (View lengths measure 0.85 mm (A) and 3.3 mm (B).)

(c) Peridotite

To many readers, peridotites may seem out of place here. Are they not igneous rocks? Some are, but the majority are thoroughly metamorphic, having undergone prolonged deformation and recrystallization. Their terminology is the same for metamorphic and igneous varieties, and details are given in section 1.7, together with a brief account of metamorphic textures.

(d) Serpentinite

Serpentinites are hydrothermally metamorphosed, often metasomatized, peridotites. The formation of serpentine is discussed in sections 2.3.1 and 3.5.4. Massive serpentinite is extremely ductile, so that it is easily squeezed tectonically away from its site of origin; serpentinites also rise diapirically in the crust. Hand specimens attest to tectonic mobility with a complex of shiny shear surfaces, but the unsheared material within may still display the pseudomorphic replacement textures of olivine (an irregular mesh of serpentinite veins) and pyroxene (a more regular replacement with visible ghosting of cleavage and/or exsolution lamellae planes parallel to the *c*-axis).

Fig. 2.13 Profuse deformation twinning in calcite in Arthur Marble, NZ. (View length measures 3.1 mm.)

(e) Eclogite

Eclogites are rocks of basaltic composition formed at $P > 10$ kbar and $T > 500°C$, and their presence defines the eclogite facies. They are found as xenoliths in kimberlites and some associated volcanic rocks, and these high-T eclogites probably derive directly from the mantle. Medium-T eclogites (550–900°C) are found in nappes within crustal gneisses where transitions between amphibolite and eclogite facies exist. The presence of coesite in some of these implies $P > 25$ kbar and transport of crustal assemblages to unusual depths and pressures. Low-T eclogites are found in metamorphosed accretionary complexes, most typically as blocks of unknown provenance in melange, or as metamorphosed pillow lava. In the latter, transitions between blueschist and eclogite facies are sometimes seen in pillows with cores of eclogite and rims of blueschist. Recent reviews include those of Smith (1988) and Carswell (1989).

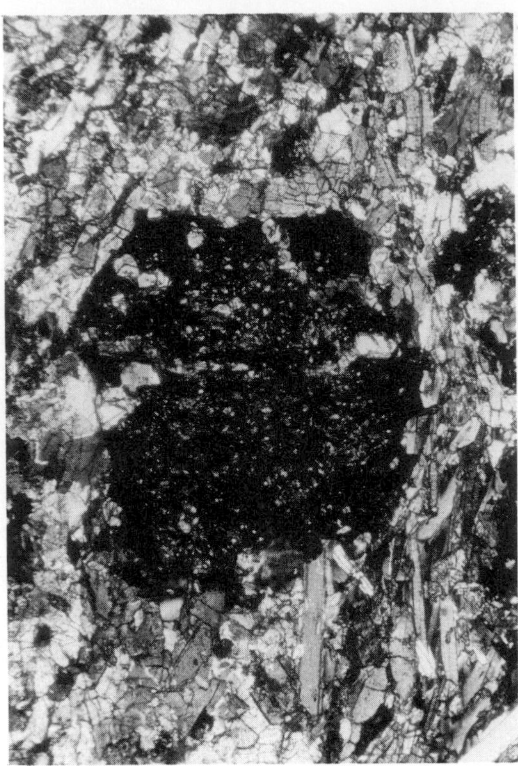

Fig. 2.14 Eclogite with subhedral pyrope garnet (with fine-grained inclusion trails) surrounded by omphacite. The very elongate crystals are mica. (View length measures 3.3 mm.)

Mineralogy: Eclogites have a distinctive essential mineralogy of green Na-rich clinopyroxene called omphacite, and pyrope, the Mg-rich garnet (Fig. 2.14). Although the garnet cannot easily be determined as Mg-rich in thin section, the omphacite is distinguished from other green clinopyroxenes by its high 2V. Plagioclase is always absent except as a retrogressive phase. Accessory minerals, some retrogressive, include hornblende and glaucophane (possibly after omphacite), rutile, titanite, kyanite, zoisite, quartz, phengite and chlorite (after garnet). The high-pressure silica polymorph coesite is sometimes present.

Textures: Most eclogites are coarse grained and many lack obvious mineral preferred-orientations. However, elongate grains of pyroxene sometimes define a foliation that is wrapped around the commonly euhedral crystals of garnet. If coesite is present, it is included within either pyroxene or garnet, and partial or complete inversion of coesite to quartz causes a characteristic set of radiating expansion cracks within the enclosing mineral (Fig. 5.16).

2.5.2 Rocks lacking obvious mineral preferred-orientations (hornfelses, and some granulites and charnockites)

Included here are well-recrystallized or crystallized rocks, lacking obvious mineral preferred-orientations, not already dealt with as a hydrothermally metamorphosed rock or as a special case in section 2.5.1. The main rock types considered come under the general heading 'hornfels', and represent products of thermal metamorphism. Also included are some granulites and charnockites.

(a) Hornfels

Hornfelses usually comprise some mixture of an equidimensional mosaic of minerals such as calcite or quartz, and randomly oriented inequidimensional minerals such as the sheet-silicates or porphyroblastic phases such as andalusite. The lack of preferred orientation helps to impart the tough, massive nature of a typical hornfels.

Hornfelses occur in the thermal aureoles surrounding igneous intrusions and, in the case of a simple single intrusive event, the temperature immediately adjacent to the contact is the mean of the combined initial temperatures of the magma and country rock, plus possibly as much as 100°C generated by the heat of crystallization (Jaeger, 1957, 1959). The rate at which rocks conduct heat is slow so that rocks, say 2 km from the contact of a 4 km-wide pluton, do not reach their metamorphic maximum until >400 000 years after intrusion, well after the immediate contact rocks have started cooling (Fig. 2.15). One can predict that pyroxene-hornfels facies rocks (*c.* 650–800°C) only occur adjacent to large basic intrusives, and that sanidinite-facies rocks (>800°C) can only occur adjacent to vents where there is a flow of high-temperature magma over a lengthy period of time.

There is no hard and fast dividing line between hornfels and schist. Many large plutons strain the country rock during emplacement, and depending on time/temperature relationships, schistose fabrics may then be imparted. This is most likely to occur during multiple episodes of intrusion where country rock has been preheated before the particular emplacement that causes strain. One must not forget either that many thermal metamorphisms are imposed on already schistose or slaty rocks. Incipient effects are signified by the 'spotting' of slates and schists, but when the effects are more pronounced, it may be difficult to decide whether hornfels or schist is the more appropriate name.

Terminology, mineralogy and metamorphic facies: The principal name is hornfels, and this is usually qualified in one of two ways. Firstly, the parent rock type can be indicated as in **pelitic** or **semi-pelitic hornfels** (derived from muddy rocks), **calcareous hornfels**

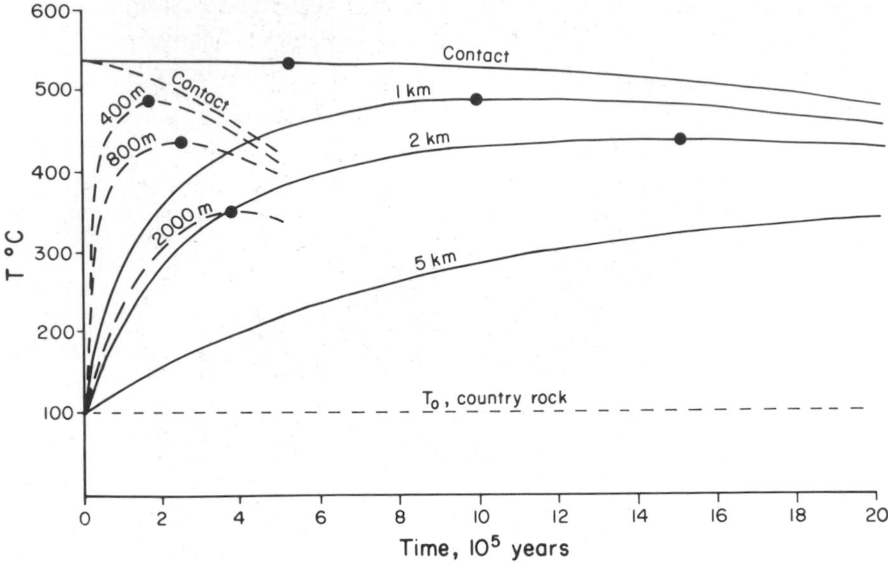

Fig. 2.15 Gradients of temperature in time at specified distances from contacts with granodiorite plutons 10 km (full curves) and 4 km (dashed curves) in diameter. Initial temperature of country rock 100°C, initial magma temperature 800°C, and heat of crystallization 80 cal g^{-1}. Reproduced from Turner (1981, Fig. 1.6), with permission of Hemisphere Publishing Corp., New York.

(from impure carbonate rocks) or **basic hornfels** (from rocks of basaltic composition). Secondly, mineral qualifiers are freely used, as in **cordierite–biotite–quartz hornfels**, for example. Both types of qualifiers can be used together. Quartz- or carbonate-rich rocks are called quartzite or marble whether they occur in a thermal or regional metamorphic setting, and the main distinguishing feature is the absence or presence of lattice mineral preferred-orientation (they are dealt with in section 2.5.1). Thermal metamorphism is a relatively short-lived affair and essentially lacking in strain energy, and therefore the effects are often incipient below 450°C (i.e., in the albite–epidote hornfels facies). Most well-developed hornfelses represent either the hornblende or pyroxene-hornfels facies. The three main types in terms of parent rock are described below.

Pelitic and semi-pelitic hornfelses: Quartz and/or feldspar form equidimensional mosaics, and tabular sheet-silicates have a random orientation (**decussate texture**), often occupying some or all of the grain boundaries of quartz and/or feldspar (Fig. 2.16A). The Al-rich pelitic rocks are characterized by the growth of the low-*P* minerals andalusite and/or cordierite, both of which tend to be poikiloblastic and sometimes

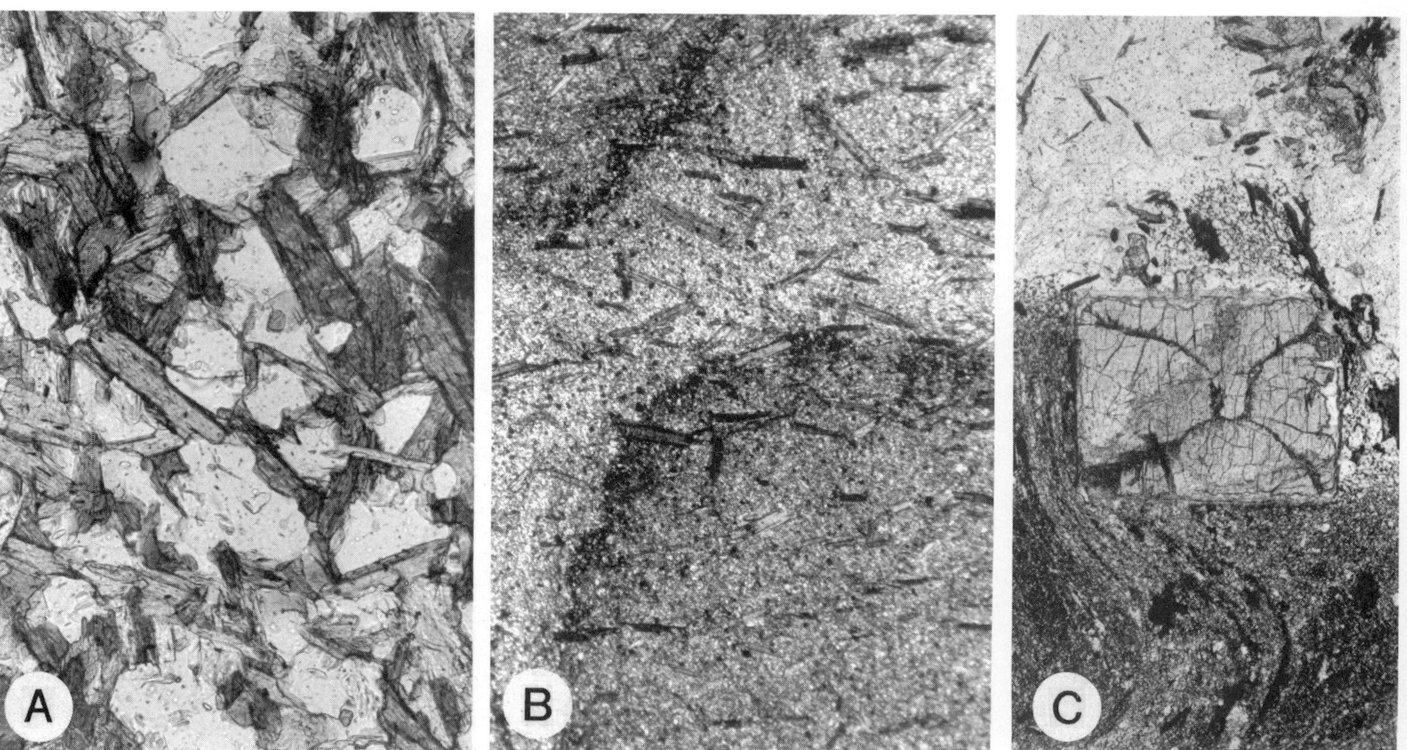

Fig. 2.16 (A): Decussate texture displayed by biotite in Greenland Group hornfels, NZ. (B): Parts of two large, spongy (poikiloblastic) cordierite crystals in hornfels, Wisconsin Range, Antarctica. Cordierite overgrows biotite crystals that already had a preferred orientation. (C): Andalusite (variety chiastolite) in slate, Buller Gorge, NZ. Discussion in text. (Lengths of view measure 0.83 mm (A) and 3.2 mm (B) and (C).)

so riddled with inclusions of quartz and sheet-silicates that one may fail to notice the large spongy crystals that exist between the inclusions (Fig. 2.16B). This problem is most acute when andalusite or cordierite is altered to sheet-silicates (frequently the case). Andalusite is frequently in the form of chiastolite (section 3.5.5), especially in graphitic slates (Figs 2.16C, 3.33 and 3.34). Apatite and tourmaline are common accessories.

The pyroxene-hornfels facies (rather than hornblende-hornfels facies) is suggested by the presence of sillimanite instead of andalusite, the presence of metamorphic K-feldspar, and the absence of muscovite. Some pelitic (rather than semi-pelitic) rocks are so Al-rich that they lack quartz and crystallize corundum and/or spinel in the pyroxene hornfels facies. Almandine garnet and kyanite are not typical hornfels minerals, but if they do occur, then relatively high-pressure thermal metamorphism is suggested.

Pre-existing slates or schists with large poikiloblasts of andalusite (chiastolite), cordierite or staurolite are called **spotted slate** or **spotted schist**. In some cases the cleavage/schistosity may represent strain connected with the emplacement of the adjacent igneous body, rather than some earlier, separate regional metamorphic event.

Pelitic hornfelses immediately adjacent to basaltic vents, or enclosed as xenoliths in basaltic rocks, usually contain corundum and spinel as well as minerals typical of the sanidinite facies (mullite, sanidine and tridymite). If melting occurs (signified by the presence of glass and pitted and corroded crystals), the rock is called **buchite**.

Calcareous hornfelses: Carbonate-rich varieties, simply named marble, are dealt with in section 2.5.1. Given that marble is very susceptible to plastic deformation, a mild strain during thermal metamorphism may produce a calcite preferred-orientation, and certainly calcite twinning. Contact- and regional-metamorphic marbles therefore may be difficult to distinguish out of the field context. Typical mineralogical changes in impure calcite and dolomitic marbles are listed in section 2.5.1. Marbles in contact aureoles are particularly prone to metasomatism, especially the addition of silica. The resulting calc-silicate rock is called a **skarn**, and typically consists of some mixture of diopside, hornblende, epidote, grossular, vesuvianite, titanite and wollastonite in the hornblende hornfels facies, and the same minerals minus hornblende and epidote and plus anorthite in the pyroxene-hornfels facies. The distinction between skarn and an impure marble must be made in the field (can the skarn be correlated with similar compositions outside the contact aureole?). Textures of calcareous hornfels are typically a mixture of equidimensional mosaics of the principal and/or finer-grained phases, and randomly oriented euhedral to subhedral crystals of minor and/or coarser-grained phases (Fig. 2.17A).

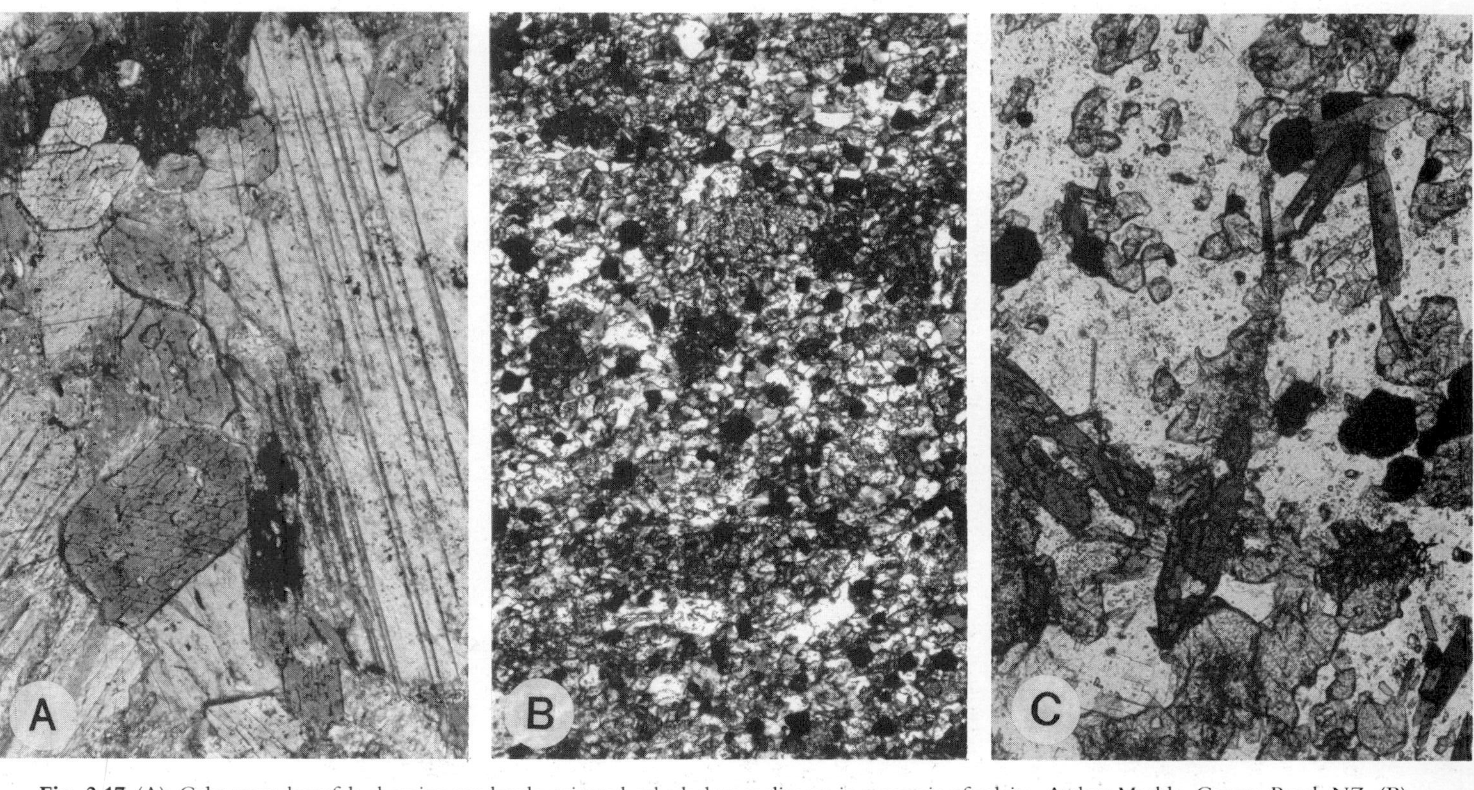

Fig. 2.17 (A): Calcareous hornfels showing randomly oriented euhedral tremolite set in a mosaic of calcite. Arthur Marble, Canaan Road, NZ. (B) and (C): Two examples of hornfels derived from basaltic tephra, Bluff, NZ. (B) is composed mainly of clinopyroxene, plagioclase and opaques, and (C) the same with the addition of subhedral hornblende crystals (dark). (View lengths measure 0.83 mm (A), (B) and (C).)

Basic hornfelses: Coarse-grained basic igneous rocks generally resist thermal metamorphism. However, fine-grained, possibly glassy volcanics may respond by a recrystallization involving removal of glass and general grain coarsening. The mineralogy of basic hornfelses of the pyroxene-hornfels facies is much the same as the original basalt, with clino- and orthopyroxene, olivine, calcic plagioclase and magnetite. In the hornblende-hornfels facies, hornblende (possibly with clinopyroxene, but not orthopyroxene) and calcic plagioclase are typical. The presence of biotite suggests an andesitic rather than basaltic parent. Common accessories include the usual apatite and titanite. The hornfelses are distinguished from their parent igneous rocks by the characteristic equidimensional mosaics of plagioclase. However, coarse igneous features (phenocrysts, amygdales, etc.) may survive texturally, especially in lower-grade hornfelses, even if the actual mineralogy has changed.

One can expect basaltic tephra to respond to thermal metamorphism more readily than basalt or andesite, and to produce similar hornfelses (Figs 2.17B and C). The only way to distinguish the two may be in the field, on the basis of relic sedimentary structures, or on the basis of compositional variations (more-or-less calcareous or pelitic) indicating mixtures of ash and background sediment. Decarbonated dolomitic marls may also approach basalt in composition (Leake, 1964), and field relationships and sedimentary compositional variations again need to be observed before a distinction can be made.

(b) Granulites and charnockites

Rocks of the granulite facies are water deficient, and metamorphosed at temperatures sufficient to cause melting in the presence of water. Not only does dry melting require a higher temperature than wet melting, but dry and water-saturated melting curves have positive and negative slopes in a $P-T$ diagram (Fig. 4.41). The granulite facies essentially lies between these curves. The necessity for dryness means that sheet-silicates and amphiboles are absent or present as minor components (there are inevitable transitions from amphibolite to granulite facies which are impossible to categorize strictly). In progressive metamorphism, the transition from amphibolite to granulite facies is marked by the removal of muscovite (which probably reacts with quartz to produce K-feldspar, sillimanite and water which is lost), and the partial or complete removal of biotite (reacts to produce orthopyroxene and K-feldspar plus water which is lost). Some hornblende may survive, but is likely to be replaced by pyroxene.

Many granulites lack obvious mineral preferred-orientations, although the mosaics of pyroxene and feldspar may have less-obvious lattice preferred-orientations not easily detectable using simple techniques with the sensitive tint plate (Chapter 6). Other granulites, particularly those with

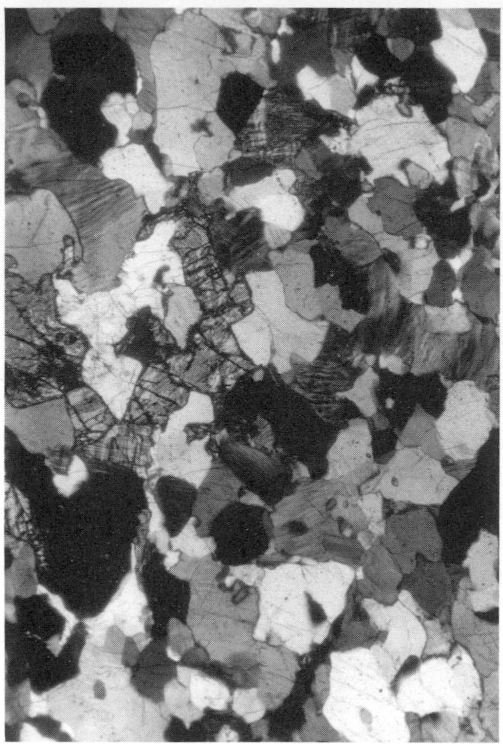

Fig. 2.18 Quartzofeldspathic granulite (with orthopyroxene, lower centre) lacking obvious mineral preferred-orientations. Charnockite from Madras, India. (View length measures 3.3 mm.)

abundant quartz and a little biotite or phlogopite, display obvious sheet-silicate preferred orientations and elongate grains of feldspar and quartz, and these are dealt with briefly in the next section (2.5.3).

Terminology and mineralogy: The name **granulite** should only be used for rocks of the granulite facies (previously the name was used more widely for 'granular' rocks of various facies, and for two-mica granites, as well as granulite-facies rocks). It can be qualified in any useful way, but the usual qualifiers are mineral names as in **quartzo-feldspathic granulite** for granitoid compositions (Fig. 2.18), and **pyroxene granulite** for gabbroic compositions. Granulite-facies rocks that contain orthopyroxene and appear to be igneous or possibly meta-igneous in the field can be called **charnockite**, and the terminology and textural characteristics of charnockites are given in section 1.6.

Mineralogically, granulites approach igneous compositions, with quartz, feldspar and pyroxene particularly common. Orthopyroxene is distinctive of the granulite (and its low-*P* equivalent pyroxene-hornfels) facies. Some hornblende and biotite may be present. In addition, Al-

and/or Mg-rich varieties may contain cordierite (low-*P*), sillimanite (medium-*P*), kyanite (high-*P*) or sapphirine; almandine garnet is another mineral commonly present. Do not assume that granulites indicate metamorphism in the lower crust. Many do, but cordierite in some granulitic terranes indicates otherwise, and indeed there is no simple boundary or distinction to be made between basic hornfelses of the low-*P* pyroxene hornfels facies and pyroxene granulites (lattice preferred-orientations are likely to exist in the latter but may not be obvious). In granitoid composition granulites, Ca-rich perthites and antiperthites are typically present.

2.5.3 Rocks with obvious mineral preferred-orientation (slate, schist, gneiss, greenschist, blueschist, amphibolite, some granulites and charnockites, and migmatites)

Included here are the most common and well-known metamorphic rocks, rocks that are well crystallized or recrystallized, and which display an obvious mineral preferred-orientation. They characterize regional metamorphic terranes in orogenic zones, although some schistose rocks also develop during localized strain around intrusive igneous rocks.

Mineral preferred-orientations are the subject of Part Three. Here it is appropriate to introduce the following general terms used in the field or during routine thin-section study:

1. **L and S tectonite fabrics:** Mineral preferred-orientations may give a linear structure to a rock (rather like the grain of wood), especially if minerals like quartz, feldspar and amphibole have rod shapes and/or sheet-silicates are highly crenulated (Fig. 2.19). Such a rock can be called an **L-tectonite**. On the other hand, if quartz and feldspar grains are platy and/or sheet-silicates have a planar preferred-orientation (Fig. 2.19), the rock is an **S-tectonite**. There is no sharp boundary between L- and S-type fabrics: a complete spectrum exists between the two. Owing to this, terms like schistosity can be used for linear or planar structures, and it is a common mistake to allow only the latter.
2. **Slaty cleavage** describes a penetrative or pervasive fabric in fine-grained pelitic or semi-pelitic rocks. Roofing slates have a slaty cleavage, but the term is used more widely to include the poorly developed or spaced cleavages in semi-pelitic rocks, provided that the grain size of the oriented sheet-silicates is fine. Often the qualifier 'slaty' is omitted but, unless the context is clear, this is inadvisable because 'cleavage' is also used for structures unrelated to mineral fabrics (as in fracture cleavage).
3. **Schistosity** describes a penetrative or pervasive fabric along which the rock splits easily. The term is used for moderate- to coarse-grained rocks of any composition.

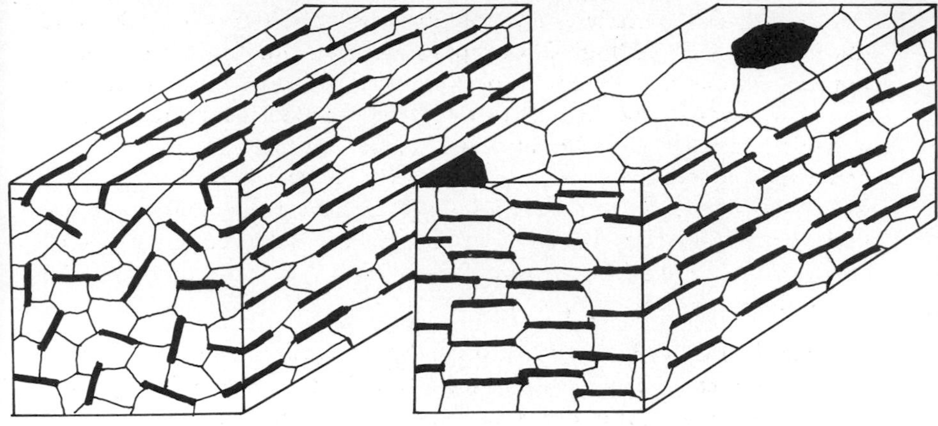

Fig. 2.19 L-tectonite fabric (left), with platy grains oriented about the lineation and mosaic grain shapes elongate parallel to the lineation, and S-tectonite (right), with platy grains and mosaic grain shapes defining a planar schistosity.

4. **Gneissosity** describes a penetrative or pervasive fabric along which the rock does not split easily. It can be applied to moderate- or coarse-grained rocks of any composition.
5. **Foliation** is a general term that encompasses slaty cleavage, schistosity and gneissosity; it also includes igneous lamination and preferred orientations.
6. Layering of metamorphic origin (section 5.3) is variously called **metamorphic layering** (less appropriately banding), **secondary layering**, or **segregation layering**. Some petrologists have used the term foliation for metamorphic layering, a practice not recommended here.

(a) Slates and semi-pelitic slaty rocks

Slates are very fine-grained rocks with a slaty cleavage, which develop from muddy sediment during regional anchimetamorphism. Slates are usually sub-greenschist facies but are not of suitable composition to develop characteristic minerals of the zeolite or prehnite or pumpellyite-bearing facies. Their mineralogy is similar to that of the original mud, namely muscovite (or 'illite'), chlorite, quartz; also present may be carbonaceous material, carbonates, feldspar and iron oxides or sulphides. Sheet-silicates undergo continuous modification during cleavage formation and increasing grade; X-ray peaks become progressively more sharply defined (a measure of the so-called 'illite and chlorite crystallinity'), allowing zoning within anchimetamorphic rocks to be established (Frey, 1987). This work is beyond the scope of this book as it requires the use of X-rays.

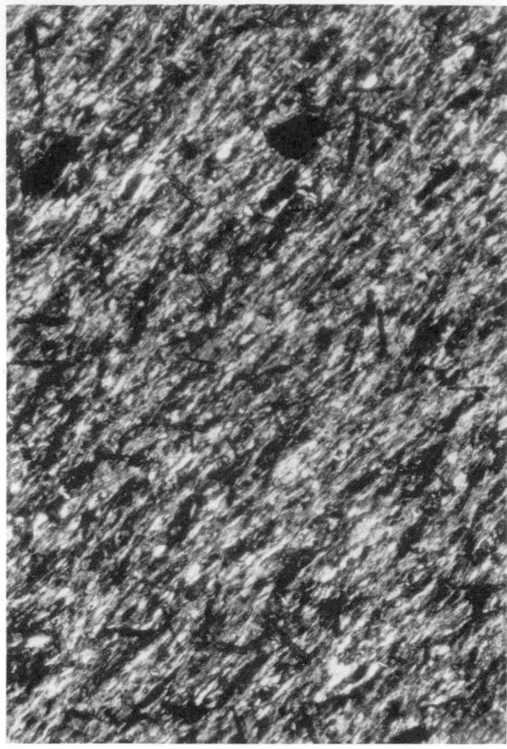

Fig. 2.20 Slaty cleavage, slate, Cumbria, UK. (View length measures 0.85 mm.)

Slate is by definition a very fine-grained rock, but similar textural and structural developments take place in the muddy matrix of semi-pelitic rocks. A detailed discussion of the origin of slaty cleavage itself is deferred until sections 5.3 and 7.2.

Structures and textures in slates and slaty rocks: The dominant texture is, of course, the alignment of sheet-silicates, often in an almost perfect planar preferred-orientation, as in quality roofing slates (Fig. 2.20). In some cases, the cleavage is 'spaced', so that planes rich in oriented sheet-silicates alternate with layers rich in quartz and feldspar, usually as a result of preferential pressure solution of quartz along the steeper limbs of crenulations (sections 5.3 and 7.2). In semi-pelitic rocks, similar structures develop, but the seams rich in sheet-silicates become wrapped around larger clastic sand grains, sometimes in an anastomosing fashion (Fig. 2.5A). Elongate, partly dissolved relics of clastic quartz may survive in sheet-silicate seams, and coarser quartz outside the seams may also be partially dissolved, especially where it is in contact with the seams. The sheet-silicate seams are often rich in iron oxides staining them yellow or red. Another feature that adds to the sum total of the cleavage in semi-pelitic rocks is the 'beards' of newly crystallized muscovite and chlorite that grow in pressure shadows on either side of clastic sand grains (Fig. 2.5B).

Slates tend to develop quartzose pressure shadows around relatively large pyrite cubes or framboidal aggregates. Relative rotations of matrix and pyrite are indicated by curved fibres of quartz in the shadows (Fig. 5.39C), a subject of great interest to structural geologists (Ramsay and Huber, 1983), and treated briefly in section 5.3.5.

(b) Phyllites, and the transition from slate to schist

Although slates have undergone substantial mineralogical change (as reflected, for example, in measures of 'crystallinity') and thorough structural transformation, grain size is not substantially increased, and associated sediments may well rank as non-metamorphic. Semi-pelitic rocks are slaty to some degree, but the most appropriate name is likely to be slaty sandstone or metagreywacke, not schist. In general, it is only in the greenschist facies (sometimes the blueschist facies) and higher-T facies that the majority of lithological types become thoroughly metamorphic, and true pelitic schists develop. Boundary problems are inevitable: how does one separate slate from schist, and when does a metagreywacke become a quartzofeldspathic schist? One approach is to invent a name to cover transitional types. Such a solution is given by the name **phyllite**, widely used for slaty rocks with a micaceous sheen on cleavage surfaces, but not coarse enough to be considered schistose. In my view, such names serve only to create two problems instead of one (how does one separate slate from phyllite, and phyllite from schist?).

(c) Schists

Leaving aside the special rock types blueschist and greenschist, considered later, most schists are pelitic or some mixture of pelite with sandy and/or carbonate material. Schists are coarse enough for crystals to be clearly visible in the field, and they split easily along directions determined by preferred grain-shape orientations.

Schist terminology: Schist can be qualified in any useful way, but common qualifiers refer to the nature of the parent rock, as in **pelitic schist**, and **semi-pelitic schist** (mud, and mud–sand mixtures), or to the mineral content, as in **biotite schist**. Some names such as **quartzofeldspathic schist** and **talc schist** seem to fall in both camps, indicating substantial quartz/feldspar and talc contents, but also indicating 'greywacke' and ultrabasic rock parents. Mineral qualifiers may constitute a complete list (with the more abundant mineral first), as in **chlorite–biotite–muscovite–quartz schist**, or the dominant or prominent or

Table 2.1 Pelitic mineral assemblages in typical Barrovian-type sequences

Zone	Possible minerals
Chlorite	Chlorite, muscovite, quartz, albite, stilpnomelane
Biotite	Biotite, chlorite, muscovite, quartz, albite, epidote
Garnet	Garnet, biotite, chlorite, muscovite, quartz, albite, plagioclase, epidote, chloritoid
Staurolite	Staurolite, garnet, biotite, muscovite, quartz, plagioclase
Kyanite	Kyanite, staurolite, garnet, biotite, muscovite, quartz, plagioclase
Sillimanite (1)	Sillimanite, kyanite, staurolite, garnet, biotite, muscovite, quartz, plagioclase
Sillimanite (2)	Sillimanite, K-feldspar, garnet, biotite, muscovite, quartz, plagioclase
Common accessory minerals	Calcite, ilmenite, magnetite, hematite, graphite, tourmaline, apatite, zircon, and in kyanite zone, rutile

Notes: 1. Biotite may form earlier in K-feldspar bearing semi-pelites.
2. Chloritoid and staurolite form only in Al-rich pelites, not in semi-pelites.
3. The first sillimanite zone results from the transformation of kyanite to sillimanite and reactions involving staurolite. The second sillimanite zone results from the breakdown of muscovite and the consequential formation of K-feldspar with sillimanite, and often coincides with migmatite formation.
4. Chlorite through garnet zones are within the greenschist facies, staurolite through sillimanite, the amphibolite facies.

most significant mineral or minerals may be referred to. The problem with this liberal system is that **garnet schist**, for example, could simply indicate the presence of garnet, or it may have been used because the rock belongs to the garnet zone.

Pelitic schist mineralogy and metamorphic zones: Pelitic rocks of very low grade are relatively insensitive to change (hence the need for X-rays to determine the subtle changes of 'crystallinity' in slate). This compares with metagreywacke and metatephra where numerous mineralogical changes affect unstable igneous material. As grade increases, roles are reversed: pelitic rocks become more sensitive to change than those of more direct igneous parentage. Hence, most mappable zones in regional metamorphic terranes are based on pelitic lithologies. This book is not the place for a full treatment of metamorphic zones, but it is important to remember the following: zone boundaries are drawn at the

Table 2.2 Pelitic mineral assemblages in sequences of lower pressure than Barrovian types

Zone	Possible minerals
Chlorite	Chlorite, muscovite, quartz, albite
Biotite	Biotite, chlorite, muscovite, chloritoid, quartz, albite
Garnet	Garnet, biotite, chlorite, muscovite, quartz, albite, plagioclase
Staurolite + andalusite	Andalusite, staurolite, garnet, biotite, muscovite, quartz, plagioclase
Cordierite	Cordierite, andalusite, staurolite, garnet, biotite, muscovite, quartz, plagioclase
Sillimanite	Sillimanite, andalusite, cordierite, garnet, biotite, muscovite, quartz, plagioclase, K-feldspar

Notes: 1. The principal difference from the Barrovian type is the absence of kyanite and the presence of andalusite and cordierite.
2. In many sequences, the first-formed low-pressure schists contain biotite.
3. Garnet and staurolite do not always appear, and the garnet is usually Mn-rich, especially the first-formed crystals.
4. Andalusite and cordierite may appear even earlier in some sequences. The presence of some minerals, and their order of appearance, when present, is variable from area to area, so that the division between greenschist and amphibolite facies (or their hornfels equivalents) is less easy to define than for Barrovian sequences.

point of entry of some distinctive new mineral; the next zone boundary does not necessarily or usually signify the disappearance of the preceding index mineral, so biotite survives from the biotite zone right through the greenschist and amphibolite facies into the sillimanite zone (although it continuously changes in composition). It is a common mistake for beginning students to regard the presence of garnet, for example, as indicating the garnet zone; in fact garnet survives through several succeeding zones.

The successive mineral changes in pelites are described in tabular form for: 1. The classic Barrovian sequence (Table 2.1); 2. Relatively high geothermal gradient (low-*P*) sequences (Table 2.2); and 3. Relatively low geothermal gradient (high-*P*) sequences (Table 2.3).

Pelitic schist textures: Textures are dominated by the platy habit of sheet-silicates and their preferred orientation, and the resulting schistosity is typically axial planar to folds (not necessarily seen on the thin-section scale). Although sheet-silicates may exhibit some degree of bending and shearing, especially in connection with second-generation

Table 2.3 Pelitic mineral assemblages in sequences of higher pressure than Barrovian types

Zone	Possible minerals
Pyrophyllite – chlorite	Chlorite, pyrophyllite, Fe-carpholite, quartz, phengite
Chloritoid – carpholite	Mg-carpholite, chloritoid, quartz, phengite plus either chlorite or pyrophyllite
Kyanite – chlorite	Chlorite, chloritoid, quartz, phengite, plus kyanite or talc or garnet
Talc – kyanite	Talc, kyanite, chloritoid, garnet, chloritoid, quartz, phengite

Notes: 1. Biotite is absent.
2. The assemblage talc – kyanite is called whiteschist (Schreyer, 1973).
3. Carpholite-bearing zones represent the blueschist facies, kyanite-bearing zones, the eclogite facies.
4. The zones noted above have been established in the European Alps (Chopin and Schreyer, 1983). Elsewhere, true pelites are absent or rare in high-pressure metamorphic sequences (the sediments in accretionary prisms are usually Ca rich, rich in igneous debris and immature).
5. In New Caledonia, relatively Ca-rich 'pelitic' schists first consist of phengite, chlorite, quartz and albite, then lawsonite enters the assemblage, followed by epidote and finally hornblende and omphacite (Black, 1977), all possibly accompanied by glaucophane. Common accessories consist of graphite and titanite, the latter replaced by rutile at high grades. Estimated conditions for the entry of hornblende and omphacite are $P > 10$ kbar and $T > 530°C$, indicating a transition from the blueschist to the eclogite facies.

microfolds or intersecting foliations (Fig. 2.21A), strain effects on the grain scale are not severe. Indeed, it is common for the sheet-silicates around microfolds to be completely annealed rather than bent, thus giving a tiled effect (Fig. 2.21B). In pelitic schists, quartz is not the dominant mineral, and quartz grains are often elongate, not due to deformation, but because of the spacing of adjacent sheet-silicates (Fig. 2.22). Quartz and feldspar may be segregated into layers parallel to schistosity and/or into veins by solution-transfer processes, leaving residual layers even more rich in sheet-silicates and opaque materials (section 5.3). Chloritoid, garnet, staurolite, kyanite, andalusite and cordierite tend to form porphyroblasts (5.1.1), feldspar, chlorite and biotite less frequently so. Porphyroblasts are often poikiloblastic (Fig. 2.23), and the nature of the inclusion trails, and the relationship of the external schistosity to the porphyroblast and inclusion trails, are subjects of great significance in determining metamorphic histories; they are discussed fully in section 5.5.

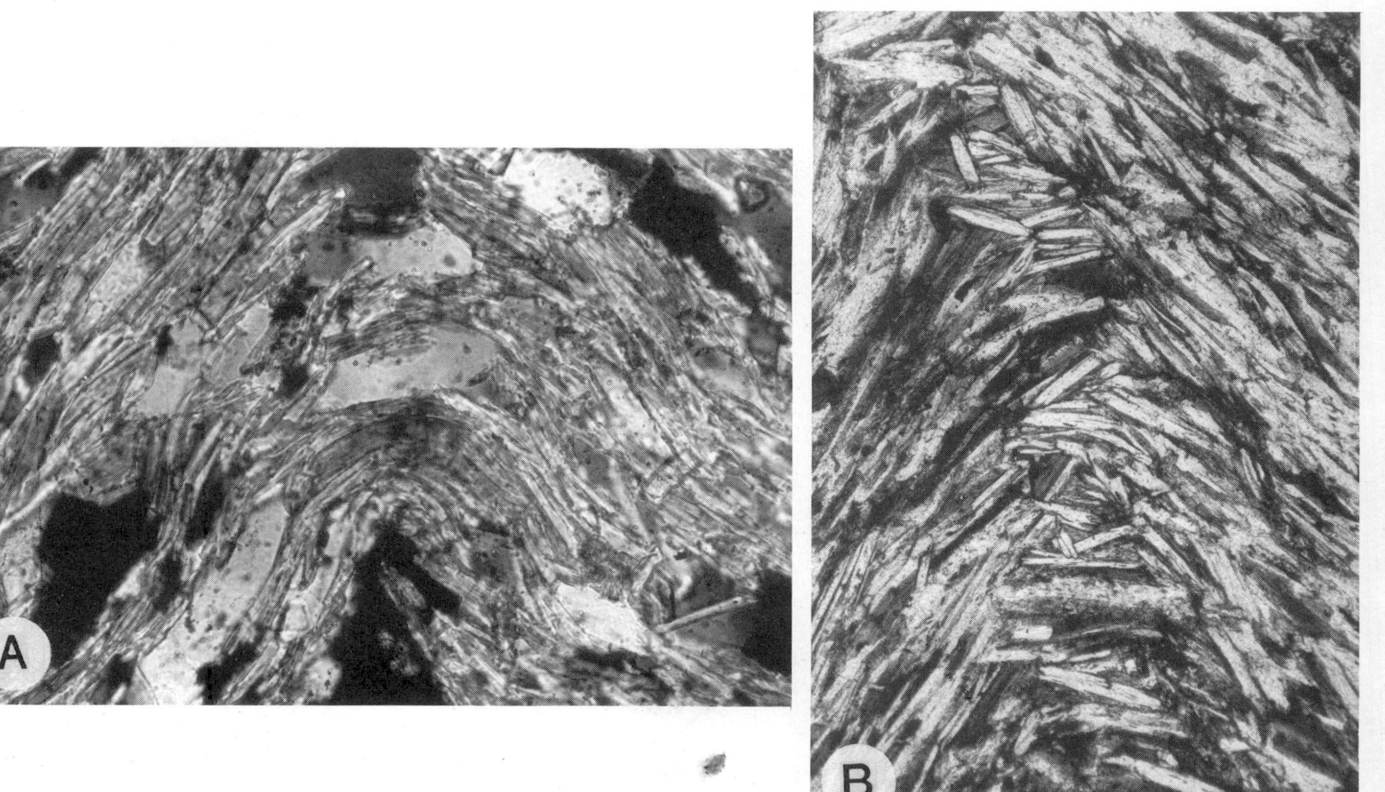

Fig. 2.21 (A): Crenulations that bend pre-existing micas. Springs Junction, NZ. (B): Crenulations where pre-existing micas are recrystallized to give a tiled effect. Brittany, France. (View lengths measure 0.35 mm (A) and 3.3 mm (B).)

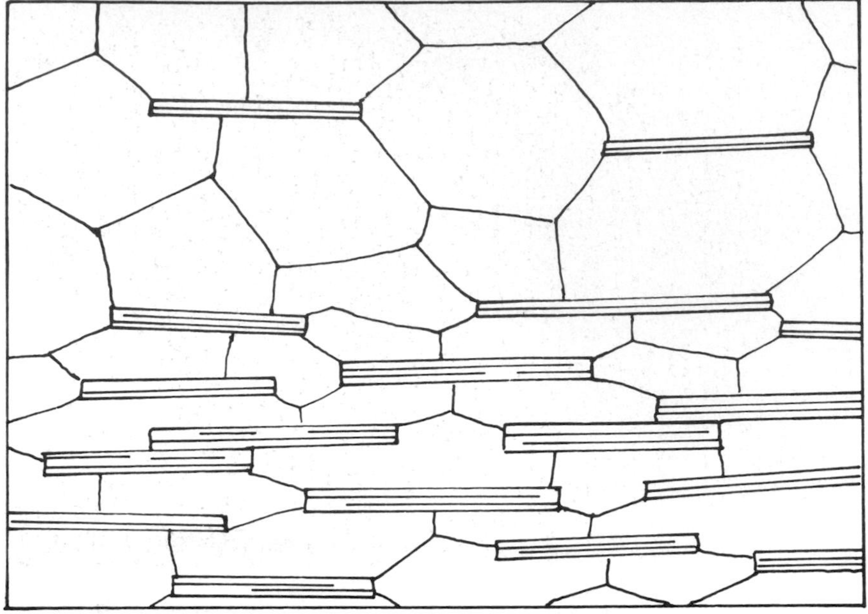

Fig. 2.22 Quartz–mica schist illustrating how aspect ratios of quartz depend on mica spacing.

Semi-pelitic and quartzofeldspathic schists: These lithologies become schistose somewhat later within a metamorphic sequence than pelitic schists, but the incipient effects are always to be seen as pressure-solution seams, partially dissolved quartz grains where they contact pressure solution seams and sheet-silicate 'beards' in the strain shadows about coarse clastic grains (Fig. 2.5). These lithologies are less aluminous than pelites, and therefore the most Al-rich index minerals do not always form (staurolite, kyanite, andalusite, sillimanite and cordierite). Thus, in the New Zealand Otago Schists (a relatively monotonous sequence of quartzofeldspathic schists), the Barrovian chlorite, biotite and garnet zones are present (garnet is seldom abundant), but the other index minerals are almost completely absent; the principal change of mineralogy from the greenschist-facies garnet zone into the amphibolite facies is that of albite + epidote to calcic oligoclase (plus residual epidote in many cases), and a garnet–oligoclase zone therefore replaces three of the classic Barrow zones.

Textural zones, widely used for quartzofeldspathic schists in New Zealand and California, and deserving of greater use elsewhere, are based on the degree of metamorphic segregation layering (section 5.3) where quartz and feldspar are segregated from the sheet-silicates by pro-

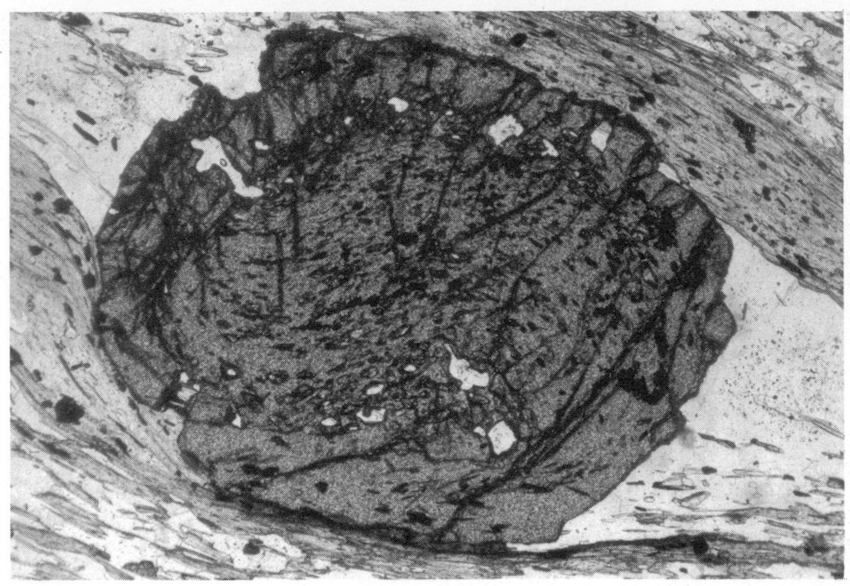

Fig. 2.23 Garnet porphyroblast (with S-shaped inclusion trails) in a mica-schist. Pikikiruna Schist, NZ. (View length measures 3.3 mm.)

cesses of solution transfer. They are designed specifically for use with medium-grained metagreywacke, and were originally mapped within the chlorite zone of the Otago Schists in New Zealand as subzones called chlorite I, II, III and IV. Later it was found that the subzones cross-cut mineralogical zones, and they are now called textural zones (Bishop, 1972b; Norris and Bishop, 1990). Definitions of the zones are:

I: Indurated metagreywacke. No foliation, no secondary layering.
IIA: Poorly developed cleavage. No secondary layering.
IIB: Well-developed cleavage. No secondary layering.
IIIA: Very well-developed cleavage/schistosity. Incipient secondary layers (1 mm–1 cm long) parallel to foliation.
IIIB: Very well-developed schistosity. Thin segregation layers (0.1–0.2 mm thick and >1 cm long) parallel to foliation.
IV: Very well-developed schistosity. Well-developed secondary layering (>0.2 mm thick).

Textures within the quartz–feldspar segregation layers vary from regular granoblastic mosaics to mosaics where the crystals are strained and elongate to some degree. In the greenschist facies, albite displays only simple primary albite twins, whereas oligoclase in the amphibolite facies may show profuse pericline deformation twinning. Sheet-silicate rich layers display a similar range of textures to those of pelitic schists, with prominent schistosity defined by mica and chlorite, elongate residual grains of quartz and feldspar, concentrations of opaque minerals, and the occasional porphyroblast of garnet, etc. Crenulation of the sheet-silicates and inclusion trails in the porphyroblasts may provide a means of determining metamorphic and structural histories (section 5.5).

Calcareous schists: There is every possible gradation between marble and pelitic schist, and schistosity defined by sheet-silicates in calcareous schists may be enhanced by platy or elongate calcite grain shapes. Zonal schemes for calcareous pelitic schists are known, one published example (Ferry, 1983) being:

1. Ankerite zone: ankerite, quartz, albite, muscovite, calcite and chlorite (equivalent to pelitic chlorite zone).
2. Biotite zone: biotite, ankerite, quartz, albite or plagioclase, muscovite, calcite, chlorite (= pelitic biotite and garnet zones).
3. Amphibole zone: Ca-amphibole, quartz, plagioclase, calcite, biotite, chlorite (= pelitic garnet or andalusite and staurolite zones).
4. Zoisite zone: zoisite, Ca-amphibole, quartz, plagioclase, calcite, biotite, K-feldspar (= pelitic andalusite and staurolite or sillimanite zones).
5. Diopside zone: diopside, zoisite, Ca-amphibole, calcite, quartz, plagioclase, biotite, K-feldspar (= pelitic andalusite and staurolite or sillimanite zones).

Accessory minerals include graphite, pyrite at low grades, pyrrhotite at all grades, apatite, tourmaline, ilmenite, titanite and at high grades scapolite and garnet. The appearance of minerals in calcareous pelitic schists is strongly influenced by partial pressures of water and CO_2, and in the case described by Ferry (1983), the rocks were infiltrated by large volumes of externally derived water. One can expect mineralogical sequences to differ markedly from region to region, and it is advisable to sample adjacent pelitic rocks to make a more confident assessment of metamorphic grade. A good summary of these problems is provided by Yardley (1989, Chapter 5).

Talc schists: Ultrabasic rocks hydrothermally metamorphose to massive bodies of serpentinite (section 2.5.1), but talc can also form, especially at temperatures >500°C, and it may be accompanied by magnesite if CO_2 was present in the fluids. Massive soapstone or magnesite–talc rock is sometimes produced, but the schistose equivalent is called a talc or talc–magnesite schist. Magnesite is often subhedral, bound by rhombs, and it is distinguished from calcite and dolomite by its total lack of twinning.

(d) Gneiss

Gneiss is the general name for coarse-grained, moderate- to high-grade metamorphic rocks that have a foliation but do not split easily along it. Foliation is such a non-specific term that it is useful to use the specific gneissosity for the fabric of a gneiss. Excluded from gneiss are special rock types such as amphibolite, quartzite and marble. The names granulite, charnockite and gneiss are to some extent interchangeable, a problem discussed later under granulites and charnockites.

Many gneisses are layered, but not all are, and it is a mistake to confuse layering with gneissosity. There are, for example, many schists with well-developed metamorphic layering.

Some gneisses represent plutonic igneous rocks deformed in the granulite or amphibolite facies, and these rocks never went through a schistose phase. On the other hand, low- to medium-grade schists may become gneissose with increasing grade, and the reasons for this include:

1. A reduction in the proportion of highly inequidimensional minerals such as mica and hornblende, responsible for schistosity (micas, for example, break down in the amphibolite facies to form K-feldspar and various other minerals).
2. The common presence of reinforcing mats of fibrolite within the sheet-silicates.
3. The aspect ratio of most minerals reduces with increasing grade, and this reduces grain boundary area per volume, and usually increases the number of interlocking grains. This applies also to the sheet-silicates (Rivers and Fyson, 1977) which may be as well oriented in gneisses as schists, but are less elongate.
4. General grain coarsening reduces the possibility of splitting on the hand specimen scale.

Terminology: The rules for qualifying gneiss are as liberal as those for schist. Thus the parent rock may be indicated, as in **pelitic gneiss, basic gneiss** and **granitic gneiss**, or a mineral list or a selection of prominent or important minerals can be used, as in **quartz–biotite–garnet gneiss** (Fig. 2.24). Structural and textural terms can also be used, as in **layered** (or **banded**) **gneiss** and **augen gneiss** (containing eye-shaped structures formed by flowage or strain around large crystals). Gneisses are coarse and relatively even-grained, and therefore many resemble plutonic igneous rocks, so it is particularly useful to indicate whether the gneiss is indeed of igneous (**orthogneiss**) or sedimentary origin (**paragneiss**). Deformed rocks that are gneissic but retain obvious igneous textures can usefully be described as **gneissic granite**, etc. Boundaries such as the one between granitic gneiss and gneissic granite have not been adequately defined.

Fig. 2.24 Hand specimen of gneiss. Wide dark layers are garnet, thin black layers are biotite, and white mainly quartz. Southern Alps, NZ. (View length is 17 cm.)

Mineralogy and textures: Many gneisses have a mineralogy that approaches that of igneous rocks of similar composition. Thus granitic gneisses contain K-feldspar, plagioclase, quartz, biotite, possibly muscovite and often sillimanite. Basic gneisses consist of plagioclase and clinopyroxene, frequently hornblende, orthopyroxene in the granulite facies (except at very high pressures), and garnet at relatively high pressures. In quartzose and micaceous gneisses, quartz and mica are often strained and wrapped around feldspar crystals to give an augen structure, similar to that in mylonites (Fig. 2.7C). Indeed, there is every possible gradation between gneissic and mylonitic foliation, with the possibility of recrystallized ribbon-like areas of quartz in gneiss. In basic gneisses, all of the minerals are relatively non-ductile, so augen-type textures are less common. Common in such rocks is a more-or-less granoblastic mosaic of recrystallized plagioclase, pyroxene and hornblende.

(e) Greenschists, blueschists and amphibolites

Within most schistose and gneissose terranes, pelitic rocks show the greatest variation and the most rapid change in mineralogy (hence their use for the purposes of zoning). Rocks of basaltic composition (including basalt, basaltic tephra and decarbonated mixtures of pelitic and calcareous material) change relatively slowly, and form three easily identifiable rock types, greenschist, blueschist and amphibolite, depending on pressure and temperature. Conveniently, the three principal facies of dynamothermal metamorphism are named after these three forms of metabasalt, and the presence of either blueschist, greenschist of amphibolite is diagnostic of those facies. The eclogite facies (less commonly associated with schist–gneiss terranes) is named after yet another form of metabasalt, discussed in section 2.5.1.

Mineralogy: Greenschists are so named because of their green colour in hand specimen, and the presence of two or three of the green minerals actinolitic hornblende, chlorite and epidote. Albite is also present, and common accessories include magnetite, titanite and apatite. Quartz, biotite, stilpnomelane and calcite are sometimes important components, particularly in layers representing greater proportions of background sediment in metatephra. **Blueschists** (Fig. 2.25A) are named after their blue (often very dark) colour in hand specimen, and the presence of the alkali amphibole glaucophane or crossite. The blue amphibole is accompanied by some combination of albite, actinolitic hornblende, chlorite, phengite and zoisite, together with pumpellyite, lawsonite, stilpnomelane and titanite at lower grades, and rutile, epidote, omphacite, almandine garnet and paragonite at higher grades. **Amphibolites** (Fig. 2.25B) consist essentially of hornblende and plagioclase (usually andesine), possibly accompanied by epidote at low grades, and almandine garnet. Quartz, carbonates and biotite may be common in layers within metatephra that are particularly rich in background sediment. Titanite, ilmenite and apatite are common accessories.

Textures and structures: Greenschists, blueschists and amphibolites all display strong mineral preferred-orientations. In general, the finer-grained greenschists and blueschists split more easily parallel to the grain fabric than the typically coarser-grained amphibolites. In all cases, the amphiboles tend to be subhedral, elongate parallel to their *c*-axes, and oriented so as to define a strong lineation (Fig. 7.29). Segregation of amphibole and feldspar is quite common, especially in amphibolites, and this helps to give the rock a planar fabric. Feldspars commonly form granoblastic mosaics. Amphiboles are among the most resistant of minerals to plastic deformation, and plagioclase is not plastic within the greenschist and blueschist facies. One must therefore conclude that most of the preferred orientations are due to growth, possibly based on some pre-growth anisotropy, but more likely as a result of anisotropic growth during strain. This is discussed in sections 7.5 and 7.7; the point of mentioning it here is to note the essential difference between pelitic

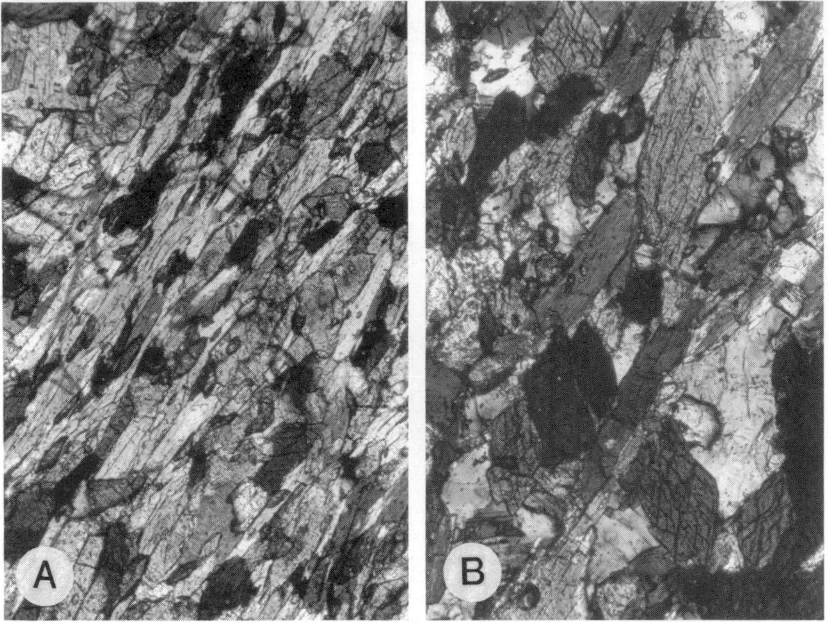

Fig. 2.25 (A): Blueschist, California, dominated by strongly oriented glaucophane, but also with epidote and titanite. (B): Amphibolite, Haast River, NZ, dominated by subhedral hornblendes together with plagioclase and epidote. (Lengths of view measure 0.82 mm (A) and (B).)

schists, quartzites and marbles, all of which deform plastically through a wide range of metamorphic conditions, and metabasaltic rocks, which even at high grades may not undergo plastic deformation easily. Textures consequently tend to be less diverse than in those other rock types, and much of the textural detail is produced by growth, not deformation.

Distinguishing metabasalt from metatephra from metamarl: In many cases one can expect to determine the nature of the parent rock from field relationships (for example, basaltic dykes are usually recognizably different from bedded tephra). However, field relations are not always clear, and neither is field information always available, and the best guide in thin section may be the nature of lithological variation. Thin, variable layers rich in quartz, carbonates and biotite, for example, strongly suggest a sedimentary origin. There may still be a problem deciding whether the rock is metatephra or a decarbonated metamarl. Compositional changes are likely to be more varied and less sharply defined in the latter. Leake (1964) has discussed the geochemistry of the problem. Do not confuse metamorphic segregation layering (where plagioclase and the amphibole are concentrated in separate layers) with primary sedimentary layering.

(f) Granulites and charnockites

Charnockites have been discussed in section 1.6 and section 2.5.2, and granulites in section 2.5.2 and briefly in the section on gneiss, above.

Terminological problems: Gneiss, granulite and charnockite are all possibly interchangeable names. However, charnockite should only be used for orthopyroxene-bearing rocks of possible igneous or meta-igneous appearance in the field (section 1.6), granulite should only be used for rocks certainly of the granulite facies (section 2.5.2), and gneiss should only be used for rocks with gneissosity. Without clear international guidelines on this subject, I recommend that the names charnockite, granulite and gneiss be used in a hierarchical fashion, in that order. In other words, a rock that can be called a charnockite should be, otherwise the name granulite should be considered, and if that seems inappropriate, gneiss should be applied. If at this stage, gneiss is inapplicable because of a lack of gneissosity, examine again the hierarchical systems given in sections 1.1.2 and 2.2, and look for some alternative name. Another approach to the problem of the three interchangeable names is to use one as a qualifier for another, as in charnockitic gneiss.

Textures: Many granulites and charnockites lack obvious mineral preferred-orientations and have granoblastic textures. However, some that contain quartz have a pronounced gneissic or mylonite-like foliation, defined by ribbons of quartz (Fig. 2.26A). In pyroxene granulites from New York, twinned ribbons of plagioclase occur (Fig 2.26B), and in the classic granulites of Saxony (Fig. 2.26C), quartz ribbons are made up of strain-free mosaics, presumably annealed at high temperature following strain.

(g) Migmatites

Migmatites are high-grade, pervasively heterogeneous rocks, that are partly metamorphic, partly igneous-like in appearance. The igneous-like part is most commonly granitic (in the field classification sense), and the metamorphic part is generally a biotite schist or gneiss. The heterogeneities are secondary, and their origin is controversial, with at least four possibilities to be considered:

1. An open system so that the igneous-like part is intruded from some neighbouring pluton.
2. An open system with the introduction of fluids that effected a significant metasomatism during the formation of the igneous-like part.

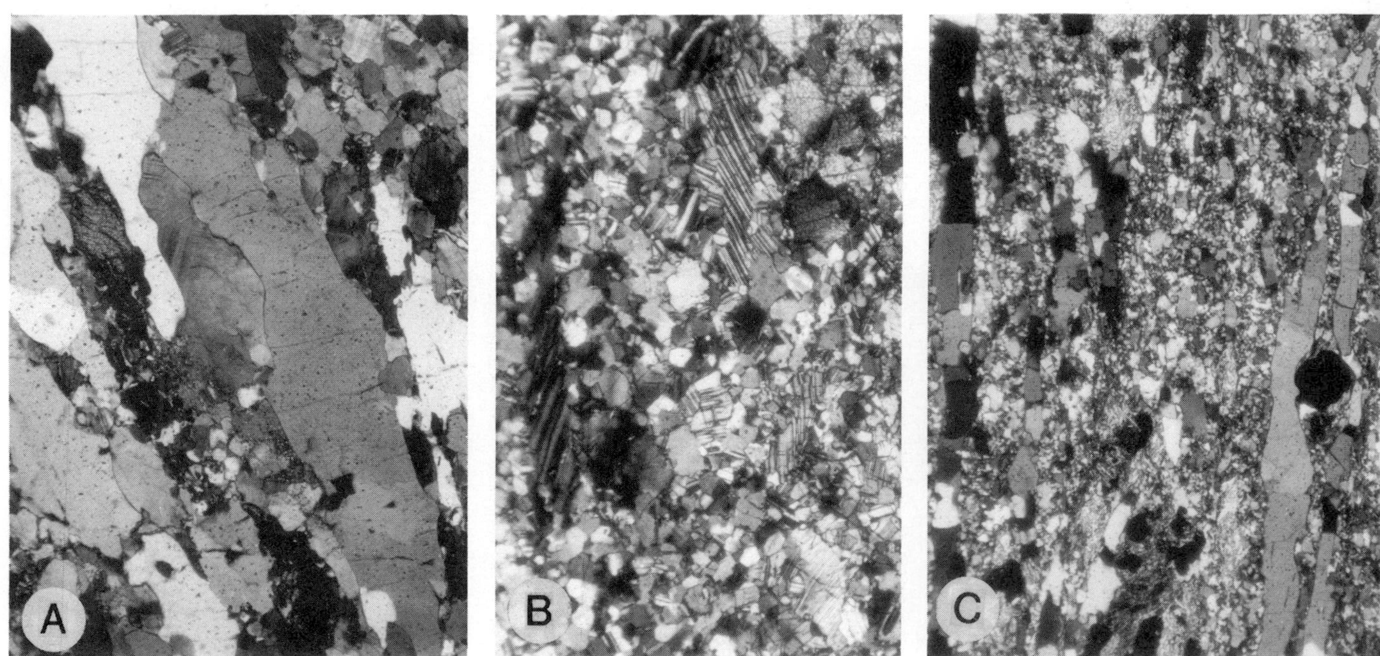

Fig. 2.26 (A): Charnockite, Brazil, with quartz ribbons. (B): Pyroxene granulite, New York, with ribbons of twinned plagioclase. (C): Granulite from Saxony with quartz ribbons consisting of strain-free mosaics that previously had been wrapped around garnets (as in centre right of view). (Lengths of view measure 3.0 mm (A), (B) and (C).)

3. A closed system. Partial melting is responsible for the migmatite structures.
4. A closed system. Metamorphic segregation is responsible for the migmatite structures.

Of course, before calling a heterogeneous rock a migmatite, one should be quite certain that the heterogeneities are secondary (this needs to be considered primarily in the field: can the same lithological variations be found in lower-grade equivalents?). In fact, many characteristic migmatitic structures, such as veins and the biotite-rich melanosomes bordering leucosomes, are not easily matched in sedimentary successions.

It is generally agreed that migmatites of type 1 are likely to occur only in the immediate vicinity of the pluton responsible. As far as type-2 migmatites are concerned, there has been a general reluctance to invoke metasomatism during migmatite formation, since the 'granitizers' lost the 'granite controversy' debate in the 1950s; however, this situation is changing, and petrologists are again considering the matter seriously (e.g., Olsen, 1985; Sorensen, 1988). Most recent debate on the subject of migmatite genesis has centred on types 3 and 4, partial melting versus metamorphic segregation and the possible ways of deciding between the two are investigated in section 5.3.6.

In terms of thin-section petrography, migmatite heterogeneities are usually on a scale such that the igneous-like and metamorphic parts have to be viewed separately, and they should be described in much the same way as any other granitoid, quartzofeldspathic gneiss or biotite-rich schist or gneiss. In terms of field description, the following terms (McLellan, 1988) are in general use:

Leucosome: Leucocratic part of migmatite – usually granitic.
Melanosome: Melanocratic part of migmatite – rich in mafics.
Mesosome: A descriptive term for an unmigmatized body within a migmatite. It may possibly be interpreted as (a) palaeosome, or (b) resister material.
Neosome: Leucosome and melanosome.
Palaeosome: An interpretative term for the precursor to neosome.
Resister: An interpretative term for rocks in migmatite that resisted migmatization (resisters should differ in composition from palaeosomes).
Restite: An interpretative term for the residual material from which mobile material has been extracted.

Structural types of migmatite include:
Agmatitic: Breccia-like.
Diktyonitic: Mesosome is extensively veined by leucosome.

Schollen: Blocks or rafts of non-leucosome in leucosome.
Stromatic: Layered.

2.6 METASOMATIC ROCKS

Whether or not a rock is metasomatic is impossible for the microscopist to determine. Field criteria are particularly important, and geochemical and isotopic data are essential to judge the matter properly. There are the usual boundary problems deciding what is or is not a significant enough chemical change to be called metasomatism. Here, it may be useful simply to list the rock types most commonly ascribed to metasomatic activity. The following list is not claimed to be exhaustive:

1. **Serpentinite** (sections 2.5.1 and 3.5.4) formation involves a major influx of water. Two extreme cases can be considered:
   (a) a very high volume increase occurs because sheet-silicates are less dense than olivine and pyroxene, Si is gained, some Mg is lost and Ca is lost from pyroxene;
   (b) no volume change takes place, in which case large amounts of Si, Mg, and Ca must be lost from the system. Evidence for Ca loss is often found in associated rodingites (shown in next section).
2. **Rodingites** are striking rocks in hand specimen, consisting of euhedral Ca-pyroxenes set in a featureless compact mass of white or pale-green grossular or hydrogrossular. They form dyke-like bodies or replace country rock *en masse* in the vicinity of serpentinite, and represent the Ca expelled during serpentinization of peridotite.
3. **Spilites** (section 2.3.1) are hydrothermally metamorphosed basalts, and significant losses in Ca and Si and gains in Mg and water may be involved.
4. **Anthophyllite–cordierite rocks** are amphibolite-facies rocks that are peculiarly rich in Mg and poor in Ca. They cannot be correlated directly with known igneous or sedimentary rocks, and a metasomatic origin is therefore suspected. One view is that they represent hydrothermally metasomatized basalts, enriched in chlorite (as are spilites), subsequently metamorphozed in the amphibolite facies (Vallance, 1967).
5. **Skarn** is a calc-silicate rock identical to calcareous hornfels (section 2.5.2) but which represents marble in a contact aureole that has been subjected to silica metasomatism.
6. **Greisen** (sections 1.7 and 4.2.3) are formed from S-type granitoids as a result of F-, B-, Li-, W- and Sn-rich fluid activity. The minerals Li-mica, fluorite, topaz and quartz are typical, and greisen are associated with **tourmaline–quartz** metasomatic rocks (Fig. 4.43), and W–Sn mineralization.

7. **Fenites** are desilicated crustal rocks formed by Na- and K-rich fluids expelled from carbonatite or highly undersaturated alkaline magmas. Fenites have the composition of syenite or feldspathoidal syenitoid or foidolite, and often mimic igneous rocks of those compositions.
8. **Granitization** is a hypothetical metasomatic process by which granitic composition rocks, including gneissic and migmatitic types, are produced. The idea was stretched to extreme lengths by some of its proponents during the 'granite controversy', but most petrologists were completely won over by the opposing magmatists after the work of Tuttle and Bowen (1958) was published. The unfortunate consequence for many years was that it became taboo for petrologists to discuss metasomatism in connection with granite or migmatite formation, despite the widespread advocacy of metasomatism in association with a wide range of other crustal and mantle processes. Fortunately, petrologists are now beginning to reintroduce some balance into the subject (Olsen, 1985; Sorensen, 1988).
9. **Mylonites** form in shear zones where considerable diffusive transfer and metasomatism may take place. In section 2.4, the metasomatic production of quartz–muscovite ultramylonites from granitic rocks in Scotland is noted.

Part Two

Textures and Microscopic Structures

3

Crystals and crystallization

3.1 INTRODUCTION

Igneous and metamorphic rocks are for the most part crystalline, but the ways in which the crystals develop are extremely diverse. Essential minerals in igneous rocks develop by some process of crystallization during cooling, but these crystals are usually accompanied by others that form during secondary processes such as deuteric alteration and weathering; in addition, xenocrysts may be present. In metamorphic rocks, the minerals often represent some mixture of premetamorphic relics and newly crystallized material, any of which may have been modified by recrystallization or alteration.

Understanding igneous and metamorphic rocks involves understanding crystals, what they are, how they form and what controls their numbers, shapes and sizes. Perhaps the single most important observation leading to modern crystallography was that of Steno, in the seventeenth century: he noted that angles between crystal faces for any one mineral are constant, regardless of how well developed those faces are, and regardless of the particular shape and size of the crystal. The best way to illustrate this is in stereographic projection, whereby all particularities of an individual crystal shape are removed, and only angular relationships remain (Fig. 3.1). Such a constant angular relationship and symmetry is now known to reflect the lattice structure.

Given that minerals are characterized by particular angular relationships, symmetries and lattice structures, it is all the more remarkable that any one mineral may display such staggering variation in form and habit. Thus crystals may be euhedral, subhedral or anhedral, they may be relatively large or small, and they may exhibit a great diversity of form ranging from regular forms with planar faces to skeletal crystals, and from highly inequidimensional to squat forms. In addition, crystals may be twinned or zoned, and one mineral may be intergrown with another. The reason for such diversity is in part found in the variety of circumstances which lead to crystallization, including the cooling of magma, devitrification of glass,

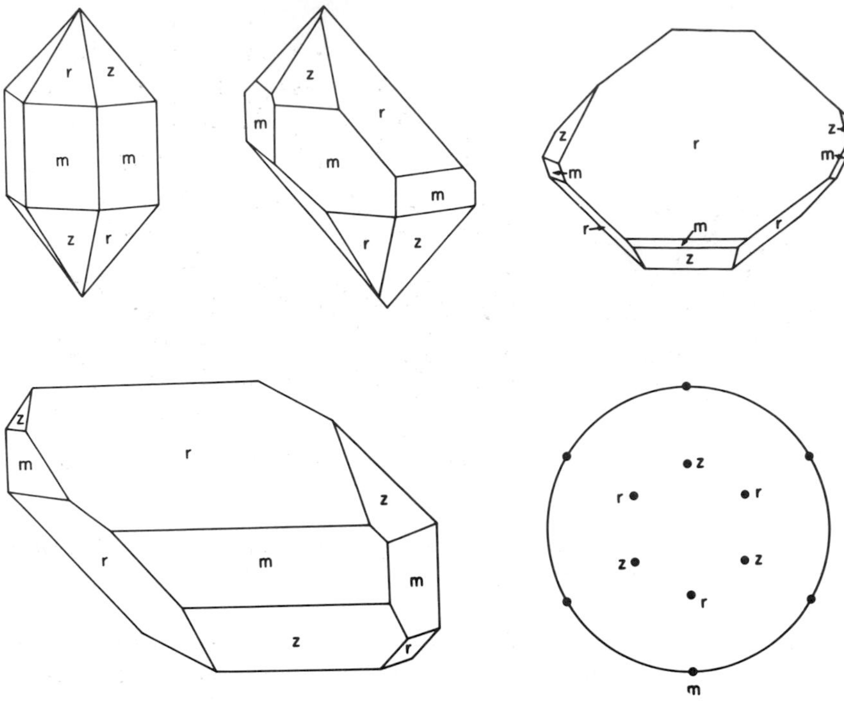

Fig. 3.1 Examples of the possible variety of quartz crystal shape, copied from Frondel (1962). All examples provide identical plots in stereographic projection (upper hemisphere).

production of new stable phases concomitant with the breakdown of unstable material and precipitation from solution. One of the principal goals of petrology, and the underlying theme in Part Two of this book, is to understand this diversity.

In this chapter, the principles of crystallization are introduced, followed by a discussion of general textural features such as twinning, zoning, mineral intergrowths, replacement textures and so on. The distinctive aspects of crystallization in igneous and metamorphic rocks are dealt with separately in Chapters 4 and 5.

3.2 CRYSTAL NUCLEATION, DIFFUSION AND GROWTH

The development of a new crystal involves three processes, nucleation, diffusion and growth, all of which are characterized by 'rates' which vary according to circumstances. Whichever rate is the slowest tends to control the way that crystallization proceeds. For example, if diffusion rates are slow relative to growth rates, the likely consequence is the development

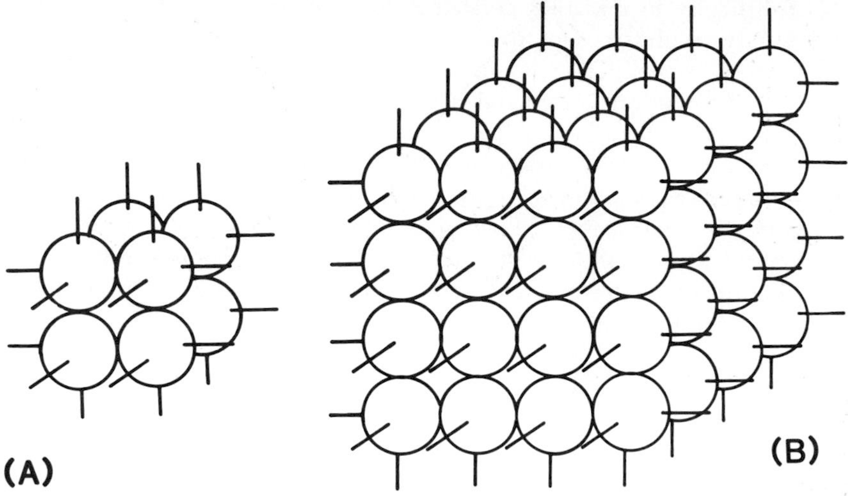

Fig. 3.2 (A): Very small embryonic cluster of eight atoms with only 50% of bonding satisfied. (B): Larger cluster of 64 atoms with 75% bonding satisfied.

of a skeletal or dendritic habit. This is because material necessary for growth is not made available at the growth site quickly enough for a stable regular crystal form to develop; instead, the crystal has to reach out towards the source of material in disequilibrium crystal form.

Nucleation

The initiation of crystallization from solution or melt is typically delayed. Thus crystals do not normally precipitate until some degree of supersaturation or undercooling has been achieved.

Although a reduction in total free energy of a system is the driving force for crystallization of a stable phase, this reduction is counterbalanced by the necessity for crystals to go through an embryonic stage. An embryo has a very large ratio of surface area to volume, and therefore it has a very high surface energy due to the large number of atoms not electrically balanced. For example, the bonding of a cluster of 8, 64 and 512 atoms respectively, may be only 50%, 75% and 87.5% satisfied (Fig. 3.2), and embryonic clusters spontaneously disassemble as easily as they assemble when conditions are otherwise right for crystallization. There is a critical size which nuclei must reach before the counterbalancing surface-energy effects are overcome; once the nuclei survive, growth can proceed. In general, simple-structured minerals will nucleate most easily. Thus in a basaltic magma, olivine and magnetite might be expected to nucleate more easily than plagioclase.

The difficulty in initiating crystallization can be termed the nucleation energy barrier, and it helps to explain many phenomena in igneous and metamorphic rocks. To overcome the barrier requires overstepping the appropriate conditions for crystallization of the stable phase, and this may involve supersaturation of solutions, undercooling of melts or an increase in temperature during progressive metamorphism. Disequilibrium in terms of texture and composition is the common result, as discussed later. The nucleation barrier may allow metastable phases to develop instead of stable phases if nucleation of the metastable phase is relatively easy.

Formation of nuclei from a melt or solution, simply by a sufficient degree of undercooling or supersaturation is termed homogeneous nucleation (Dowty, 1980a). Nucleation may happen more easily by heterogeneous nucleation which involves some other factor such as a 'seed' crystal or an impurity; the wall of a magma chamber may induce nucleation, and in the case of metamorphic crystallization, existing crystal boundaries or other defects may be nucleation sites. All these potential sites of nucleation are characterized by relatively high surface energy, and it is the balancing of that with the high energy of the embryonic new phase that allows the nucleation energy barrier to be overcome. Seed crystals are the most easily understood sites of heterogeneous nucleation: their lattices have certain structural elements in common with the nucleating phase. Closely related minerals, such as two species of feldspar, or the chain silicates pyroxene and amphibole, are sufficiently similar for one mineral to nucleate on the other, but one also finds unexpected seed relationships between sillimanite and muscovite, for example. The ability of one mineral to nucleate on the structure of another is called epitaxy; when the two crystals have an essentially parallel crystallographic orientation it is called syntaxy.

Seed crystals may sometimes induce the nucleation of an unstable phase. In the Al_2SiO_5 polymorphic system, for example, there is only a small difference in free energy between the polymorphs, and the fact that sillimanite may be seeded by muscovite may energetically favour its nucleation and growth outside its stability field.

Diffusion

Crystals cannot grow unless the appropriate ions are able to move towards the crystal for attachment. In general, diffusion is easiest in fluids, less easy in glass, and slowest in crystals. Hofmann (1980) shows that in silicate liquids and glass the rates of diffusion are related to ionic radius and charge, so that elements with a large ionic radius (such as K^+) or high charge diffuse relatively slowly. In metamorphic rocks, diffusion takes place most readily along grain boundaries and most quickly by solution transfer.

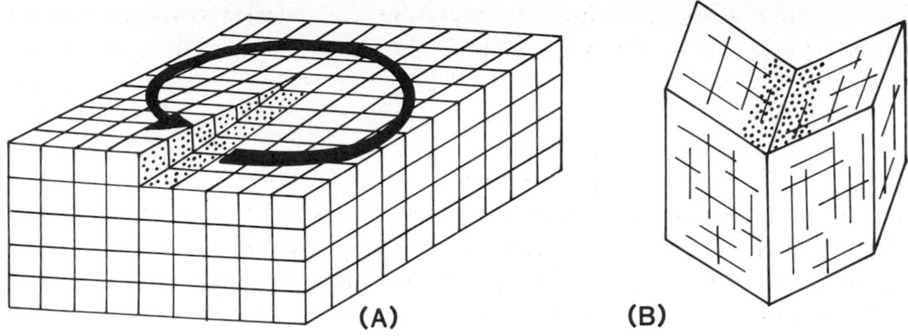

Fig. 3.3 (A): Screw dislocation which provides steps to which new atoms can be attached. Growth proceeds as a layer that forms a spiral around the locus of the dislocation. (B): Twin re-entrant angle amplifies the potential of screw dislocations in providing suitable growth sites.

In the simplest of solutions and melts, ions may be totally free to diffuse towards crystals, but this is not the case in viscous magma, glass or metamorphic rocks. Polymerization is general in magma, which means that Al, Si and O are already linked, although without the degree of order present in crystals (Mysen, 1988). For these elements to diffuse requires the polymers to be disassembled, and the introduction of water and alkalis may help achieve this. In metamorphic rocks, unstable minerals must react before the various elements are free to diffuse to sites of nucleation and growth.

Crystal growth

Crystal growth is dependent on nuclei developing, and on diffusion moving the appropriate elements to the growth site. Crystals are never perfect, but are characterized by defects or dislocations such as illustrated in Figs 3.3 and 5.21. In fact, crystals would not grow at all easily were it not for defects, and particularly favoured attachment sites for atoms are the steps on crystal faces associated with screw dislocations (Fig. 3.3A). Re-entrant angles associated with twin boundaries enhance the effects of screw dislocations (review of Sunagawa, 1987), making these especially attractive sites for growth (Fig. 3.3B).

The most stable crystal shapes are simple faceted forms, such as are commonly developed by phenocrysts. Such equilibrium shapes are produced if diffusion rates are faster than crystal growth rates, and this is generally the case in low-viscosity magma with only slight undercooling. However, diffusion rates slow as viscosity increases with higher degrees of undercooling; at the same time, nucleation and growth rates may be increasing, and the net result is that disequilibrium textures develop. These include skeletal, dendritic and spherulitic forms, and they will be discussed in detail in section 4.1. Dendritic and skeletal forms are also efficient at releasing the latent heat of crystallization.

3.3 TWINNING

Twinning in igneous and metamorphic minerals is so common and conspicuous that it is essential to understand its nature and genesis. In fact, twinning is even more common than at first appears, because in a number of minerals such as quartz and the sheet-silicates, twinning is present but invisible in thin section.

A twin is an intergrowth of two crystals of the same species with some special crystallographic relationship between the two parts. In terms of atomic structure, the special relationship is a lattice plane or row that is common to both parts; in terms of geometry, the relationship is such that one part can be rotated (usually by 180°) about an axis (the twin axis) into the orientation of the other; often the one part is also a mirror image of the other across a plane (the twin plane). The plane that joins the two parts is called the composition plane. The composition and twin planes may be the same, as in albite twinning, but that is not always the case, as for example in Carlsbad twinning (see Smith and Brown, 1988 or most standard mineralogy texts for details of feldspar twin geometry).

The various mechanisms that produce twinning can be listed under two main headings, primary and secondary, and these are considered in turn below. Terms used to describe twins include simple twins – just two crystals joined together, and multiple or polysynthetic twinning – an intergrowth of numerous crystals in twin orientations.

3.3.1 Primary twins

Attachment of atoms during crystal growth is not necessarily easy, and indeed attachment to a perfectly smooth crystal face may require a nucleation step called ‘surface nucleation’. Atoms are most easily attached at sites of dislocations in the crystal structure (Fig. 3.3A). The lowest energy state is attained if atoms are attached in the same orientation as the host crystal. However, the twin orientation has certain features in common with the host, and it is also a favoured orientation (although second best) which may develop occasionally ‘by accident’. Such accidental developments are most likely to happen during rapid growth, immediately following nucleation, for example, when the high concentrations built up during supersaturation are quickly reduced.

It is not surprising, therefore, that growth twins are often simple twins, as commonly seen in feldspars (Fig. 3.4A). It is clear that the twin

Fig. 3.4 Primary twins and synneusis. (A): Simple growth Carlsbad twin in orthoclase in granite, Siberia Bay, NZ. (B): Stepped, irregularly spaced twin-lamellae in plagioclase in gabbro, Onawe, NZ. (C): Simple albite twin in plagioclase showing evidence for synneusis. Olympus Granite, NZ. (D): Cluster of oscillatory zoned plagioclase crystals in rhyolite joined by synneusis. In (C) and (D) note the independent zoning in the attached crystals, and the late stage zones that join the various parts. (View lengths measure 2.9 mm (A), 0.8 mm (B), 3.2 mm (C) and 3.0 mm (D).)

boundary was initiated at or near the start of growth, and that the paired crystal then continued to grow as a twin. A re-entrant angle at a twin boundary enhances the effect of dislocations making the boundary itself a favourable attachment site for further atoms during growth, and this may be a factor in the survival and development of simple twins (Fig. 3.3B).

Growth twins may continue to develop by accident during crystallization, and according to Vance (1961) are characterized by irregular spacing and steps along the composition planes (Fig. 3.4B). Vance (1969) also advocates the idea that simple twins may develop by 'synneusis', or the floating together of two crystals in magma, and their attachment in a twinned orientation. The idea is based on the fact that the {010} composition plane for albite and Carlsbad twinning is also the principal crystal face in feldspars and therefore a likely attachment surface. The principal thin-section evidence for synneusis is independent zoning in each twin part (Fig. 3.4C). Even more likely to be formed by synneusis are clusters of crystals in the same orientation, well illustrated by Fig. 3.4D. Dowty (1980b) expressed doubts about synneusis, pointing out that apparently independent zonal patterns could be an artefact of oblique sections of either simply zoned crystals or twins resulting from epitaxial nucleation of one crystal on the other (Fig. 3.5). Nevertheless, the experiment of Schaskolsky and Schubnikov (1933) with alum crystals showed that two growing crystals are likely to become cemented together if they accidentally contact one another in parallel or twinned orientation. Further work is needed to establish the relative frequencies of crystals truly joined by synneusis and those which are mere artefacts of the thin section, as suggested by Dowty (1980b).

3.3.2 Secondary twins

Subsequent to crystal growth, twins may develop by twin gliding during strain or during the transformation of a mineral from one crystal system to another, for example during cooling.

In **twin gliding** the crystal lattice is rotated during strain into the twin orientation. Familiar examples are albite and pericline twinning in feldspar and the twinning on *e* of calcite (Fig. 7.21). According to Vance (1961), deformation twinning in plagioclase is characterized by regularly spaced lamellae which may bend and wedge out in a systematic way (Fig. 3.6A). Deformation twins probably initiate as very thin wedges which progressively thicken as they spread through the crystal. The density of such twinning can be related to the differential stress, so that Rowe and Rutter (1990) show for calcite that the log twin density is proportional to stress and independent of grain size. For a particular grain size, the volume of twinning increases with differential stress (Fig. 3.7).

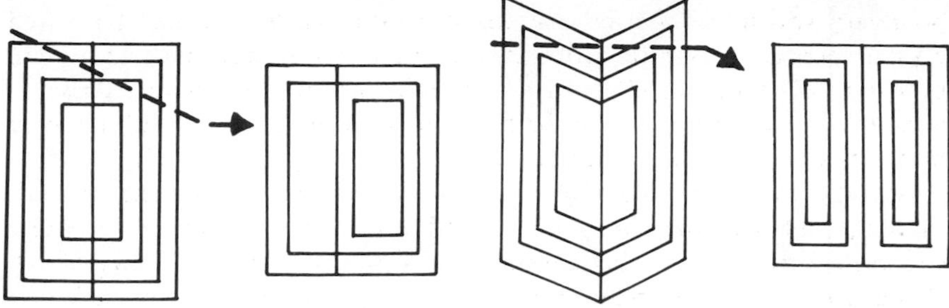

Fig. 3.5 Sections through simply zoned and twinned crystals along the dashed lines give the false appearance of independently zoned crystals joined through synneusis. After Dowty (1980b).

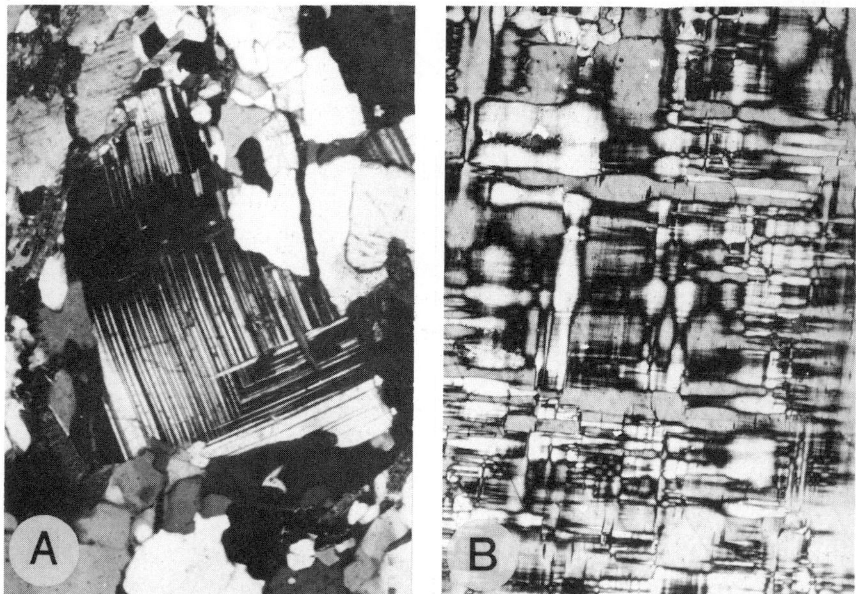

Fig. 3.6 Secondary twins. (A): Typical bending and wedging associated with deformation albite and pericline twinning in plagioclase, Constant Gneiss, NZ. (B): Transformation twinning in microcline, granite, Bavaria. (View lengths measure 1.8 mm (A) and 0.7 mm (B).)

Twin gliding operates with a unique glide line and sense of gliding. For the calcite twin in Fig. 7.21 the ideal principal compressive stress direction is indicated, and the fact that twin gliding only operates in response to a particular stress system allows one to determine palaeostress orientations. The method for calcite is well established (Turner and Weiss, 1963) and will be discussed more fully in Part Three. Lawrence (1970) has used

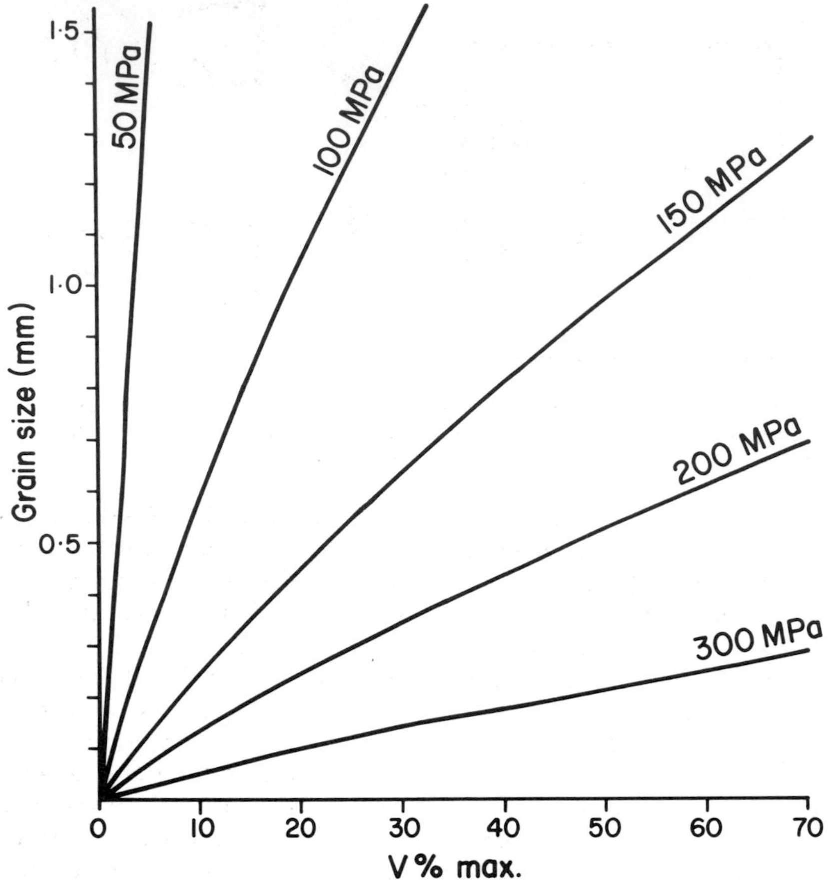

Fig. 3.7 Values of differential stress in MPa for various grain sizes and maximum volume % twinning (*V*). Based on equation 4 of Rowe and Rutter (1990).

a similar method for palaeostress determinations from deformation twinning in plagioclase.

It is not geometrically feasible for all twin laws to develop by deformation. For example, Carlsbad twinning can only develop by primary processes. On the other hand, albite and pericline twinning, and *e*-twinning in calcite can develop either by growth or deformation.

Transformation twinning occurs as a crystal changes from a higher to lower symmetry. The best-known examples are in low-temperature pseudo-cubic leucite, which is truly cubic at high temperatures, and in triclinic microcline (Fig. 3.6B) which develops from monoclinic feldspar during cooling. The result is a complex intergrowth of twins which can be

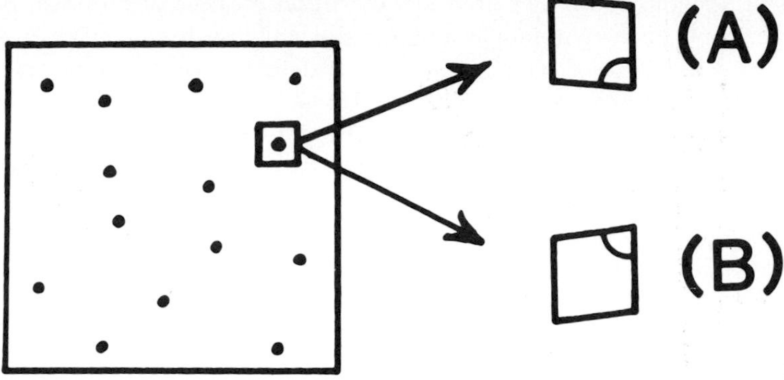

Fig. 3.8 A series of random points within the *b–c* section of monoclinic K-feldspar. On transformation to the triclinic state (microcline) there is an equal chance of these points adopting either of the orientations (A) or (B). Further discussion in text.

explained with the aid of Fig. 3.8. Consider a monoclinic K-feldspar, and the *b–c* section. The angle between *b* and *c* is 90°. On transformation to the triclinic state (at around 500°C) that angle departs progressively from 90°, but there are two equally valid options, A or B in Fig. 3.8, which happen to represent the two orientations of an albite twin. The change from monoclinic to triclinic will occur randomly throughout the crystal, and at any point within the crystal there is an equal chance of either orientation A or B nucleating. The result is a complex anastomosing albite-twin intergrowth. The same crystal transformation also produces the possibility of pericline-twin formation. The sum result of the transformation is therefore a complex anastomosing intergrowth of pericline and albite twinning. In pseudo-monoclinic orthoclase, the twinning is submicroscopic, producing a 'tweed' texture in thin section. Coarsening of this to the more familiar cross-hatched or tartan pattern typical of microcline results from tectonic strain or deuteric alteration (Brown and Parsons, 1989). The intimate twin intergrowth also obviates the strain that would result if a single triclinic crystal were developed. Anorthoclase undergoes a similar transformation, although its pericline twin plane (an irrational plane) has a different orientation from that in microcline.

3.3.3 The occurrence of common feldspar and carbonate twins

Carlsbad twinning only forms during growth, and may form in either monoclinic or triclinic feldspar. It is very common in igneous rocks (Gorai, 1951), especially in calcic plagioclase, but is otherwise rare.

Albite and pericline twins are restricted to triclinic feldspar, can be of primary or secondary origin and occur in both metamorphic and igneous rocks; in igneous rocks they are most common in sodic plagioclase. Albite and pericline glide-twinning is not possible in low-temperature ordered albite, and this is confirmed by the lack of deformation twinning in albite in the greenschist-facies New Zealand Haast Schists, for example; however, simple primary albite twins are present (Shelley, 1977, 1989a). In the amphibolite facies, secondary pericline twinning becomes abundant as the feldspar composition changes to calcic oligoclase. In many high-grade metamorphic rocks, both albite and pericline twinning are equally well developed. Much albite twinning in igneous plagioclase results from minor high-temperature deformations associated with igneous emplacement.

In calcite, *e*-twinning is possible from room temperature up through the entire range of metamorphic conditions (Turner and Weiss, 1963). It is discussed further in Part Three. Dolomite twins on *f*, but not easily at low temperatures. Experimental work indicates a minimum temperature of 300°C for twinning, most recently confirmed by Barber *et al.* (1981), and this explains why dolomite twins are not found in sediments or very low-grade metamorphic rocks.

3.3.4 Twins invisible in thin section

The most familiar examples are in quartz and the sheet-silicates. Quartz is very commonly twinned, and the twin laws involve a parallelism of the *c*-axes. Since the *c*-axis (optic axis) is the only optically identifiable direction in thin section, the twins are invisible. In sheet-silicates, the twin-plane is {001}, and the *c*-axis is 'reflected' across this plane (Fig. 3.9). However, in thin section there is no trace of the *c*-axis, and extinction is parallel to the visible {001} cleavage, so twinning can be seen only in euhedral sheet-silicates displaying re-entrant angles.

3.4 ZONING

3.4.1 Concentric compositional zoning in igneous crystals

Compositional zoning in minerals implies that complete equilibrium was not reached during crystallization, and this happens most frequently during rapid volcanic crystallization. It is least common in metamorphic rocks.

The following discussions will often focus on plagioclase zoning which is particularly common and conspicuous. It is common because equilibration of plagioclase crystals with a melt would involve the difficult interchange of Si and Al; it is conspicuous because plagioclase is triclinic, and optical directions rapidly change position relative to crystal axes with changes of composition, the basis of the extinction-angle methods of determination.

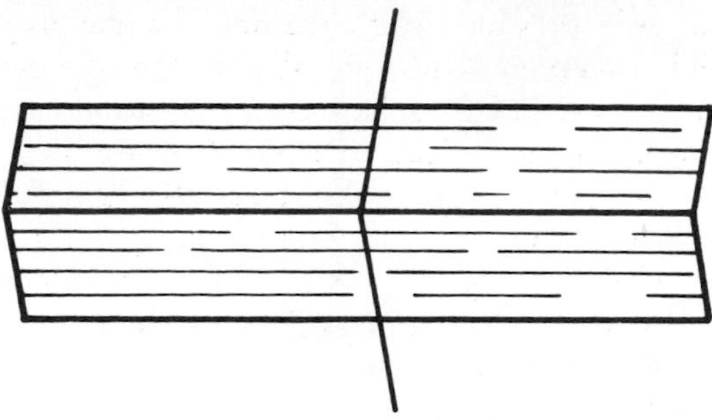

Fig. 3.9 Euhedral twinned mica in an *a–c* section, and with the *c*-axis position in both twin parts shown. Discussion in text.

Some of the more complex zonal patterns involving sieve textures and mantling of perhaps extensively resorbed phenocrysts will be discussed later in section 4.1.6.

(a) Normal zoning in igneous plagioclase

Albite and anorthite form a complete solid-solution series at high temperatures, as represented by the familiar solidus and liquidus curves given in Fig. 3.10. It is convenient to use this diagram here even though in natural melts the curves always lie at lower temperatures due to the effects of other components; the principles remain the same. In the pure system, a melt of composition *X* crystallizes plagioclase of composition *M* which causes the remaining melt to become enriched in the albite component and move towards and beyond *Y*. Crystallization proceeds with falling temperature, and the plagioclase crystals become increasingly albite rich (from *M* towards and beyond *N*). Such a variation in composition from more An-rich core to more Ab-rich rim is called normal zoning, and it is very common in igneous rocks although often interrupted by oscillations in composition, as described below.

If the plagioclase system were in perfect equilibrium, crystals would react continuously with the melt to produce unzoned plagioclase with the same composition as the original melt. The fact that normally zoned crystals are common shows that equilibration is generally much slower than the rate of crystallization, especially in volcanic rocks. Plagioclase does not react readily with the melt because that would involve a change in the Al/Si ratio.

(b) Reverse zoning in igneous plagioclase

This is the opposite of normal zoning with more Ab-rich feldspar towards the centre of a crystal. Loomis (1982) shows how a reduced undercooling

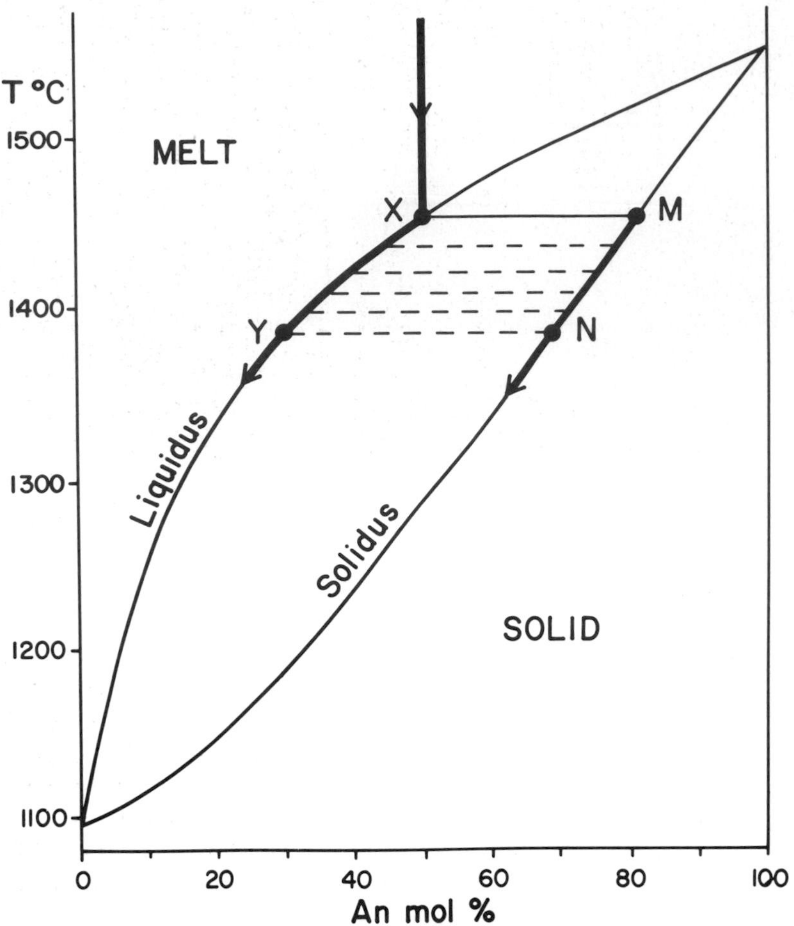

Fig. 3.10 Crystallization of the plagioclase solid-solution series (after Bowen, 1913 with permission of the *American Journal of Science*). Discussion in text.

or a build-up in volatiles may produce reverse zoning in igneous plagioclase, but it is rare except as an integral part of oscillatory zoning, as described below.

(c) Oscillatory zoning in igneous plagioclase

The compositional changes across a plagioclase crystal are seldom perfectly regular, and in igneous rocks are frequently oscillatory in fashion, perhaps involving many tens of reversals or abrupt changes in An % (Figs 3.11 and

Fig. 3.11 Oscillatory zoning in plagioclase in andesite from White Island, NZ. (A): Concentric pattern of zones in both halves of a Carlsbad twin displaying a re-entrant angle. Crystal is part of a glomeroporphyritic cluster. (B): Complex pattern of zones; some boundaries cross-cut earlier zones. (View lengths measure 1.4 mm (A) and 2.2 mm (B).)

3.4D). The two main features that characterize oscillatory zoning are: abrupt changes which usually mark a sudden outwards increase in An % and which may feature corrosion of the underlying crystal surface; gradual changes of An % (either reverse or normal) in the plagioclase deposited between successive sharp breaks. The An % change at sharp boundaries varies from small to large (33% is reported by Nixon and Pearce, 1987, for one crystal); the gradual changes are always small (less than 10%). Examples of oscillatory zoning profiles taken from the recent literature are shown in Fig. 3.12.

Abrupt compositional changes require abrupt changes in the dynamic conditions of crystallization, and most researchers now advocate magma mixing. Thus Nixon and Pearce (1987) show that repeated injection of fresh, hot, basic magma into a chamber of already differentiated and cooled magma repeatedly caused resorption of already crystallized plagioclase. This will be discussed further in section 4.1.6.

The more gradual changes in oscillatory zoning are best interpreted in terms of local effects of disequilibrium crystallization (Loomis, 1982). In this context it is important to understand that liquidus–solidus curves for pure anhydrous plagioclase (Fig. 3.10) are depressed by the presence of other components. Water alone can depress the curves by several hundreds of degrees, and the shape of the curves changes too. Consider then the local environment of a crystal during magma cooling. As growth

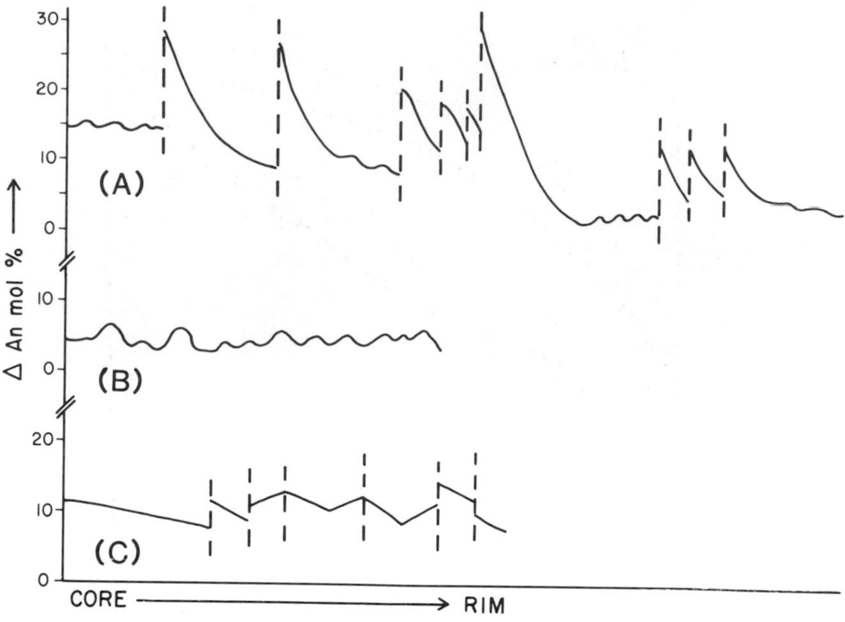

Fig. 3.12 Schematic illustration of typical oscillatory zoning profiles from core to rim. (A): Repeated major sharp reversals due to magma mixing caused by injection of fresh basic magma, as described by Nixon and Pearce (1987). (B): Small irregular oscillations due to local disequilibrium crystallization, as described by Loomis (1982). (C): Complex oscillatory zoning due to combinations of magma mixing and disequilibrium crystallization, as described by Stamatelopoulou-Seymour *et al.* (1990).

proceeds, the components not required by the plagioclase crystal increase in concentration immediately adjacent to the crystal (such chemical gradients are well known from microprobe studies of crystals immersed in glass). These components will include H_2O. The solidus and liquidus curves for the immediate environs of the crystal become depressed (Fig. 3.13), and at a fixed temperature (T_1) the crystallizing plagioclase composition shifts from A to B. If, on the other hand, undercooling increases (T_1 to T_2), perhaps as the result of an increased cooling rate, composition moves from A to C.

Reverse zoning forms if undercooling decreases or residual components adjacent to the crystal face increase; normal zoning forms if undercooling increases or the residual components decrease. The changes are likely to be small and transient. Loomis (1982), for example, shows that the build-up of volatiles next to the crystal eventually leads to a density gradient that will in turn cause local convection. This convection will readjust the chemical composition of the magma next to the crystal; repeated convection will cause oscillations in plagioclase composition. Anderson

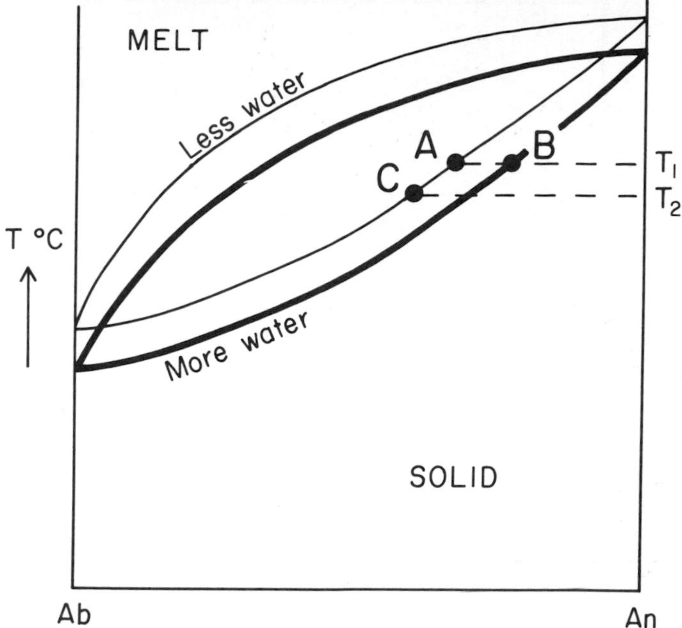

Fig. 3.13 Schematic illustration of how local disequilibrium crystallization can cause oscillatory zoning, as described by Loomis (1982). Explanation in text.

(1984) suggests the alternative mechanism of liquid shearing for interrupting the build-up of chemical gradients next to the crystal. He proposes that the crystal–liquid interface is sheared twice daily by tidal pulses.

Oscillatory zones are usually euhedral in character and the resorption surfaces associated with abrupt changes in composition cut across previous crystal faces and zoning patterns. However, not all embayments are the result of resorption: some result from dendritic growth, as demonstrated by convolute oscillatory zone patterns that parallel embayment surfaces (Fig. 3.14).

(d) Concentric compositional zoning in other minerals

Few minerals show the degree of zoning that plagioclase does. This may be because equilibration was relatively easy as, for example, in the olivine ($(Mg,Fe)_2SiO_4$) solid-solution series that requires only a redistribution of Mg and Fe (cf. the difficulty of redistributing Al and Si in plagioclase). On the other hand, many minerals are compositionally zoned but simply do not display it in thin section. Thus Helz (1987) describes reverse and normal zoning in olivine (in Hawaiian basalt), but, because the extinction

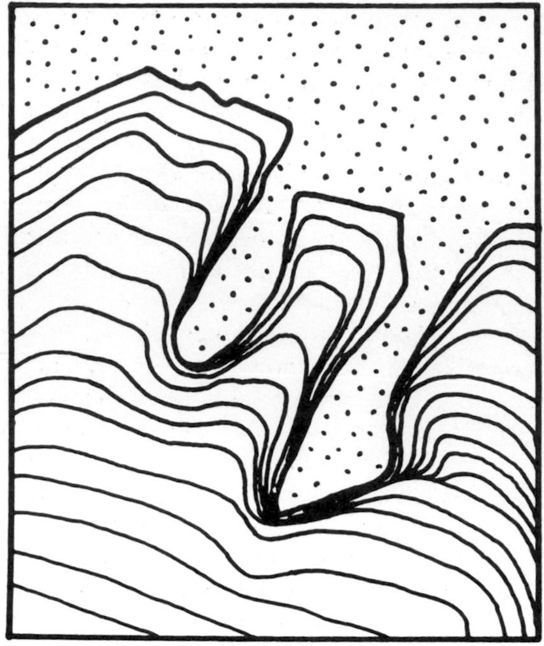

Fig. 3.14 Convolute oscillatory zoning around embayments of glass (stippled) in phenocryst of plagioclase in basaltic andesite. Field of view width is 0.6 mm. Based on a photograph in Blackerby (1968).

positions do not change, it is necessary to use an electron microprobe to see it. Major changes of composition in olivine (or epidote), may be made visible by changes in birefringence. In some minerals, colour variations make zoning conspicuous, as, for example, in the sodic pyroxenes.

The only mineral other than plagioclase that commonly displays obvious oscillatory zoning is augite, particularly titanian augite, where it may be developed in conjunction with sector zoning (discussed below).

3.4.2 Concentric compositional zoning in metamorphic minerals

Zoning is relatively rare or weakly developed in metamorphic minerals, a consequence of the slower development rate of metamorphic textures, and the lesser degrees of solid solution possible at lower temperatures. Plagioclase, for example, is conspicuous for its lack of zoning in metamorphic rocks especially in the greenschist facies; broad patterns of zoning do appear increasingly commonly at higher grades as a wider range of compositions becomes possible. Oscillatory zoning is particularly rare, and in fact the first question to be asked if it is found is whether or not the crystal could be a premetamorphic igneous relic. Indisputable metamorphic oscillatory zoning does occur, however, as described, for example, in vein pyroxene in eclogite facies rocks (Philippot and Kienast, 1989). The reasons for its development are not well explored, and it is usually ascribed to fluctuating supplies of fluid to the growth site.

Most minerals are members of various complex solid-solution series, and most reactions in metamorphic rocks are therefore continuous rather than discontinuous. That is to say, there is an element partitioning between coexisting minerals which changes progressively with changes in *PT* conditions or with the production of new phases. For example, garnet growth is accompanied by a reduction in the An content of plagioclase as it reacts and fractionates into garnet as grossular. The reaction causes some plagioclase grains to be resorbed; others grow with a decrease in An outwards (Spear *et al.*, 1990). Interestingly, in other rocks, epidote breakdown in the amphibolite facies provides a supply of the An component which leads to an increase in An outwards. Zoning in a metamorphic mineral usually represents growth over a range of *PT* conditions, and therefore indicates a degree of disequilibrium; diffusion was evidently not fast enough to change the composition of early-formed grains which are progressively protected from reaction by new growth. Spear and Selverstone (1983) and Spear *et al.* (1990) developed a quantitative approach to determining *PTt* histories (discussed in section 5.5.2) from the progressive complex compositional changes that develop between coexisting zoned minerals in metamorphic rocks. The theory presumes that local equilibrium was maintained along the grain boundaries of all coexisting phases at all times, and that diffusion was not effective in changing chemistry after the crystals grew, presumptions generally thought to be valid unless temperatures are well into the amphibolite facies when homogenization of minerals becomes common.

However, disequilibrium crystallization may be common if growth rates are fast, due perhaps to delayed nucleation of phases, and Hickmott and Shimizu (1990) show that trace elements are more sensitive indicators of disequilibrium partitioning in garnet than major elements. Obviously considerable care needs to be exercised in deciding the significance of mineral zoning in metamorphic rocks.

Some metamorphic mineral zoning is obvious in thin section (for example, the blue amphibole cores to green actinolitic hornblende in low-grade metamorphic rocks in New Zealand, as described by Yardley, 1982), but much of the zoning present is invisible in thin section without the aid of the electron microprobe (as in most garnets in pelitic rocks). It is clear that the level of sophistication of methods such as those of Spear and Selverstone (1983) far exceeds the capabilities of the polarizing micro-

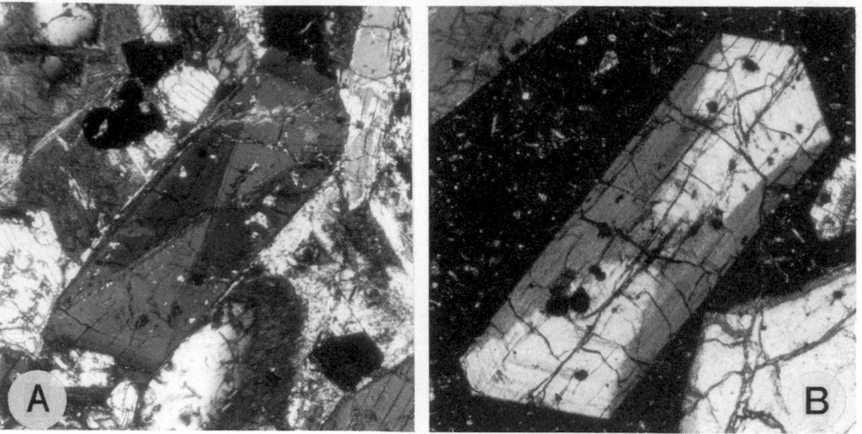

Fig. 3.15 Sector zoning in titanian augite. (A): In teschenite, Teschen, Czechoslovakia. (B): In limburgite, Kaiserstuhl, Germany. View lengths measure 2.26 mm (A) and (B).

scope. The electron microprobe, even the ion microprobe, has become an essential tool. Nevertheless, the preliminary step in all such investigations is the detailed examination of textural relationships, using the ordinary microscope.

3.4.3 Sector zoning

Sector zoning (sometimes called hour-glass zoning because of its appearance in thin section) is developed when different faces of a crystal grow with differing compositions. It may be combined with concentric zoning.

Perhaps the best-known examples of sector zoning are found in titanian augite (Fig. 3.15), but the phenomenon is widespread and may develop during igneous crystallization, metamorphism or diagenesis. As with all other kinds of mineral zoning, it is only conspicuous when revealed by colour variations, or changes in extinction position or birefringence.

Consider a growing pyroxene crystal. Each face exposes a different aspect of the crystal structure (Fig. 3.16), and {100} faces expose half-filled sites of sixfold (M_1) and eightfold (M_2) co-ordination. These sites (called proto-sites) are particularly flexible in terms of geometry, and are likely to attract highly charged small cations in non-equilibrium proportions (Nakamura, 1973; Dowty, 1976). Other faces do not expose the same half-empty proto-sites, and so do not attract the same cations. Thus a different composition develops on {100} and growth perpetuates the difference in

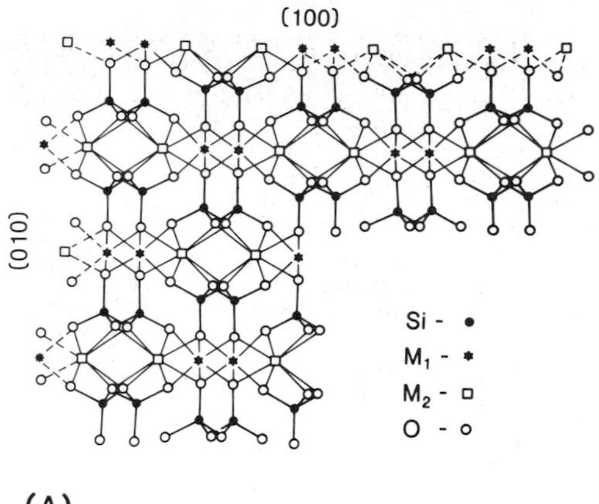

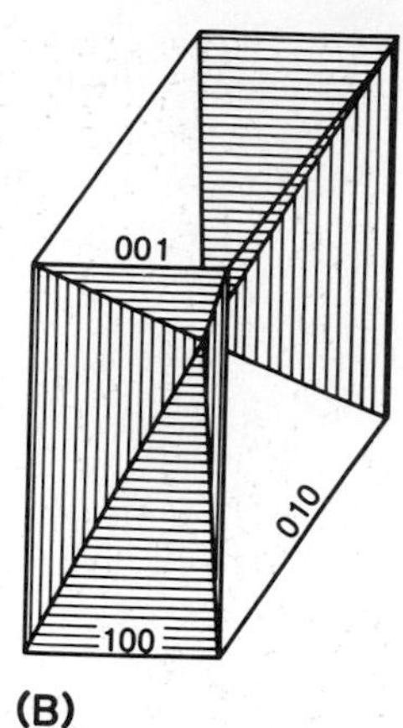

Fig. 3.16 Explanation of sector zoning in pyroxene (discussion in text). (A): Pyroxene structure looking down [001] showing half-filled M_1 and M_2 sites on (100) but not on (010). Based on Dowty (1976, Fig. 2) with permission. Copyright: Mineralogical Society of America. (B): Drawing of a simple pyroxene crystal to show typical pyramidal compositional sectors.

pyramidal sectors (Fig. 3.16B). In the case of titanian augite, the {100} sector may attract a disproportionately large amount of titanium.

Other factors are also important. Thus Carpenter (1980) and Philippot and Kienast (1989) describe sector-zoned omphacitic pyroxenes from veins in blueschists and eclogites. Omphacite is preferentially developed along {100} faces due to the attraction of Al and Ca to the {100} proto-sites; in turn, these elements produce a charge imbalance that is corrected by absorbing appropriate numbers of Mg and Na ions as well. In the example described by Philippot and Kienast, the zoning is more complex, so that {100}-based sectors contain omphacite, {010} sectors aegirine augite and {001} sectors acmite augite. Not only are proto-sites and charge balancing required to explain the degree of complexity, but immiscibility gaps may also be important. Thus two or more immiscible pyroxene compositions precipitate from a common solution on to different faces according to whichever face exposes the best sites for each particular composition.

Sector zoning also involves the degree of order/disorder, and the omphacite sectors described by Carpenter (1980) are well ordered compared with the sodian augite sectors. The familiar case of variable optics (from microcline to sanidine type) in sectors of vein adularia (found in a wide variety of igneous and metamorphic rocks) is a reflection of the different degrees of order in those sectors.

Sector zoning develops most commonly when crystal growth is relatively fast as described, for example, by Carpenter (1980) for vein pyroxene, and by Smith and Lofgren (1983) for experimentally grown plagioclase when Ca is attracted preferentially to {010}. However, the earlier idea that different growth speeds for each crystal face contribute to sector zoning is not supported in the recent literature (e.g., Reeder and Grams, 1987).

3.5 INTERGROWTHS AND TEXTURES WHERE MINERAL GRAINS ARE SET WITHIN EACH OTHER

There are few aspects of thin-section petrology more confusing for the student than the plethora of textures where two or more minerals are intergrown or set within each other. Consider a crystal cloudy with inclusions. Is the texture due to exsolution of the included material? Or could it be the result of deuteric alteration? Or is it a poikilitic texture? And how does one decide on these alternatives? Students can at least be comforted by the fact that researchers often have been just as confused. Myrmekite, for example, has been variously ascribed to processes of eutectic crystallization, unmixing, replacement or reaction, as well as a special variety of poikiloblastic texture. In some cases it is accepted that a texture could possibly have arisen in a variety of ways, as with K-rich/Na-rich feldspar intergrowths (eutectic crystallization, unmixing or replacement).

In Table 3.1, the causative mechanisms for these textures are categorized under four main headings:

1. Coupled mineral growth.
2. Unmixing.
3. One mineral grain overgrown or replaced by one or several other grains.
4. One mineral grain encloses several other mineral grains.

It is useful to subdivide further according to whether the minerals have coherent (more-or-less parallel lattice orientations) or incoherent relationships.

The table embraces textures with a very wide variety of origins. Nevertheless, the products of this wide variety may mimic each other. From the student point of view, sericitized feldspar looks much the same as other textures ascribed to unmixing or poikilitic growth. Of course, the student should learn, as soon as possible, what processes the common textures are normally ascribed to; however, there are many situations when a consideration of diverse options is more appropriate, and there lies the justification for putting all this material in the one table.

It would be impossible in any ordinary-sized text to be fully comprehensive, so the following discussions are necessarily selective. I at least try to mention the more common examples, and a few are discussed in detail to illustrate the principal causative mechanisms.

3.5.1 Products of coupled growth: eutectic/cotectic crystallization

Solders are familiar examples of eutectic systems, where two metals mixed together reduce each other's melting point. Eutectic-type systems are common in nature, although they may become exceedingly complex when multi-phase systems exhibiting varying degrees of solid solution are involved. In the simple albite–quartz eutectic system (Fig. 3.17), a melt of composition *X* first crystallizes quartz at *S*°C. With falling temperature the remaining melt becomes enriched in the feldspar component, but continues to crystallize quartz until the eutectic point *E* is reached. Feldspar and quartz crystallize simultaneously at the eutectic point, and all the melt is consumed at that temperature. If the melt is richer in feldspar than *E* (say composition *Y*), feldspar crystallizes first and the melt is enriched in silica; with falling temperature, feldspar continues to crystallize until eutectic point *E* is reached.

Crystal fractionation in igneous rocks naturally leads to magma compositions of an essentially eutectic or cotectic nature; similarly, the first-formed melts during melting are likely to have eutectic compositions.

(i) Granophyric intergrowth

The three-component system Qz–Ab–Or (Tuttle and Bowen, 1958) can be used to model granitic and rhyolitic magma behaviour, and it consists of two eutectic systems, Qz–Ab and Qz–Or, and a cotectic line joining the two eutectic points and separating the fields of crystallization of quartz and feldspar (Fig. 3.18). The cotectic line is rather like a valley with a deep depression at its centre; granite and rhyolite compositions plot in the vicinity of this temperature low point.

Most granites consist of separately grown crystals of quartz and two feldspars, a consequence of slow near-equilibrium growth at water pressures high enough to restrict solid solution in alkali feldspar (Fig. 3.19A). However, granites emplaced near the surface, sometimes in association with basaltic volcanism, crystallize more quickly and at low water pressures so that complete solid solution in alkali feldspar exists (Fig. 3.19B); hence quartz and one feldspar crystallize instead. If volatiles are lost (along ring-fracture systems, for example), the liquidus/solidus curves are raised, producing a rapid undercooling and freezing. Under these conditions, independent crystals do not develop; instead the simultaneous growth of quartz and alkali feldspar is coupled to produce granophyric intergrowth

Table 3.1 Textures due to coupled growth, overgrowth, unmixing, alteration, replacement or reaction, and inclusion of existing grains

Coupled mineral growth	Due to eutectic/cotectic crystallization	Graphic intergrowth (in granite pegmatites)
	Mainly incoherent	Granophyric intergrowth (in granophyre, and interstitially in tholeiitic rocks)
		K-rich/Na-rich feldspar intergrowths (in some syenitic rocks)
	Due to reactions Mainly incoherent	Symplectites (in a wide variety of rocks, but particularly along grain boundaries in metaplutonic rocks)
		Myrmekite (in granites and granitic gneisses)
Unmixing	Due to spinodal decomposition Mainly coherent	Most but not all perthitic intergrowths (in alkali feldspar in plutonic and metamorphic rocks)
		Pyroxene/pyroxene exsolution lamellae (in pyroxene-bearing plutonic and metamorphic rocks)
	Nucleation required Mainly incoherent	Fe and/or Ti oxides from a wide variety of Fe–Mg minerals
		The 'clouding' in many minerals

One mineral grain overgrown or replaced by one or several other grains	Usually coherent	Chloritization of biotite
		Various mantling relationships (e.g., in feldspar phenocrysts due to magma mixing)
	↓ ↓ ↓ ↓	Some perthitic intergrowths
		Uralitization (deuteric alteration or greenschist-facies metamorphism of pyroxene to amphibole)
	Decreasing or variable coherency	Various alterations of olivine (serpentinization, etc.) and other Fe–Mg minerals
	↓ ↓ ↓ ↓	Saussuritization of plagioclase
		Sericitization/clay mineral alteration of feldspars and other Al-rich minerals
Complete replacement may produce a pseudomorph	Usually incoherent	Corona textures due to reactions between adjacent grains
One mineral grain encloses several other mineral grains	Nearly always incoherent	Poikilitic texture involving a wide variety of minerals in a wide variety of igneous rocks
		Poikiloblastic texture involving a wide variety of minerals in metamorphic rocks, including pre-, syn- and post-tectonic fabrics enclosed within porphyroblasts
		Crystallographically arranged inclusion patterns, as for example in chiastolite

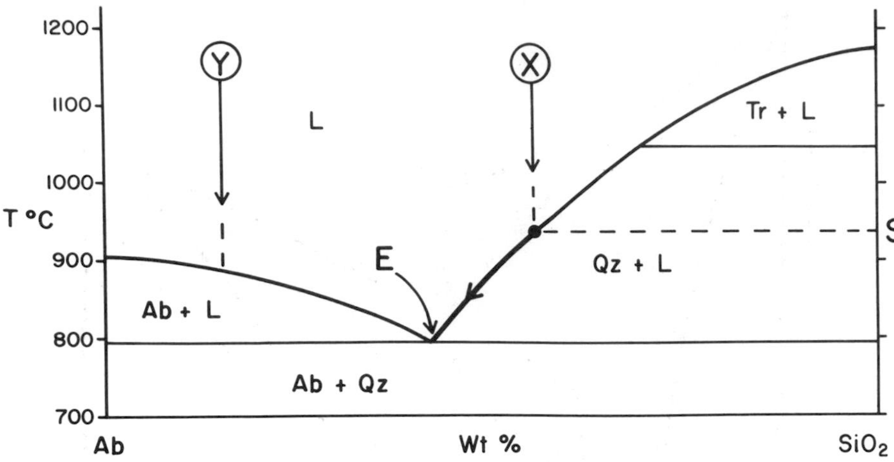

Fig. 3.17 Eutectic system $NaAlSi_3O_8$ (Ab) and SiO_2 at 1 kbar pressure in the presence of water. L = liquid, Qz = quartz, Tr = tridymite. Crystallization of two liquids (*X* and *Y*) is discussed in text. Data from Tuttle and Bowen (1958).

(Fig. 3.20). Granophyric intergrowth often nucleates on pre-existing phenocrysts which result from earlier slow cooling and/or pre-cotectic crystallization. The intergrowth appears as radiating or branching interconnected quartz grains set in a single crystal of feldspar. The feldspar itself may be euhedral, and the quartz grains actually form one or two crystals so that neighbouring branches extinguish simultaneously. The quartz grains are often angular, triangular or cuneiform in section (Fig. 3.20), and they vary in grain size from fine microscopic branches to grains sufficiently large to be seen in hand specimen. The feldspar may have unmixed to mesoperthite.

Granophyric intergrowth is the principal constituent of granophyre, a variety of granite, and it also occurs as a final interstitial product of crystallization in other rocks, especially tholeiitic basalt, dolerite and gabbro. Very fine-grained examples are also found in rhyolite in the outer parts of coarse spherulites (Fig. 4.9B).

Some writers (e.g., Lofgren, 1971; Smith and Brown, 1988, Fig. 20.3; Swanson *et al.*, 1989) extend the term granophyric intergrowth to include other less-co-ordinated mixtures of silica and feldspar crystallized in rhyolite. This seems to me to be unnecessary and confusing, and I restrict the term here to the intergrowth of a branching crystal of quartz in a single crystal of feldspar. If the growth consists of silica with radiating or otherwise separate feldspar crystals of variable lattice orientation, then it is best named spherulitic, poikilitic or snowflake (poikilomosaic) texture, as discussed in section 4.1.

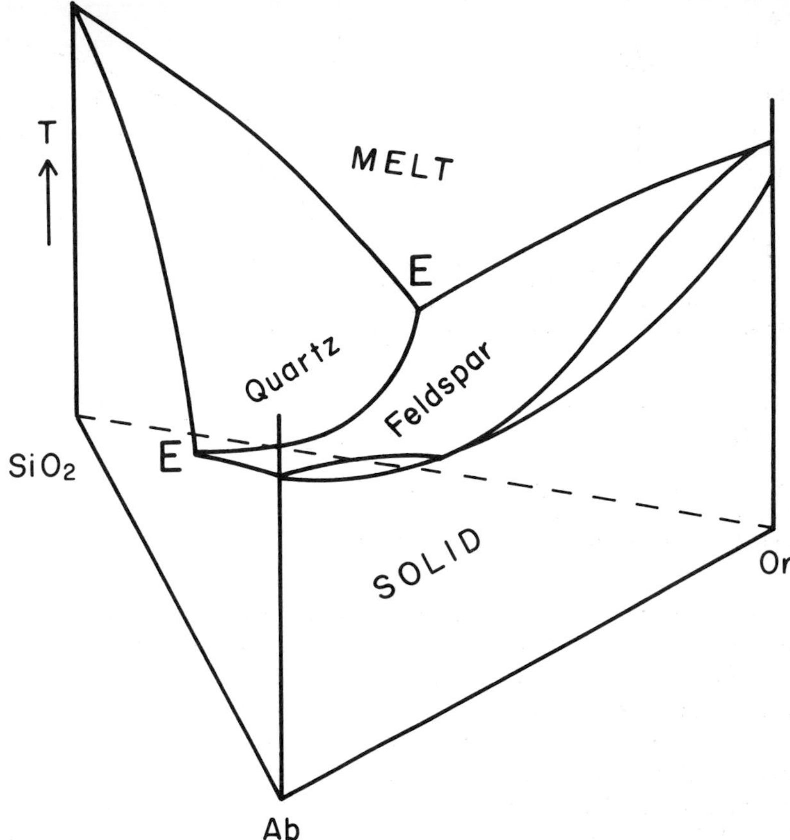

Fig. 3.18 Crystallization in the water-saturated system Qz–Or–Ab. Actual temperatures vary with pressure, and are depressed as pressures are raised. E = eutectic point. Compare with Figs 3.17, 3.19, and 4.19. Based on Tuttle and Bowen (1958).

(ii) Graphic intergrowth

Graphic intergrowth is similar to granophyric intergrowth. The name graphic (Greek for writing) derives from the angular hieroglyphic or cuneiform nature of the interconnected branches of quartz. Graphic intergrowth is coarser than granophyric, and consists of K-rich and Na-rich feldspar rather than intermediate compositions (Barker, 1970); it is therefore a product of subsolvus crystallization (section 4.2.3) at higher water pressures (Fig. 3.19A) in the Qz–Ab–Or system (at either point *E* in Fig. 3.18). It occurs principally in granite pegmatites, and its formation may be connected to the presence of abundant fluids. The composition is often reported as being too feldspathic for a true eutectic, but this may be

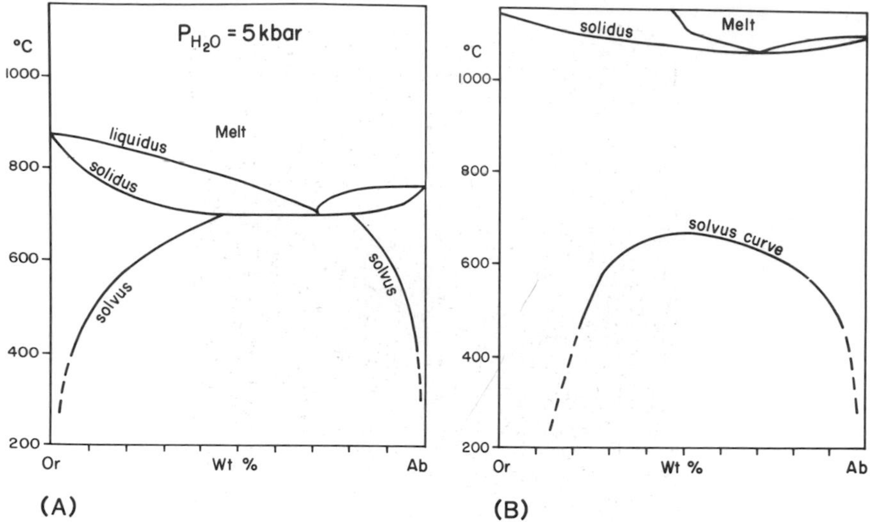

Fig. 3.19 Crystallization of the alkali feldspars, (A) at 5 kbar water pressure, and (B), at zero water pressure. (A) is from Morse (1970) by permission of Oxford University Press, and (B) from Tuttle and Bowen (1958).

a result of feldspar crystallizing first, and the influence of kinetics of growth on the system, as demonstrated experimentally by Fenn (1986). Thus, feldspar and quartz both require a degree of undercooling to start growing, and feldspar nucleates more easily than quartz. The initial feldspar growth causes the boundary liquid layer to become supersaturated in SiO_2 and enriched in H_2O; this produces a skeletal growth habit in the feldspar, so that when the quartz eventually crystallizes it does so in the spaces of the cellular feldspar. The situation is complicated by the facts that growth rates speed up significantly during coupled growth, and increases in fluid content shift peak growth rates to lower temperatures.

(iii) Other eutectic/cotectic mixtures

Eutectic/cotectic systems are common in geological situations, and potentially they may all result in coupled intergrowths, especially if crystallization is rapid. Plagioclase and pyroxene often crystallize as a cotectic in basaltic rocks (Fig. 4.12), and the granophyric-like intergrowth of the two minerals, illustrated in MacKenzie *et al.* (1982, Fig. 90), probably results from coupled cotectic growth.

Plagioclase and alkali feldspar may crystallize as a cotectic, and Petersen and Lofgren (1986) describe experimentally grown intergrowths of oligoclase and sanidine that mimic those found in some of the well-known

Fig. 3.20 Granophyric intergrowth in granophyre, NZ. (View length measures 3.3 mm.)

larvikite building stones from the Oslo region. Just as Fenn (1986) reported for graphic granite, variable nucleation and growth rates do not always allow for perfectly regular eutectic proportions in all the intergrowths, despite the essentially eutectic nature of the crystallization. The feldspar intergrowths are similar in many ways to the more common perthites (described below), but differ in the fact that the intergrown feldspar often shows faceted interfaces and branching radiating forms reminiscent of granophyric intergrowth.

3.5.2 Products of coupled growth: symplectites (including myrmekite)

Symplectite is the general term for the fine-grained products of coupled growth during secondary reaction and replacement. Occasionally eutectic crystallization is proposed. The products are usually microscopic and characteristically found along grain boundaries of primary minerals. The reactions may be promoted by the concentration of water-rich fluids along grain boundaries during cooling of an igneous rock, or infiltration of fluids during a secondary metamorphic event, and in gabbroic rocks, symplectites develop in essentially dry rock because of very slow rates of cooling at great depth. The very fine-grained intergrown nature of symplectites and their restriction to the grain boundaries of specific minerals indicate a general immobility of the material involved: diffusion is the rate-controlling factor. By far the most common variety of symplectite is myrmekite.

(i) Myrmekite

Myrmekite (Fig. 3.21) is an intergrowth of branching rods of quartz set in a single crystal of plagioclase; neighbouring quartz rods have the same lattice orientation, and extinguish together. It is almost ubiquitous in granites and granitic gneisses, and it most commonly occurs at grain boundaries of K-rich feldspar. Myrmekites appear to have grown inwards from grain boundaries, invading and replacing K-feldspar; the quartz rods branch in that direction, and the plagioclase may be euhedral or have a bulbous appearance representing a minimum surface area to volume configuration for the plagioclase.

Myrmekite has provoked a very considerable literature on its origins, but by far the most accepted hypothesis goes back to Becke (1908) who proposed the following reaction of Na- and Ca-bearing fluids with K-feldspar:

$$KAlSi_3O_8 + Na^+ = NaAlSi_3O_8 + K^+, \text{ in combination with:}$$
$$2KAlSi_3O_8 + Ca^{2+} = CaAl_2Si_2O_8 + 4SiO_2 + 2K^+$$

Silica is released because the Al/Si ratio is different in K- and Ca-feldspar; because of the immobility of Al and Si, quartz forms a microscopic intergrowth with the feldspar in quantities that are directly in proportion to the An % of the plagioclase (Fig. 3.22).

Myrmekite and the partly replaced K-feldspar have an incoherent relationship which avoids the strain associated with coherent interfaces which get pinned because of similar but not identical crystal lattices. Myrmekite along the grain boundary between two adjacent K-feldspars may have a 'swapped-rim' relationship indicating that myrmekite nucleated coherently on one grain but then grew preferentially into the adjacent grain with an incoherent interface (see also 'swapped-rim' perthite below).

Myrmekitic quartz and plagioclase are also in incoherent relationship, and it often seems that quartz nucleates on some pre-existing quartz grain at the K-feldspar boundary. Interestingly, Stel and Breedveld (1990) describe myrmekitic quartz which nucleated on groundmass quartz with a

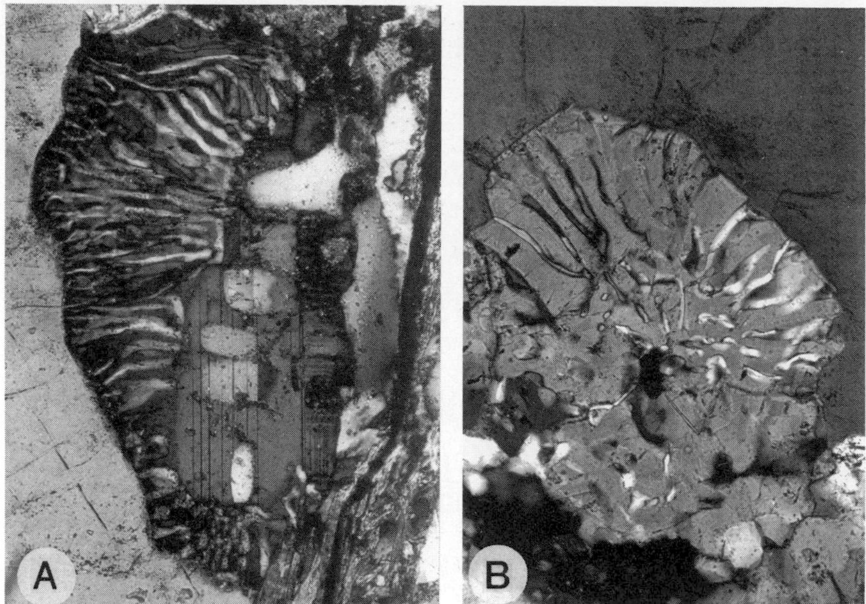

Fig. 3.21 Examples of myrmekite in Constant Gneiss, NZ, which have grown at the boundary of large K-feldspar crystals at sites of deformation and recrystallization. In (A) the plagioclase is slightly bulbous and one can see the adjacent severely deformed matrix (right). In (B) the plagioclase is subhedral. (View lengths measure 0.80 mm (A) and (B).)

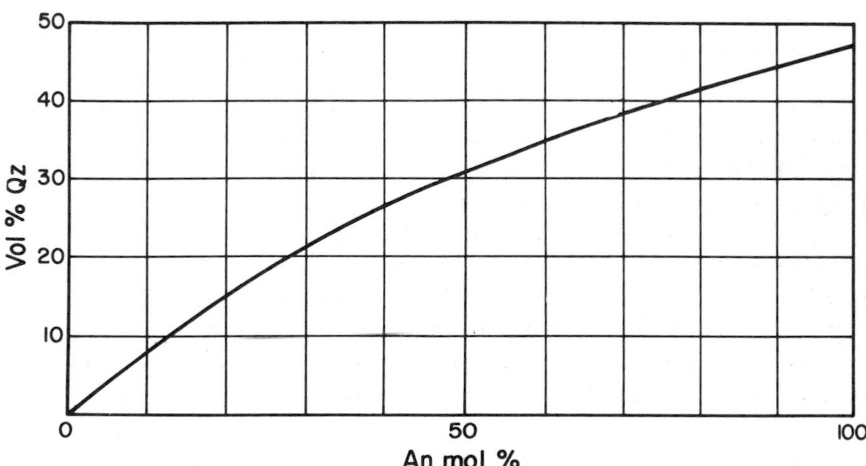

Fig. 3.22 Volume % quartz versus plagioclase An %, as expected according to Becke's hypothesis. Curve from Phillips and Ransom (1968, Fig. 1) with permission. Copyright: Mineralogical Society of America.

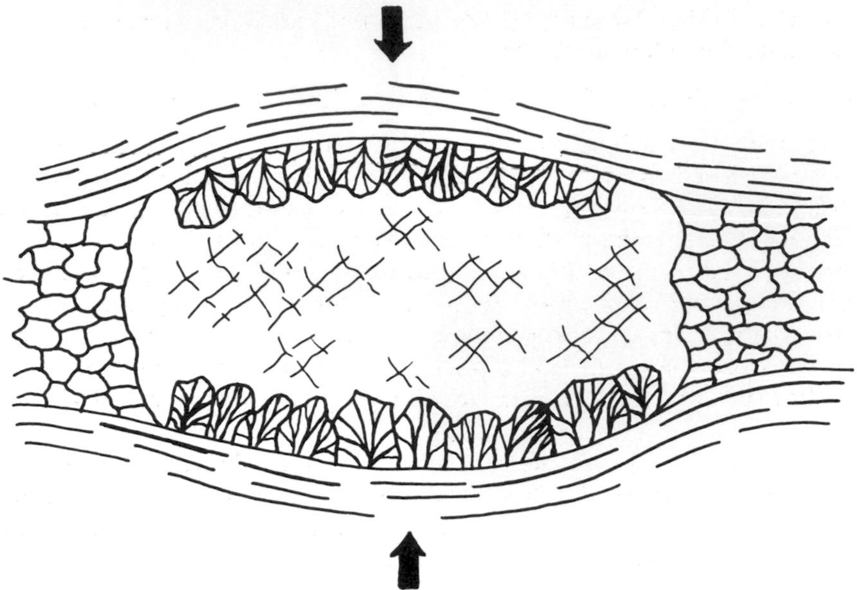

Fig. 3.23 Schematic illustration of K-feldspar porphyroclast in a deformed granitoid with recrystallized 'tails' and myrmekite growing on sides facing the shortening direction. Based on Simpson and Wintsch (1989).

tectonically-induced *c*-axis preferred orientation; that preferred orientation is preserved in the myrmekite but eliminated in the rest of the rock by further deformation and recrystallization.

Myrmekites are often associated with strain features in deformed granitic rocks, and Simpson and Wintsch (1989) show that they may form only on sides of K-feldspar facing the shortening direction (Fig. 3.23). Myrmekite forms at those sites because of the volume decrease associated with myrmekite formation, and because of the greater pre-existing strain which helps drive the replacement reaction. Myrmekites may display secondary twinning, a response to progressive strain or subsequent deformation.

Myrmekite is commonly associated and intergrown with similar intergrowths of muscovite and vermicular quartz (Phillips *et al.*, 1972) formed in the following reaction:

$$3KAlSi_3O_8 + H_2O = KAl_2(AlSi_3)O_{10}(OH)_2 + 6SiO_2 + K_2O$$

All the above-mentioned examples of myrmekite form during retrogression or secondary deuteric alteration. Myrmekite also develops independently of K-feldspar when albite changes to Ca-oligoclase in pelitic schists at the boundary between the greenschist and amphibolite facies

(Shelley, 1973). Ashworth (1986) calls this 'prograde myrmekite' and advocates the following reaction:

$$(1 + x)\,\mathrm{NaAlSi_3O_8} + x\mathrm{Ca^{2+}} = \mathrm{Na_{1-x}Ca_xAl_{1+x}Si_{3-x}O_8} + 4x\mathrm{SiO_2} + 2x\mathrm{Na^+}$$

He suggests it will be preserved only if formed at the thermal climax of metamorphism. In the New Zealand Haast Schists, Shelley (1973) shows that 'prograde myrmekite' only develops in quartz-rich segregation layers. Oligoclase that replaces albite in mica–feldspar layers does not develop myrmekite, hence the alternative idea that myrmekite is a poikiloblastic growth that includes pre-existing quartz drawn out along the feldspar growth directions as it recrystallizes. If one were to accept the reaction proposed by Ashworth (1986), then it would be necessary somehow to explain the removal of quartz from the oligoclase in the mica–feldspar layers.

Other occurrences of myrmekite include very An-rich examples in anorthosite, described by Dymek and Schiffries (1987). The myrmekite is labradorite–bytownite and develops along grain boundaries of the dominant andesine. They propose the following reaction:

$$\begin{aligned} &1.2857\ \mathrm{(Na_{0.6}Ca_{0.8})(Al_{1.4}Si_{2.6})O_8} + 0.2857\ \mathrm{Ca^{2+}} = \\ &\qquad \mathrm{(Na_{0.2}Ca_{0.8})(Al_{1.8}Si_{2.2})O_8} + 0.5714\ \mathrm{Na^+} + 1.1428\ \mathrm{SiO_2} \end{aligned}$$

Andesine is albitized due to the release of Na^+.

(ii) Other symplectites

Gabbroic rocks, subjected to very slow cooling at great depth in the crust, develop reaction rims between the primary igneous crystals. Olivine crystals become surrounded by a mantle of reaction products called kelyphitic rims or corona texture. The products of reaction typically include ortho- and/or clinopyroxene, amphibole, and spinel or garnet; the products are often arranged in a series of concentric zones around olivine, and spinel or garnet may be in symplectitic intergrowth with amphibole or pyroxene (Fig. 3.24, also Fig. 2.2).

Corona textures are discussed further in section 3.5.4. The salient point to note here is that the reaction zones are restricted to the boundaries between certain specified mineral pairs. This is clear evidence for metamorphic reaction rather than igneous crystallization, and shows too that diffusion was the rate-controlling process; the fine-grained nature of the symplectite itself is further evidence that diffusion was too slow to allow segregation of the reaction products into an equilibrium texture. Grant (1988), for example, has shown with the aid of chemical profiles across coronae that Al and Si were particularly immobile.

Temperatures and pressures that prevail during typical gabbroic

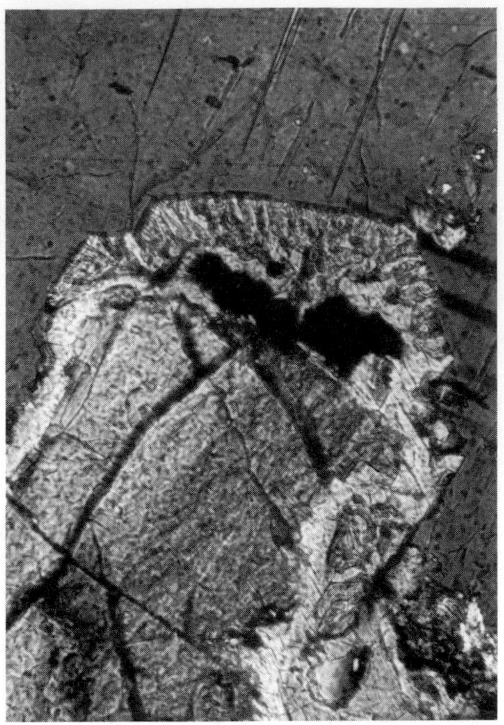

Fig. 3.24 Symplectite of amphibole and spinel at the boundary of olivine and plagioclase, Bluff, NZ. (Length of view measures 0.35 mm.)

symplectite formation are summarized in Fig. 3.25. Spinel generally indicates a higher temperature and possibly lower pressure than does garnet.

More superficially, oxidation of olivine may cause formation of an orthopyroxene–magnetite symplectite. Johnston and Beckett (1986) discuss examples from gabbros, metamorphic rocks and lherzolite xenoliths which reacted with basaltic magma.

Symplectites are not restricted to basic igneous rocks. Thus Lonker (1988) describes cordierite–quartz symplectites produced by the reaction of biotite and sillimanite in high-grade pelitic gneisses. And intergrowths of vermicular nepheline in K-feldspar and plagioclase are well known from undersaturated alkaline igneous rocks: explanations include the breakdown of leucite to nepheline–K-feldspar (the so-called pseudoleucite), the replacement of plagioclase by nepheline, and the eutectic crystallization of nepheline and K-feldspar (Smith and Brown, 1988).

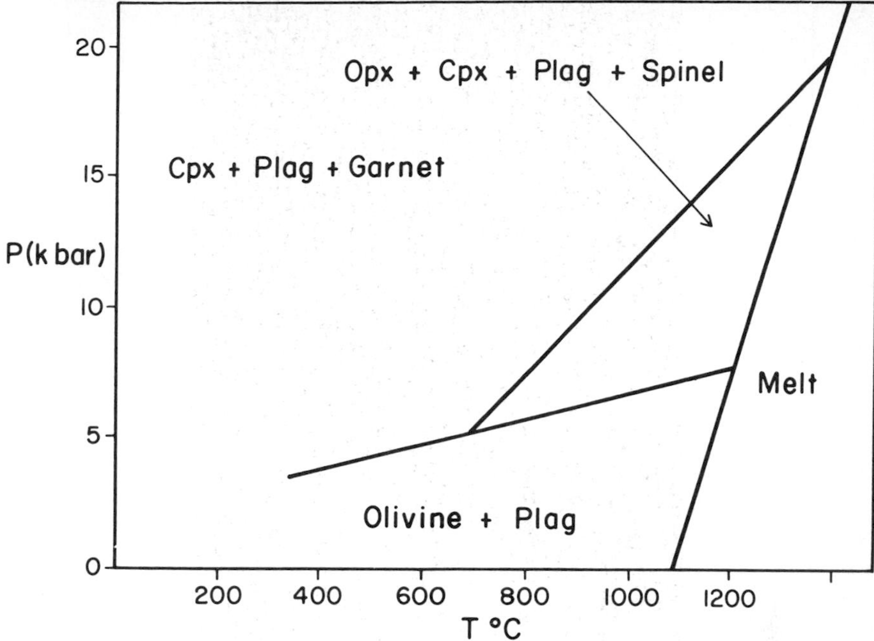

Fig. 3.25 *P–T* diagram to show the approximate stability relationships of spinel and garnet-bearing mineral assemblages in corona textures. With permission based on Griffin (1972).

3.5.3 Products of unmixing

A solvus curve defines the limits of solid solution on a phase diagram (Fig. 3.19). If the initial crystallization takes place above a solvus, the solid solution may be metastably preserved by rapid cooling; hence the occurrence of metastable intermediate-composition alkali feldspars in volcanic rocks. Slow cooling may enable unmixing into two intergrown phases to occur, although this depends on the nature of the material, and the details of the cooling history.

In the case of alkali feldspar, unmixing produces two closely similar minerals, and the unmixing involves primarily a redistribution and partitioning of K and Na, not the strongly bonded Al and Si. Therefore unmixing can take place easily and progressively in coherent intergrowth (i.e., the lattices of the two phases are more-or-less parallel).

The unmixing of plagioclase produces closely similar minerals in coherent intergrowth, but redistribution of the strongly bonded Al and Si is necessary as well as Ca and Na. Unmixing is therefore not easy, and the results are usually too fine grained to be seen in thin section.

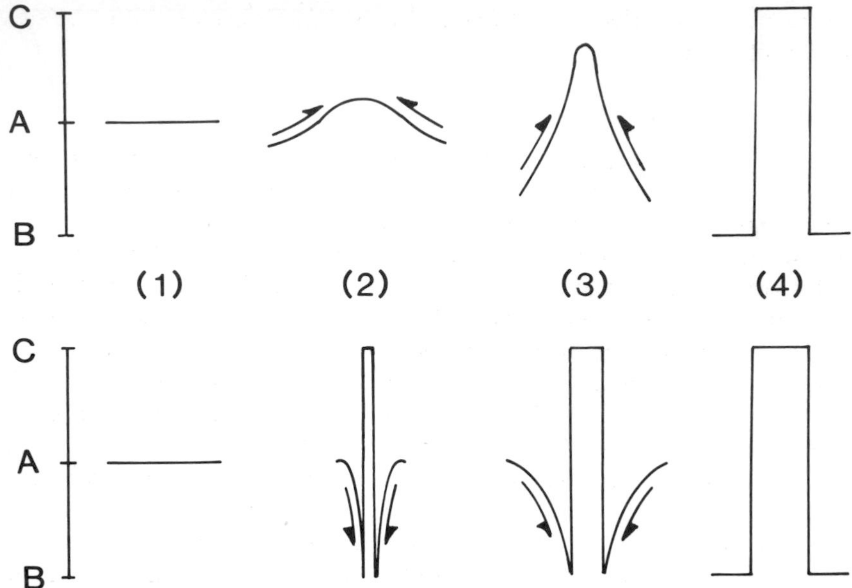

Fig. 3.26 Formation of an intergrowth of (B) and (C) from an initial composition (A). Top: Spinodal decomposition with (1), original state, (2) and (3), gradual change in composition, (4), final state. Bottom: Discontinuous transformation involving nucleation of phase C, (2), and its progressive growth, (3) and (4). Note that compositional gradients differ in the two processes. Spinodal decomposition involves so-called 'uphill' diffusion, discontinuous transformation, 'downhill' diffusion with steeper gradients. With permissions based on Putnis and McConnell (1980).

(a) Spinodal decomposition versus unmixing involving nucleation

The formation of intergrowths by unmixing requires an initial creation of the phase difference, and then growth or coarsening; the composition may change progressively during unmixing. If the unmixing takes place as the result of a gradual increase in the amplitude of compositional fluctuations, this is called spinodal decomposition. The resulting intergrowth is usually coherent. If a sudden and abrupt change in composition occurs at the initiation stage, the process requires a distinct nucleation step. The result may be a coherent or incoherent intergrowth. The two processes are illustrated diagrammatically in Fig. 3.26.

Feldspars probably unmix by spinodal decomposition, although after coarsening and maturing it may be impossible to ascertain this. Certainly, most fine-grained feldspar intergrowths produced by unmixing are coherent and involve a gradual partitioning of some cations. On the other hand, the exsolution of iron and titanium oxides from ferromagnesium minerals must involve a nucleation step because the two phases are so utterly different.

(b) Three conflicting factors

1. The primary driving force for unmixing is a decrease in bulk free energy, but this is opposed by the increase in grain-boundary energy consequential to the creation of two phases in intergrowth.
2. The initial fine intergrowth requires only short-range diffusion, but this advantage is counterbalanced by the high surface energy involved.
3. A coherent interface involves the least disruption to the lattice and may enable a gradual spinodal decomposition to take place easily; on the other hand, the coarsening of a coherent intergrowth is more difficult than an incoherent one. This is because severe strains develop at interfaces between two slightly different but parallel lattice structures. Grain boundaries that are incoherent are more mobile, and fine coherent intergrowths tend to be increasingly incoherent on coarsening.

   Undercooling is usually necessary to produce unmixing; a lower cooling rate is more likely to produce unmixing. Unmixing and/or coarsening may also develop during subsequent metamorphism and reheating.

(c) Unmixing of plagioclase

Owing to the immobility of Al and Si, intergrowths due to unmixing in plagioclase are fine grained and usually submicroscopic, though evidence of their presence may be given by iridescence in hand specimen. Peristerites are intergrowths due to unmixing in sodic plagioclase, Bøggild intergrowth is the term used for intergrowths in andesine–labradorite, and Huttenlocher intergrowth occurs in Ca-labradorite–bytownite (Fig. 3.27A). The iridescence associated with the first two types is familiar in numerous facing stones used on city buildings world-wide. The phenomenon of unmixed plagioclase may not be directly observable in thin section, but the effects of the increasingly restricted range of composition with lowering of temperature is manifest in metamorphic plagioclase. In pelitic rocks throughout the greenschist facies, for example, plagioclase is always nearly pure albite; intermediate plagioclase compositions are unstable under those conditions.

(d) Perthitic intergrowths

Unmixing of K-rich and Na-rich phases from alkali feldspar is relatively easy, and the resulting perthitic intergrowths are common. If the dominant phase is K-rich, the intergrowth is called perthite; if the dominant phase

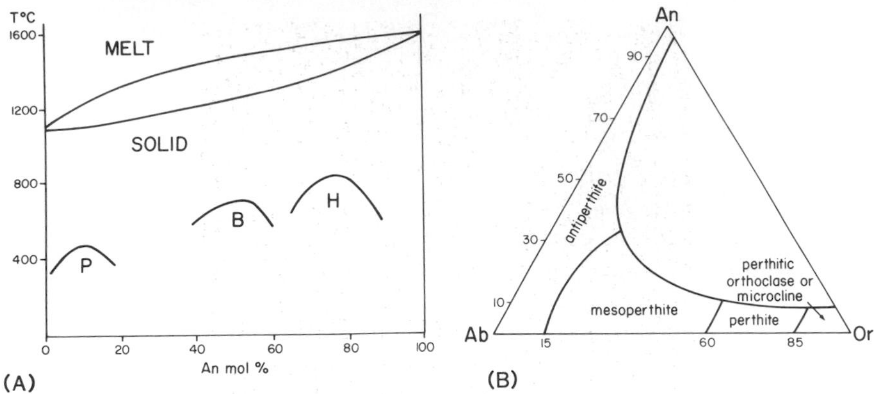

Fig. 3.27 (A): Three possible solvus curves to show limits of solid solution in plagioclase. The regions of peristerite (*P*), Bøggild intergrowths (*B*) and Huttenlocher intergrowths (*H*) are indicated. (B): Terminology for perthitic intergrowths. Based on Smith and Brown (1988).

is Na-rich then it is antiperthite; intermediate compositions produce mesoperthite (Fig. 3.27B). If cooling is very rapid, the intergrowth may be submicroscopic; it is then called cryptoperthite. Most perthitic intergrowths are coherent, especially those that are fine grained; the fact that the lattices of the two phases are parallel is confirmed by observations that cleavage runs across phase boundaries without interruption (Fig. 3.28). In effect, the intergrowth represents a partitioning of K and Na rather than movement of Al and Si. The principal morphological and other characteristics of common perthite types are given in Table 3.2.

Most workers advocate unmixing as the mechanism that forms fine perthitic intergrowths. However, coarser varieties, particularly vein and patch perthite, are usually associated with areas of turbid alteration; this was ascribed by Parsons and Brown (1984) to deuteric coarsening at temperatures below 400°C but other workers have ascribed coarse perthites to processes of replacement, as discussed later. In addition, tectonic strain may induce unmixing and influence the preferred orientation of perthitic lamellae. And as reported earlier, some other coarse coherent intergrowths, similar to perthite, may develop by eutectic crystallization of K-rich and Na-rich feldspars from a melt.

(e) Unmixing in pyroxenes

The result of unmixing in the Ca–Mg–Fe-pyroxenes is commonly visible in thin section as sets of thin exsolution lamellae. The miscibility of Ca-poor (pigeonite or orthopyroxene) and Ca-rich (augite) reduces with

Fig. 3.28 Perthite. Note that the (001) cleavage runs without interruption through both the K-rich (darker) and Na-rich (lighter) phases. (View length measures 3.3 mm.)

lowering temperature, and each may exsolve the other, probably by spinodal decomposition. Crystallographically the situation is interesting in that Ca-poor pyroxene can be either monoclinic (pigeonite) or orthorhombic. The lamellae are always strictly parallel to {100} if either lamellar or host pyroxene is orthorhombic. The orientation of lamellae for two intergrown clinopyroxenes (augite and pigeonite) is much more complex. One or two sets may develop, possibly simultaneously, one within 20° of {001}, the other within 20° of {100}. The precise orientation (Fig. 3.29) depends on the relative values of the *a* and *c* cell dimensions of host and lamellae (Robinson *et al.*, 1971; Jaffe *et al.*, 1975), and although the lamellae are usually on irrational planes, those planes nevertheless optimize the coherency of the two phases. The *b* and *c* crystal axes of all intergrown lamellae are parallel, and the *a*-axes are always within 3° of each other.

Exsolution lamellae in pyroxene may record the *PT* history of the rock in the following ways:

Table 3.2 Characteristics of various perthite types (based mainly on information in Smith and Brown, 1988)

Film perthite (forms mesoperthitic areas)	Films lie at +107° from *a* in (010) sections	Fine grained and coherent
Braid perthite (forms mesoperthitic areas)	Two sets of cross-cutting and anastomosing lamellae symmetrically disposed at about 30° to *b* in (001) sections	Mostly result from unmixing
String perthite (in K-rich perthite or Na-rich antiperthite)	Rods with the elongation direction in (010) sections at a moderately high angle to *a*. May possibly develop from film perthite by a reduction in surface area.	May coarsen in turbid areas affected by deuteric alteration
Plate perthite (in K-rich areas)	Plates that lie at about 10° to the trace of (001) in (010) sections (i.e., at a high angle to films and strings), and at about 15°–30° from the trace of (001) in sections perpendicular to *a*.	

Vein perthite	Vein-like bodies *c.* +113° from *a* in (010) sections and *c.* 30° to *b* in (001) sections. The veins are often twinned on the albite law.	Coarser-grained and may be incoherent
Patch perthite	Irregular-shaped bodies, often blocky, and the plagioclase is often twinned.	May have a very irregular distribution
Swapped-rim perthite (at boundary between two adjacent alkali feldspars)	Exsolved material at the grain boundary has an incoherent relationship with host but the same orientation as the adjacent grain on which it nucleated.	May be due to unmixing or replacement and coarsening may be deuteric

Different perthite types may be found in the one crystal. Exsolved areas often thicken towards the boundary of the host crystal.

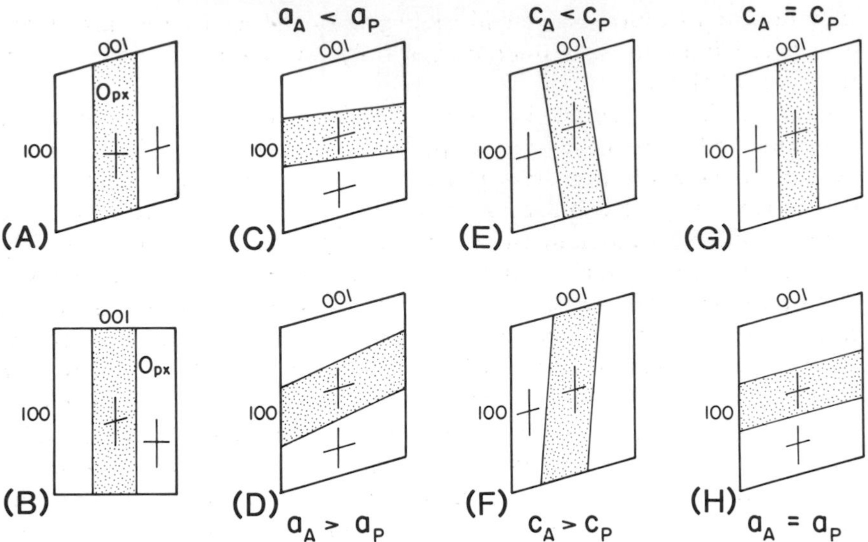

Fig. 3.29 Orientation of exsolution lamellae (stippled) in pyroxenes: (A) and (B): Orthopyroxene–clinopyroxene pairs. (C)–(H): Clinopyroxene–clinopyroxene pairs. The orientation of *a* and *c*-axes are shown for host and lamellae by the crossed lines, and the relative values of *a* and *c* cell dimensions for augite (*A*) and pigeonite (*P*) are shown. Based on Robinson *et al.* (1971) and Jaffe *et al.* (1975).

1. The *a* and *c* cell dimensions are functions of changing *P* and *T*; so therefore must be the orientations of paired clinopyroxene lamellae. Robinson *et al.* (1971) describe multiple generations of lamellae on slightly different planes which clearly reflect changing conditions during the course of exsolution.
2. Pigeonite may initially exsolve augite on irrational planes close to either or both {001} and {100}, but if it later inverts to orthopyroxene it will continue to exsolve augite strictly on {100}. Similarly, augite may initially exsolve pigeonite on two sets of irrational planes, but on cooling will exsolve orthopyroxene strictly on {100}. Orthopyroxene is significantly more Ca poor than pigeonite, so the inversion of pigeonite may produce a sudden increase in the amount of exsolved material.

(f) Other common unmixing phenomena

Exsolution phenomena are widespread and diverse in textural and mineralogical character, and this is a consequence of the general fact that minerals engage in solid solutions (at high temperatures in a relatively disordered state) that are unstable at lower temperatures. The factors that may impede exsolution are rapid cooling rates and nucleation difficulties. The possibilities are so numerous that only a few selected examples are mentioned below.

Clinoamphiboles unmix into two amphiboles in a similar way to the feldspars and pyroxenes; the amphibole exsolution lamellae grow in host amphibole on irrational planes close to {100} and {001}, as in clinopyroxenes (Robinson *et al.*, 1971). More generally, minerals exsolve material of totally different character. Thus, pyroxenes and other Fe–Mg minerals exsolve a great diversity of Fe- and Ti-rich oxides on cooling, and pyroxenes have even been observed to exsolve plagioclase; in all such cases, nucleation of the exsolved material (rather than spinodal decomposition) is required, and epitaxial relationships between host and lamellae may facilitate nucleation and control the orientation of exsolved material.

A familiar example of exsolution of a disparate mineral phase includes the red coloration in feldspars. In some cases this is due to the exsolution of hematite, in others, deuteric alteration may be the cause. Similarly, the general cloudiness in feldspars and many other minerals is usually contentious, being ascribed either to exsolution or deuteric alteration.

In any attempt to elucidate the PT history of a rock it is important to determine the nature of exsolved material: it provides a vital clue to the conditions that formerly allowed solid-solution to occur. More sophisticated tools than the polarizing microscope may be required to identify and properly examine the often fine-grained material.

3.5.4 Products of overgrowth or replacement of a mineral grain, including pseudomorphs

A normal consequence of geological events is that conditions of P and T and the presence or absence of magma or certain fluids change continuously in space and time. In the case of low-temperature sedimentary assemblages undergoing a thorough progressive metamorphism, the increase in T results in a thorough reconstitution of the rock. Original grain shapes and textures are destroyed in response to reaction rates that increase rapidly with T, aided by fluids progressively expelled from the rock. The corollary is that total reconstitution of a rock is less likely when a high-temperature mineral assemblage is subject to lower temperatures; primary grain shapes and textures are often preserved, although individual mineral grains may be altered, especially if fluids gain access to the rock.

The replacement of single pre-existing grains is therefore most characteristic of processes such as deuteric alteration, weathering, retrogressive metamorphism or the metamorphism of igneous rocks. If the replacement is more-or-less complete without changing the overall shape of the original grain, then the product is called a pseudomorph (Fig. 3.30A and B). Overgrowths strongly resemble some incomplete grain-boundary replacement products (Fig. 3.30C), but have a different origin, usually in magmatic crystallization or progressive metamorphism.

A coherent interface typifies replacement or overgrowth when the two minerals are species of the same mineral (e.g., feldspar and feldspar) or when they have closely related atomic structures (pyroxene and amphibole – both chain silicates). Replacement where the two minerals have a coherent relationship is called topotaxy; overgrowth with a coherent relationship is called epitaxy. An incoherent interface is the more likely result if the two minerals are unrelated species (Fig. 3.30D), although semi-coherent epitaxial arrangements are sometimes found as, for example, in some replacements of muscovite by sillimanite (Kerrick, 1987). Coherent topotactic and epitaxial relationships obviate or ease nucleation problems by seeding; the fact that replacement products are often confined to rather specific sites indicates an absence of widespread diffusion in the rock, and in the common hydration reactions, growth rates depend on the rate of fluid supply.

A comprehensive listing of replacement and overgrowth phenomena would be lengthy indeed. Below I discuss a few selected examples to illustrate some principles involved.

(a) Chloritization

Biotite readily alters to chlorite at low–moderate temperatures, and the product is often present as a complex sandwich of chlorite layers coherently disposed between relic layers of biotite. It is tempting to view such a replacement mainly in terms of the addition of water with the sheet structure of the original mica relatively unchanged and immobile during the replacement process. In fact the process is a great deal more complex than that, and the behaviour of K^+ and other ions need to be considered. Thus, Eggleton and Banfield (1985) advocate the following coupled reaction for chloritization in a granite (at 340°C):

$$\begin{aligned}&1\text{ biotite} + 0.21\text{ anorthite} + 1.64\text{ water} + 1.46\text{ H}^+ + 0.08\text{ O}_2\\&= 0.46\text{ chlorite} + 0.11\text{ titanite} + 0.10\text{ epidote} + 0.56\text{ muscovite}\\&+ 0.63\text{ quartz} + 0.02\text{ magnetite} + 1.46\text{ K}^+\end{aligned}$$

The K^+ released from biotite sericitizes plagioclase, and the calcium released from the plagioclase is used to produce epidote and titanite, mixed in with the sheet-silicates. The reaction conserves Mg in the chlorite. In terms of the sheet structure, two layers of biotite become one layer of chlorite, so that one-half of the tetrahedral sheets is lost whereas the other half is inherited intact. The *a* and *b* cell dimensions of the chlorite and

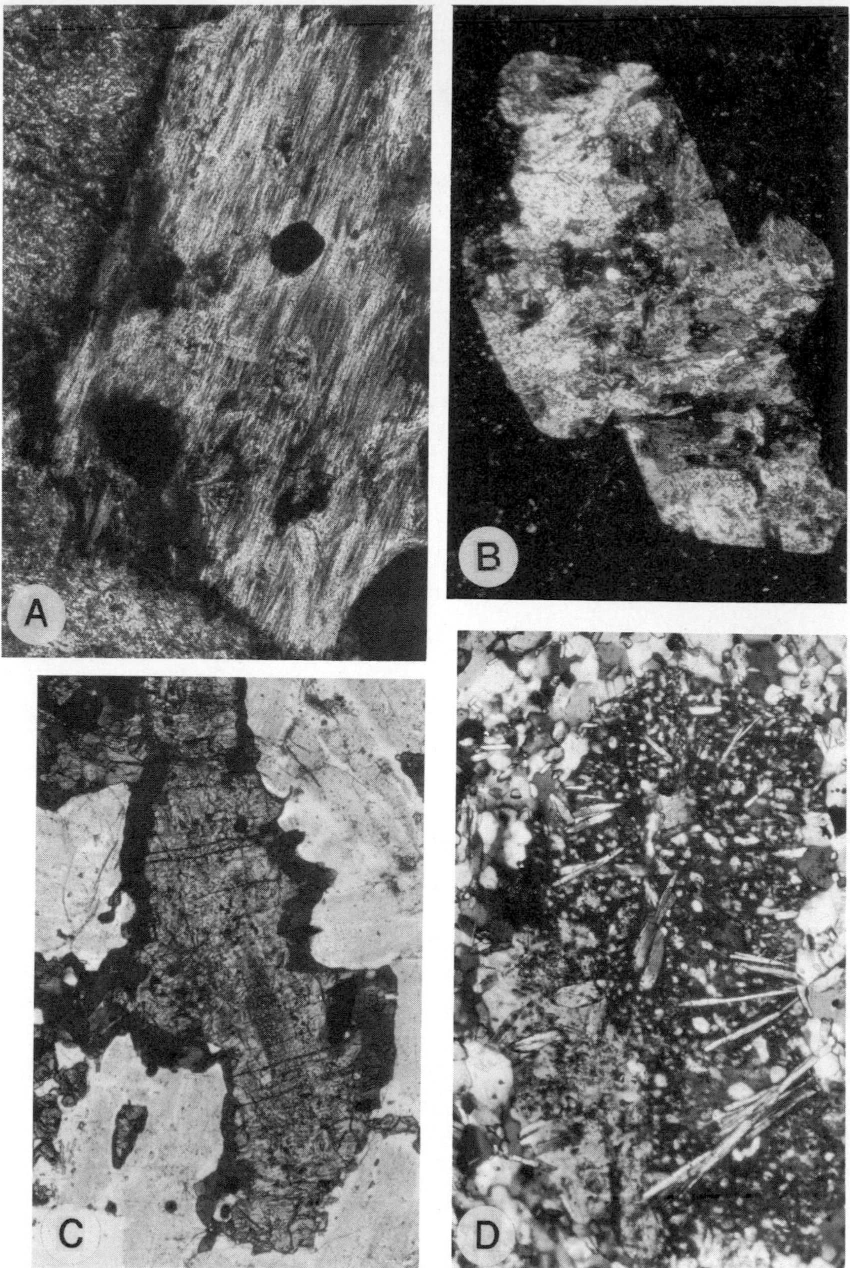

Fig. 3.30 Replacement phenomena. (A): Pseudomorph of pyroxene totally replaced by fibrous amphibole (uralitization) in diorite. (B): Pseudomorph of feldspar phenocrysts in a volcanic rock totally replaced by zeolites. (C): Hornblende mantling pyroxene in gabbro. (D): Incoherent partial replacement of large plagioclase (in extinction by amphibole in basic metatuff. (View lengths measure 0.85 mm (A) and (B), 1.9 mm (C), and 0.8 mm (D).)

biotite are within 0.1% which allows an almost perfect strain-free coherent fit.

Chlorite is one of the principal stable minerals in sedimentary assemblages, and forms at low to moderate temperatures as an alteration product, not only of biotite, but of pyroxene, amphibole and garnet, among others.

(b) Perthite

The development of perthite is rather simpler chemically than chloritization of biotite. Although fine-grained coherent perthite almost certainly results from exsolution (section 3.5.3), the coarse, less-coherent examples listed in Table 3.2 result from fluid-induced coarsening. The evidence for this is twofold: first, vein, patch and swapped-rim perthites are usually cloudy with alteration products and fluid inclusions; secondly, it is well established experimentally that K^+ and Na^+ ions are easily exchanged between K-(or Na)-feldspar and a Na-(or K)-rich fluid (Orville, 1963). The balance of opinion, according to Smith and Brown (1988), is that most coarse perthites represent a mutual replacement in an essentially closed system; in other words, a coarsening of existing perthite without significant overall change in chemistry. In an open system, however, circulating solutions may well effect a gross chemical change, and Smith and Brown (1988) believe this could be the case for some coarse perthite in pegmatite. As discussed in section 3.5.3, coarse perthites become increasingly incoherent, and swapped-rim perthite is always incoherent.

(c) Uralitization

A dry, massive, totally crystalline igneous rock is very resistant to metamorphic change not only because it lacks the fluids and fluid channels necessary for mineral reactions, but because it is difficult to deform, certainly less so than surrounding sediments. Nevertheless, the water-rich fluids that are the final products of igneous crystallization may alter already solidified, cooling igneous masses, particularly near the borders of a pluton, and along joints and fractures. Such alteration is called deuteric rather than metamorphic if it is an intrinsic part of the igneous process.

One characteristic product of deuteric alteration is uralite (Fig. 3.30A), a semi-coherent replacement of pyroxene by amphibole. In the case of thin rims of common hornblende mantling pyroxene (Fig. 3.30C), it is not always easy to decide whether they are due to late-magmatic overgrowth or secondary grain-boundary reaction involving water-rich fluids. However, when entire crystals of pyroxene are replaced by tremolitic or actinolitic hornblende it becomes clear that secondary processes are responsible. Evidence is found in the overall shape of the original pyroxene (the octagonal section is particularly characteristic and relic patches are commonly preserved). And although there is a more-or-less coherent relationship between pyroxene and hornblende with the chain length and *c*-axes parallel, the occasional incoherent area where tremolitic material diverges in orientation into a jumbled mass of acicular grains clearly reveals the secondary nature of the product. Not only that, but tremolitic hornblende typically develops at metamorphic temperatures.

The same replacement process takes place during regional metamorphism of a basic igneous rock, provided that fluids have access to the rock. Thus the rock epidiorite retains essentially igneous features but with pyroxene replaced by amphibole.

(d) Sericitization and saussuritization

Sedimentary rocks abound in muscovite and related clays and sheet-silicates, much of which derive from the alteration of Al-rich minerals in igneous and metamorphic rocks. Thus feldspars, the Al_2SiO_5 polymorphs, cordierite and staurolite are commonly speckled with or completely pseudomorphed by fine-grained white micas or related minerals called 'sericite'.

The growth of sericite most notably requires the addition of water and K^+, and sericitization can only proceed if water-rich fluids are available. One important source of K^+ may be found in the chloritization of biotite (discussed above): the K^+ released reacts with the anorthite component of plagioclase to free Ca^{2+}, and zoned plagioclase is most easily sericitized in the An-rich parts. In vigorous hydrothermal systems, there may be a general flux of ions so that intense sericitization is accompanied by a general K-metasomatism.

Unlike coherent uralite and chloritization of biotite, most sericite grows incoherently. However, in some feldspars large muscovite plates are aligned more-or-less parallel to cleavage traces, probably a response to the easy access of fluids along those directions. Diffusion is the rate-controlling factor in sericitization: reaction depends on the supply of fluid and K^+, yet the products grow only at the local sites of reaction (e.g., particular zones within plagioclase), presumably due to the relative immobility of Si and Al. Despite a lack of coherency, nucleation seems not to be a problem as numerous fine grains develop.

Saussurite is another common alteration product of plagioclase, where the addition of water changes the An-component to epidote (or clinozoisite or zoisite), and the residual plagioclase is left as albite. Saussurite is formed under the same conditions that characterize the greenschist (and equivalent low-P) facies. Sericite and calcite often accompany the epidote and albite. Common saussurite illustrates again how hydrothermal reaction products tend to grow first at localized sites; thus An-rich zones in plagioclase selectively develop the epidote. And it illustrates the contrasting behaviour of plagioclase, which changes composition to albite in a perfectly coherent manner (no fresh nuclei formed), with epidote which nucleates easily and grows incoherently in random orientation within the altered feldspar.

Marzouki *et al.* (1979) describe various stages of alteration of diorite so that following on from saussuritization there is extensive epidotization involving the addition of Ca and the release of Na and Si. Albite produced during the initial saussuritization is later transformed to more epidote by the addition of calcium and water.

(e) Pseudomorphs of olivine

Olivine is one of the most susceptible of minerals to alteration; alteration is often complete so that pseudomorphs are common, and the alteration products are rather diverse. A comprehensive review is provided by Deer *et al.* (1982).

The principal products include serpentine, iddingsite, bowlingite and chlorophaeite, and they form under a wide range of conditions. Iddingsite is a red mineraloid containing ferric oxide. It is found principally in lavas, and the observation of iddingsitized olivine apparently overgrown by fresh olivine (Fig. 3.31) suggests it may form at the vapour-rich top of the magma column before extrusion. It also forms during post-extrusive deuteric alteration, and possibly during weathering. Iddingsite develops coherently within the olivine, so that the iron-oxide completely (the sheet-silicates to a lesser extent) inherits the disposition of oxygen atoms from the olivine. Even so, the lack of strong structural anisotropy in olivine means that alteration spreads evenly across the crystal. The less-oxidized mineraloid is called bowlingite (or chlorophaeite in its isotropic form), and some lavas contain both bowlingite and iddingsite, depending on the state of oxidation. In terms of chemistry, these alterations involve the addition of iron and water and the loss of Mg; Si is immobile.

Serpentinization is the most common pervasive alteration product of olivine (Fig. 1.18). In the case of altered peridotite, entire rock masses may be converted to serpentinite, and the alteration may involve pyroxene as well as olivine. The temperatures at which serpentinization takes place may be $<100°C$ in the case of lizardite and chrysotile formation, and as high as 500°C in the case of antigorite. Iron in the olivine is often deposited as magnetite, and brucite is also commonly mixed in with the serpentine alteration products. The lack of cleavage or strong structural anisotropy in olivine has an important influence on the way that serpentinization proceeds. The alteration starts along irregular cracks to produce an anastomosing network of veinlets with fibrous or platy serpentine.

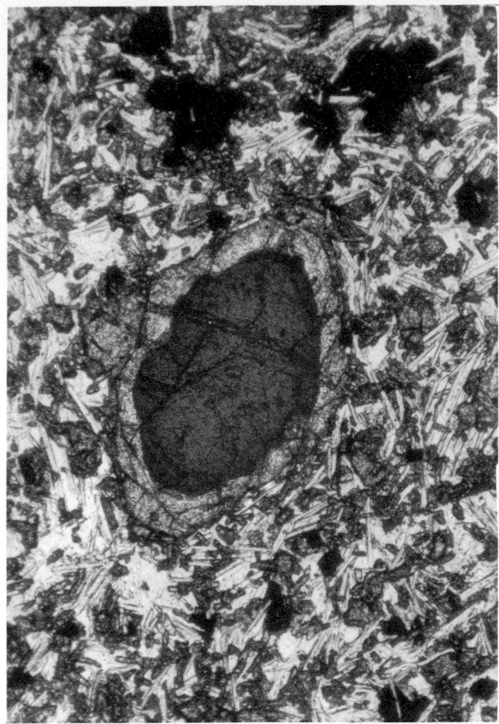

Fig. 3.31 Iddingsitized olivine (dark area) apparently overgrown by fresh olivine in basalt, Banks Peninsula, NZ. (View length measures 2.1 mm.)

Isotropic areas of serpentine may occur between the veinlets. Unlike iddingsite alteration, serpentinization proceeds incoherently.

Pseudomorphs of serpentine after olivine are common and can be recognized by the distinctive shapes of the original crystals, but since serpentine is soft and easily deformed, pseudomorphs and original textures are frequently destroyed in massive serpentinite. Serpentinization clearly requires the addition of water (copious amounts to produce large bodies of serpentinite), and often Si is added while Mg is lost in solution; the process involves a volume increase of as much as 50% for complete alteration. As a result of this expansion, in gabbroic rocks one commonly finds cracks in plagioclase, filled with serpentine, radiating outwards from serpentinized olivine (Fig. 1.18).

(f) Coronae

Among the most common mantling textures are those involving pyroxene bordered by amphibole. If the amphibole is tremolitic, then the process is certainly one of replacement (see uralite above), but mantles of common hornblende may also result from late-magmatic crystallization.

The most spectacular coronae are those developed around olivine in gabbroic rocks; symplectites in such coronae are discussed in section 3.5.2. They represent high-temperature grain-boundary reactions between anhydrous minerals in a dry, completely crystalline rock. In gabbroic coronae, the reaction products are developed mainly where olivine and plagioclase join (Figs 2.2 and 3.24), although Grant (1988) showed that material can be transferred between olivine and plagioclase through intervening crystals of pyroxene. The products of reaction are arranged in concentric rims, and common sequences include: (olivine) opx, cpx + spinel symplectite, (plag); (olivine) opx, hornb + spinel symplectite, garnet, (plag); (olivine) cpx, hornb, (plag); (olivine) opx, cpx, garnet + cpx symplectite, garnet, (plag). The products of reaction may radiate outwards perpendicular to the host-mineral grain boundary (Fig. 3.24). Incoherent interfaces are general.

It is possible to write balanced equations which seem to explain corona reactions, as for example:

$$2Mg_2SiO_4 + CaAl_2Si_2O_8 = 2MgSiO_3 + CaMgSi_2O_6 + MgAl_2O_4$$

However, Johnson and Carlson (1990) show that coronae actually evolve in an open system, not in strictly isochemical fashion. They describe the following evolution in composition of olivine–plagioclase coronae, so that plagioclase and clinopyroxene formed in the first stage are both progressively eliminated:

(opx–cpx–plag–garnet)
olivine (opx – cpx –garnet) plag + spinel
(opx – garnet)

The changes are related to continuous reactions involving a change in composition of reactant plagioclase rather than varying *PT* conditions. In the same gabbros, coronae also developed between oxides and plagioclase with (oxides), oxides, biotite, amphibole, plagioclase + spinel (plagioclase).

Coronae represent reactions that proceeded with difficulty due to declining temperatures and an absence of abundant fluid. Slow diffusion rates are the rate-controlling process, and growth proceeds only at sites close to the reactants; Grant (1988) showed that Al and Si are particularly immobile. Corona development ceases when diffusion paths between the reactants become too long to sustain growth.

(g) Overgrowths

Mantle overgrowths are formed in the following ways:

1. Crystallization from residual magma in an igneous rock. Some hornblende mantles form around pyroxene in this way; the minerals may be in coherent or incoherent relationship. A change in mineralogy in the final stages of crystallization is a common response to the concentration of elements not incorporated into the principal minerals of the rock. Residual fluids in a gabbro, for example, may be enriched in H_2O and K^+, producing the common mantling sequence of pyroxene, hornblende, biotite. Early-formed crystals may cease to be in equilibrium with late-stage magma, and mantles often grow on corroded crystal boundaries, a subject developed later in section 4.1.6.
2. Mantling, particularly of feldspar by feldspar as a result of magma mixing, and discussed later under phenocrysts (section 4.1.6).
3. Progressive or polymetamorphism, where incomplete reaction allows early-formed crystals to be mantled by secondary overgrowths, a subject more appropriately dealt with in Chapter 5.

3.5.5 When one mineral grain encloses several other grains: poikilitic and poikiloblastic textures

Minerals in igneous and metamorphic rocks sometimes enclose other mineral grains during growth. A mineral with inclusions has a larger surface area for its volume than a grain lacking inclusions, so these textures do not represent the most stable configuration. Factors that promote the enclosure of foreign material include rapid growth (perhaps with an irregular growth front), and the mutual reduction in surface energy that results when foreign material is adsorbed on to a crystal face (although the reduction is not as great as for an ideal crystal growth). The solid-state environment in which metamorphic porphyroblasts grow may make inclusion of non-reactants inevitable (section 5.5.1).

In igneous rocks, poikilitic texture is widely used to determine order of crystallization, the argument being that if one mineral species is enclosed by another then the enclosed grain must have been the first to crystallize. This may sometimes be true, but it is certainly not always so. Thus McBirney and Noyes (1979), for example, show that ophitic texture (the enclosure of plagioclase in pyroxene) may be due to simultaneous crystallization. Cited as evidence is the increase in grain size of feldspar from the centre of the texture outwards. In other words, the texture originates as a result of differing nucleation and growth rates, so that a single pyroxene nucleates and grows to a large size (low nucleation rate) in contrast to the several feldspars (higher nucleation rate) which necessarily remain

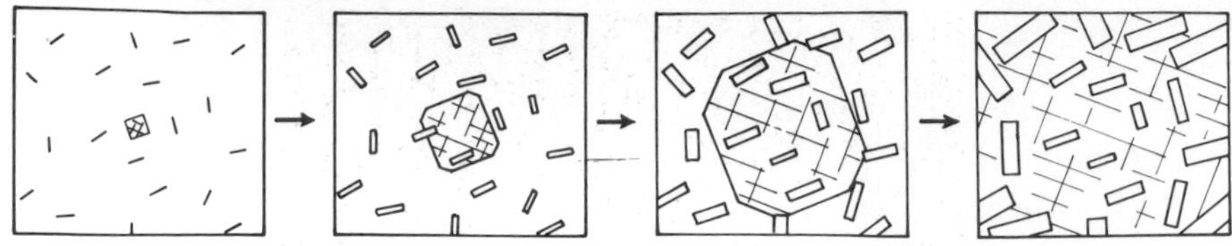

Fig. 3.32 Development of ophitic texture by simultaneous growth but different nucleation rates of pyroxene and plagioclase. Plagioclase grain size increases towards pyroxene margins.

relatively small and become successively entrapped in the pyroxene (Fig. 3.32).

In many poikilitic textures, the enclosed grains are randomly arranged. In others, the enclosed grains may be concentrated in zones, either representing times when grains were available to be included, or times when the inclusions were most easily adsorbed on to the surface of the host crystal. In some cases, included grains have a specific crystallographic relationship to the host, and this is discussed below.

In metamorphic rocks, not only is poikiloblastic texture very common, but it is also extremely important in providing evidence for complex structural, textural and mineralogical sequences. A full discussion of such inclusion textures is found in section 5.5. However, one must also be on guard for the possibility that some poikiloblasts represent simultaneous growth of two or more minerals with differing nucleation or growth rates rather than a distinctive sequence of events.

Inclusions arranged in particular crystallographic directions

The included grains in some poikilitic and poikiloblastic textures are arranged in a systematic way relative to the crystallography of the host mineral. Perhaps the most familiar example of all is the cross of carbonaceous inclusions found in the variety of andalusite called chiastolite (Figs 3.33 and 2.16C). The characteristics, as seen in thin section are: inclusions concentrated in four zones running diagonally inwards from the four edges of the prism faces; a central (often rectangular) area bordered by inclusions which marks the inner termination of the four radiating inclusion zones; a feathery pattern in the inclusion zones; possibly re-entrant angles at the four corners of the section. Explanations, summarized in Spry (1969), include the brushing aside of unwanted carbonaceous material to the edges of the crystal as it grew, although no realistic reason or mechanism for such a process has been put forward. Alternatively, dendritic growth at the edges of the crystal, trapping inclusions there, was followed by layer growth of the prism faces in the spaces between the dendrites. There is, however, no substantial evidence to support this, and

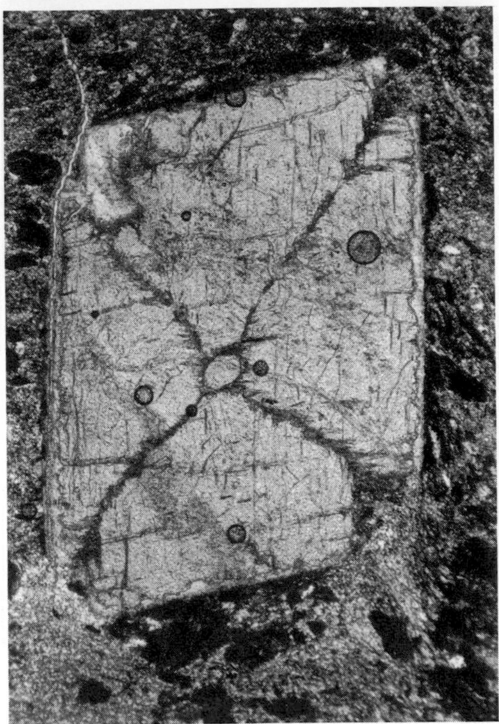

Fig. 3.33 Chiastolite (slate, Buller Gorge, NZ) illustrating the central sector, the feathery pattern of four inclusion zones, and the re-entrant angles. (View length measures 3.3 mm.)

it seems inconsistent with the common re-entrants at the crystal edges. Probably the best explanation derives originally from Frondel (1934), who noted that many crystals preferentially adsorb foreign material on to specific crystal faces and edges. In some ways the phenomenon is similar to sector zoning which originates because different crystal faces have different atomic arrangements exposed at the surface, affecting which atoms are attached there. Crystal edges and corners will always be preferred sites of attachment of atoms or foreign particles, because of their relatively high surface energy; in addition, even in simple euhedral crystals, the edges represent the positions of fastest growth and the sites most likely to trap inclusions. The development of the chiastolite pattern is shown sequentially in Fig. 3.34. The selective attachment of inclusions at the corners effectively slows growth there (called a poisoning effect) so that a re-entrant develops. The re-entrant is an even more potent high-energy site for the attachment of inclusions than the original edge, and so it is perpetuated with further growth. The feathery inclusion pattern

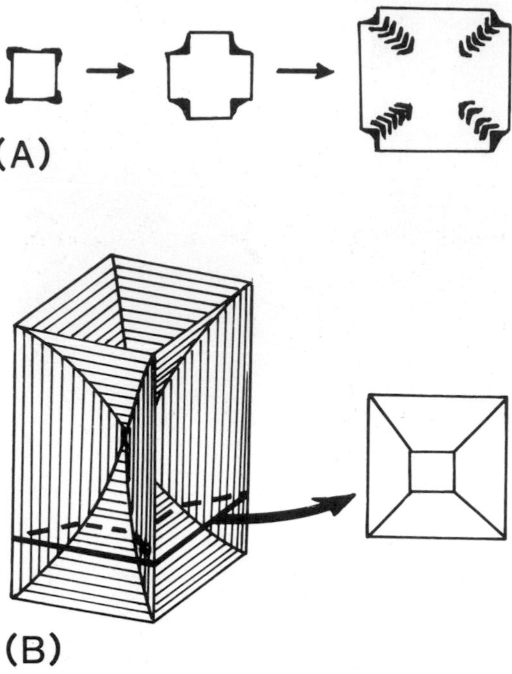

Fig. 3.34 Chiastolite formation. (A): Carbonaceous impurities concentrated at edges and corners lead to a feathery inclusion pattern, as described in text. (B): Three-dimensional view of planes rich in impurities. The crystal cross-section intersects eight planes to give the typical chiastolite pattern. See also Fig. 2.16.

represents the perpetuation of the re-entrant. The rectangular pattern in the centre of the chiastolite is explained simply as an artefact of a two-dimensional view of a three-dimensional texture. Thus it can be seen in Fig. 3.34B that the concentration of inclusions at all the edges of the growing crystal creates a complex of planes that define six pyramidal sectors. Almost any slice through the crystal cuts eight inclusion planes: four are responsible for the central rectangular pattern, the other four for the radiating inclusion zones.

Similar patterns are found in other metamorphic minerals, particularly garnet, as described by Burton (1986), for example. Garnet develops 12 pyramidal sectors, all bounded by graphitic inclusions, probably by a mechanism similar to chiastolite. In addition, within each sector, enclosed quartz rods extend from the sector boundary outwards and perpendicular to the crystal faces (Fig. 3.35). According to Burton, the presence of graphite reduces the solubility of silica at the site of garnet growth, with

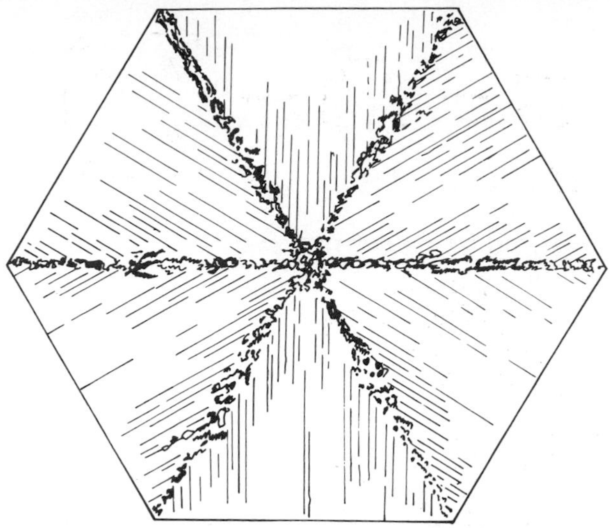

Fig. 3.35 Cross-section of euhedral garnet showing sectors based on dodecahedral faces {110} and defined by impurities along sector boundaries. Quartz rods grow perpendicular to crystal faces, as described in text. Reproduced with permission from Burton (1986).

excess silica precipitated at the site of growth spirals. Quartz continued to grow as rods perpendicular to the crystal face in order to minimize the garnet–quartz interfacial energy. Although the direction of growth is crystallographically controlled, the quartz and garnet crystal structures are in random and incoherent relationship to one another.

Poikilitic textures in igneous rocks may display similar crystallographically controlled relationships. One of the best known is the systematic arrangement of plagioclase (and mica) enclosed in K-rich feldspar phenocrysts (e.g., Fig. 1.15B), as discussed by Schermerhorn (1956), for example. The included minerals may be concentrated at sector boundaries, similar to the inclusion pattern of chiastolite, and most commonly the (010) faces of plagioclase and (001) of mica are aligned more-or-less parallel to the principal faces of the K-feldspar. Such textures probably originate by a combination of the following processes: selective adsorption of foreign material on to certain crystal faces and edges, as discussed above for chiastolite; synneusis so that plagioclase of a certain preferred orientation adheres most easily to the host crystal face (refer to previous discussion of synneusis in twinning); and nucleation of plagioclase (in a preferred orientation seeded by the K-feldspar) on the K-feldspar.

Some of the enormous variety of poikilitic and poikiloblastic textures are discussed in other relevant parts of this book.

4

Igneous crystallization, textures and microstructures

We are all familiar with crystallization from a melt in the system ice–water. This system is relatively simple in that it involves only two phases, ice and water, the two phases have the same composition, and the temperature at which ice melts to water or ice crystallizes from water is always very close to 0°C. Such a system is in marked contrast to natural silicate melts which are very complex indeed. Not only do they crystallize through a wide range of temperature of several hundred degrees, but the number of phases may be numerous, disequilibrium is common, and in some situations crystallization does not take place at all.

Numerous factors are involved. Thus, crystallization of a phase that belongs to a solid-solution series takes place over a temperature range producing zoned crystals as already discussed for plagioclase (Fig. 3.10). Differentiation over a range of temperature may be represented by more than the zoning in individual crystals. Thus, in hawaiite dykes of the Miocene Lyttelton volcano, New Zealand, phenocrysts are labradorite, microphenocrysts are andesine, and the groundmass contains anorthoclase (Shelley, 1988). Crystallization in eutectic and cotectic systems also takes place over a range of temperature, as discussed in section 3.5.1.

The range of temperature for crystallization is widened by the necessity for some degree of undercooling before nucleation and growth take place. Nucleation and growth rates first increase, then decrease with undercooling; diffusion rates only decrease. If cooling is fast, as in volcanic situations, the consequential interplay of fluctuating or changing rates of nucleation, growth, and diffusion leads to great diversity and complexity of texture.

Igneous rocks fall naturally into two main textural categories:

1. Plutonic rocks that crystallize below the Earth's surface in bodies sufficiently large for the magma to cool slowly.
2. Volcanic rocks (and minor intrusives) that crystallize from rapidly

cooled magma, either at or near the surface, or in very small intrusives at greater depth.

Plutonic rocks are always completely crystallized with textures resulting from near-equilibrium crystallization; undercooling was slight, nucleation and growth rates low, and diffusion rates high enough not to be a limiting factor in crystallization. In contrast, crystallization in volcanic rocks is often incomplete. Textures are diverse and represent disequilibrium crystallization, varying degrees of undercooling, fluctuating (from zero to high) nucleation and growth rates with diffusion rates a limiting factor on growth. We shall discuss the volcanic situation first.

4.1 VOLCANIC ROCKS

4.1.1 Crystal shape and size

Crystal nucleation and growth both take time. A melt with no degree of undercooling may take an infinite time to nucleate crystals; the nucleation rate is zero. With the beginnings of undercooling, nuclei start to form at random, and at low initial rates. As undercooling increases, the nucleation rate reaches a maximum, then declines to zero at high degrees of undercooling. Growth rates exhibit a similar increase and decrease, whereas diffusion rates simply decrease with undercooling. The rates vary with magma composition, and they may be very sensitive to the presence or absence of water and alkalis.

If magma is quenched in water it forms glass, and no crystals develop; in this situation the nucleation rate was too low to initiate crystallization during those very brief moments during quenching when the temperature was ideal for nucleation and growth. But there is every gradation in speed of cooling between quenching in water and the relatively slow cooling in the centre of a volcanic dome, and the timing of crystallization in any one volcanic rock is a complex function of nucleation, growth, diffusion, and cooling rates.

Growth and nucleation rates are difficult to quantify. Apart from anything else, nuclei are too small to observe directly under most conditions, crystal growth rates differ along different crystal axes, vary with growth mechanism, and are sensitive to changes in diffusion rate. Many workers document nucleation density rather than rates. Dowty (1980a) discusses these problems and provides theoretical curves for nucleation and growth of alkali feldspar from hydrous melts (Fig. 4.1). The curves show maximum rates for growth at undercoolings of approximately 100°C; the nucleation rates are particularly sensitive to changes in H_2O content, and may peak at widely different degrees of undercooling.

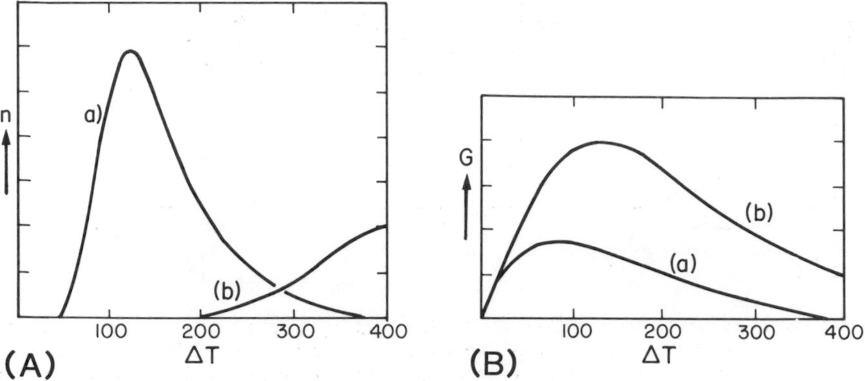

Fig. 4.1 Theoretical curves for feldspar nucleation (A) and growth rates (B) plotted against degrees of undercooling (ΔT). Curves (a), 9.5 wt. % water, curves (b) 1.7 wt. % water. Reproduced with permission from Dowty (1980a).

4.1.2 Crystal-size distribution (CSD) curves

Another approach to determining growth and nucleation rates is to plot the population density of particular grain sizes in the form of a crystal-size distribution (CSD) curve. Samples of basalt from a lava lake on Kilauea are used by Marsh (1988) and Cashman and Marsh (1988) to illustrate the principles involved (the specimens crystallized in a steady and continuous manner over known periods of time). The simplest form of curve, a plot of crystal length (L = the greatest diameter in thin section) against frequency (Fig. 4.2A), shows that there are many more small crystals than large ones. The same data can be plotted as crystal length against N, the cumulative number of crystals per unit volume (Fig. 4.2B), and the slope of the curve at any point is given by dN/dL (= n). Borrowing from well-established methods for synthetic materials, Marsh (1988) and Cashman and Marsh (1988) show that a plot of $\ln(n)$ against crystal length produces a straight line (Fig. 4.2C), and such log-linear relationships typify steady continuous crystallization marked by more-or-less constant growth rates, independent of crystal size.

Salient points derived from CSD curves include:

1. The curved inverse relationship between grain length and frequency (Fig. 4.2A) indicates an exponentially increasing nucleation rate with time (Marsh, 1988).
2. The slope of the log-linear curve is $-1/G\tau$ where G is the average growth rate and τ the growth time.
3. The intercept point on the population density axis (where $L = 0$) is equal to J/G, where J is the nucleation rate when crystallization ceased

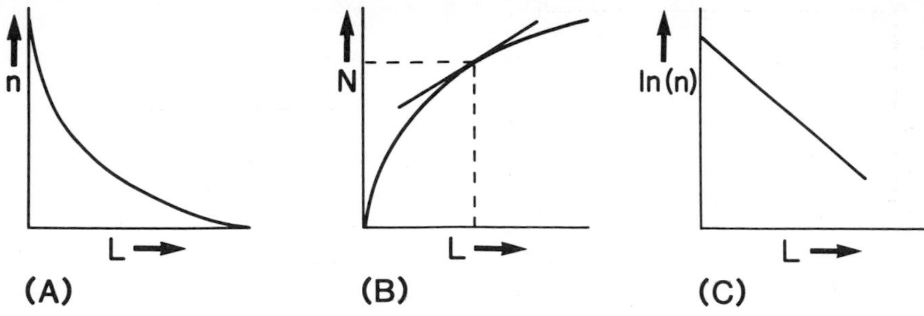

Fig. 4.2 Principles of igneous CSD analysis. (A): Typical frequency (n) against crystal length (L) curve. (B): Slope of curve of cumulative number of crystals (N) against length equals n. (C): L plotted against ln (n) gives a straight line. Further explanation in text. With permission based on Marsh (1988).

(this nucleation rate is related to a unit volume of partially crystallized rock, and Cashman and Marsh (1988) point out that it may be more appropriate to recalculate a higher rate related to the volume of residual magma).

In the particular case of the Kilauean lava lake basalt, growth times are known, so that growth and nucleation rates can be determined. For plagioclase, the average growth and nucleation rates (at $L = 0$) range from 5.39 to 9.89×10^{-11} cm sec^{-1} (*c.* 0.02 mm per year), and 0.0016–0.0339 cm^3 sec^{-1} or 5 to 100×10^4 cm^3 per year respectively. Nucleation was heterogeneous, as already noted by Kirkpatrick (1977), thus enabling growth to proceed at a very low degree of undercooling.

Several practical problems are posed when measuring data for a CSD curve from two-dimensional thin sections. Firstly, the measured population density per unit area is converted to population density per unit volume simply by raising it to the power of 3/2, but strictly this is only appropriate for crystals with uniform shapes (spheres, for example). This is obviously not the case in natural rocks, but Cashman and Marsh were unable to overcome the practical difficulties in correcting for non-uniform shapes. The error affects the determination of nucleation density. Secondly, it is assumed that crystals have a random orientation; this may be true for Kilauean samples but it is often not so in other igneous rocks. Thirdly, skeletal crystals and glomeroporphyritic clusters pose particular problems in deciding what dimension to measure. And fourthly, crystal dimensions measured in two-dimensional thin sections are not usually the actual dimensions; thus a crystal cut through its corner will seem to be much smaller than a similar-sized grain cut through its centre. A considerable literature exists on this problem, but Cashman and Marsh (1988) conclude that there is no satisfactory way to correct the data. They show that at

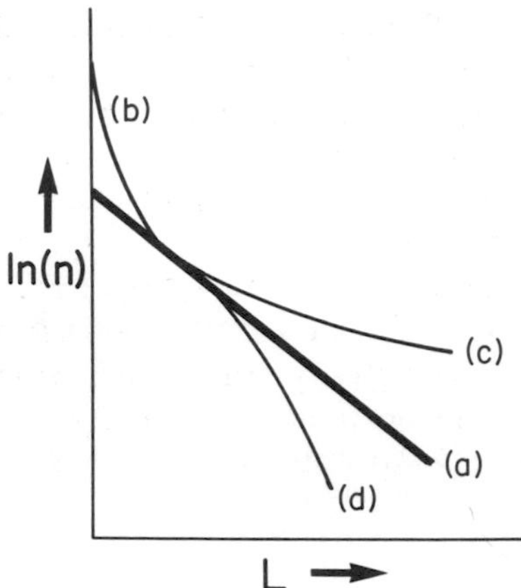

Fig. 4.3 Ideal straight-line igneous CSD (a), modified by quenching (b), crystal accumulation (c), or removal of crystals from the liquid (d). With permission based on Marsh (1988).

worst the uncorrected data give curves that are 10% in error in slope and intercept position on the vertical axis.

Crystallization in igneous rocks is often more involved than that in a Hawaiian lava lake, and CSD curves may commonly become non-log-linear (Fig. 4.3). For example, a marked increase in undercooling will cause the $\log(n)/L$ curve to steepen for smaller grain sizes. Crystal settling will preferentially remove the largest crystals (they sink more rapidly than smaller ones, as discussed in section 4.2.2): cumulate layers so produced should have CSD curves that flatten out with increasing grain size, whereas residual material will concentrate finer grain sizes with a CSD curve that steepens to the right. Concave or convex CSDs could, however, also develop as a consequence of non-linear growth laws. Kinked or irregular CSD curves are likely to be produced by the mixing of magmas with crystal populations already characterized by different CSD curves.

Marsh (1988) reports CSD curves for a variety of basaltic volcanics, and the most important finding is that they all demonstrate a smooth variation of grain size. This means that there is no sharp boundary between phenocryst and matrix grain size, despite first appearances of bimodality.

The analysis of quantitative grain-size population data using CSD curves is a new and little-tested field of igneous petrology. The subject lags behind similar work in sedimentary petrology and the study of synthetic materials, and it remains to be seen how useful CSD curves will be in routine petrology. Despite the cautionary note above about the status of phenocrysts versus matrix, the reality of many volcanics is that they consist of a complex of minerals, some derived from the subterranean magma chamber, some formed at the surface, and the quenching and release of volatiles at or near the surface can produce marked changes in undercooling and the rate of crystallization. The usefulness of CSD analysis for such complex rocks must be left for future research to decide.

In the next two sections we examine post-phenocryst textural developments in volcanic rocks. The first section deals with crystallization from basaltic magma, characterized by high fluidity and relatively high diffusion rates. By way of contrast, the second section deals with the rate-limiting effects of low diffusion rates on crystallization in viscous rhyolitic magma. Phenocrysts provide important clues to magma evolution prior to its emplacement at or near the surface, and they will be discussed separately in section 4.1.6.

4.1.3 Basaltic crystallization

Basalt magma is typically very fluid, and the potential mineralogy is quite diverse, involving pyroxenes, feldspars, olivine and various opaque materials like magnetite. Complete, or nearly complete, crystallization is common, and basaltic glass is only abundant in the most rapidly cooled rocks such as chilled margins, pyroclasts, and rocks formed in explosive underwater eruptions.

The most common texture among groundmass microphenocrysts in well-crystallized basalts is intergranular (Fig. 1.31B), this largely being a consequence of the relative grain size and shape of three mineral phases. Thus the relatively elongate plagioclases form a framework, rather like a jumble of books or cards with some on end, and the more equidimensional and typically smaller pyroxene and magnetite grains fit in between. If glass is present between the laths the texture becomes intersertal. There is every possible gradation between intergranular and intersertal texture, and also trachytic texture where the laths are in strong alignment due to flow (Fig. 1.31A), a subject discussed further in section 8.1.3.

The groundmass microphenocrysts grow during the final rapid stage of magma cooling accompanying emplacement of hypabyssal rocks, or extrusion. Consider a crystal, bounded by faces, growing from an undercooled magma. The lowest-energy-state equilibrium form is one with regular planar faces, but this requires that diffusion is able to move atoms to appropriate attachment sites as quickly as the crystal can grow. In cooling magma this is not always the case: diffusion rates are decreasing whereas growth rates are increasing. Evidence that diffusion is a limiting factor in crystallization is found in various so-called quench textures, most commonly seen in basalt as skeletal, dendritic and swallow-tailed forms. Material for crystal growth can be supplied most efficiently at the ends, edges and corners of a crystal because these sites are surrounded by a

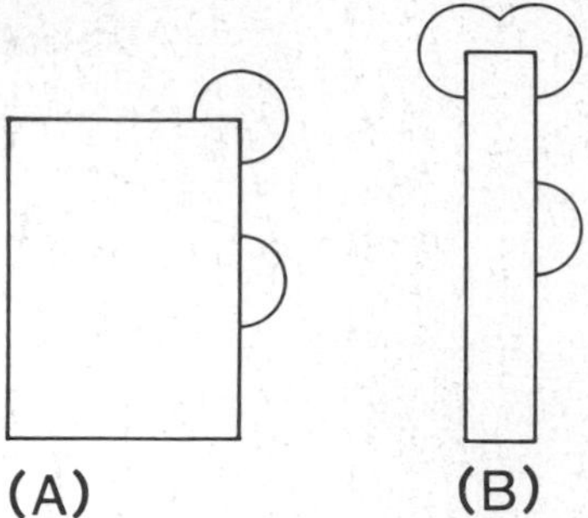

Fig. 4.4 (A): The volume of liquid in the immediate vicinity of a crystal edge or corner is greater than next to a face. (B): The volume around the ends of an elongate crystal is greater than next to the larger faces.

greater volume of magma than the centres of crystal faces (Fig. 4.4). The ends, corners and edges therefore grow more rapidly than the rest of the crystal if diffusion is unable to keep pace with equilibrium texture development, and elongate or skeletal forms may result (Fig. 4.5). The fact that atoms at crystal corners and edges have their bonding requirements less satisfied than at other sites also contributes to the preferred growth at corners and edges. A particularly common quench texture, and incipient skeletal form is swallow-tailed plagioclase formed by relatively rapid growth of the terminal edges of a microphenocryst (Fig. 4.6). Rapid quenching may also produce a network of unusually elongate feldspar laths (Fig. 4.5C), and the change from normal crystallization rates to rapid rates associated with quenching may be marked by an abrupt change in morphology. One microphenocryst in Fig. 4.5D, for example, has a plagioclase 'whisker' with an aspect ratio of >35/1 grown due to a sudden onset of rapid quenching. The quench textures represent a higher energy state than simple euhedral forms as a result of their higher surface area/volume ratio.

The dependency of these textural developments on the degree of undercooling is well established through observations on natural rocks and in experiments. Helz (1987), for example, describes skeletal olivine microphenocrysts in air-quenched pumiceous basalt from Hawaii but not in slower-cooled material. And Lofgren (1980) documents the experimental development of textural forms in plagioclase from tabular to skeletal to dendritic and finally spherulitic forms with progressive undercooling. However, most crystallization in basalt takes place at a low level of undercooling (Kirkpatrick, 1977), growth rates do not peak, and diffusion rates do not become an overwhelmingly important rate-limiting factor. Regular crystal shapes are the norm, and the more extreme quench den-

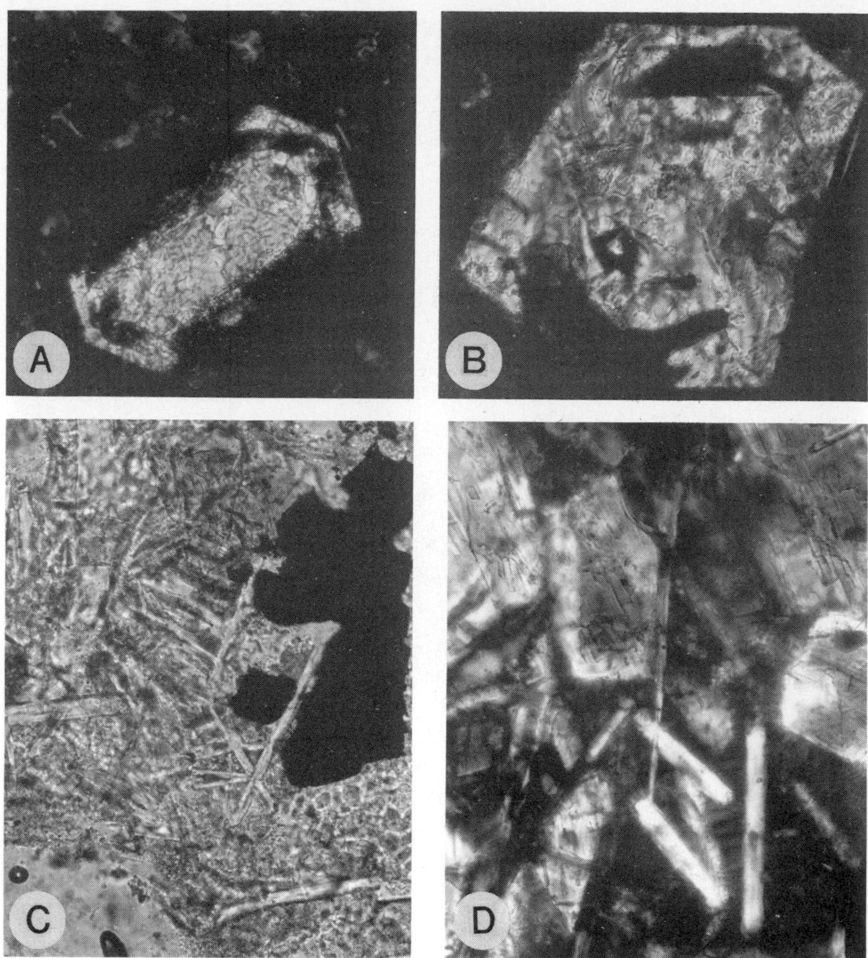

Fig. 4.5 (A) and (B): Skeletal olivine in chilled margin of basalt in pyroclastic deposits, Taupo, NZ. (C): Network of unusually elongate plagioclase laths in basaltic xenolith within ignimbrite, Rotorua, NZ. (D): Plagioclase 'whisker' with aspect ratio >35/1 grown from the corner of a microphenocryst. Same specimen as (C). (View lengths measure 0.35 mm (A), 0.24 mm (B), 0.3 mm (C) and 0.19 mm (D).)

dritic and spherulitic forms of feldspar are not common in natural basaltic rocks. The ultimate quench product is, of course, glass, and given an infinity of time one might expect it to crystallize. In practice it is stable on the geological time-scale, although particularly prone to metamorphic change and weathering to palagonite.

Crystal size is to some extent a consequence of the ratio of nucleation and actual growth rates. Thus, a mineral that nucleates with difficulty but

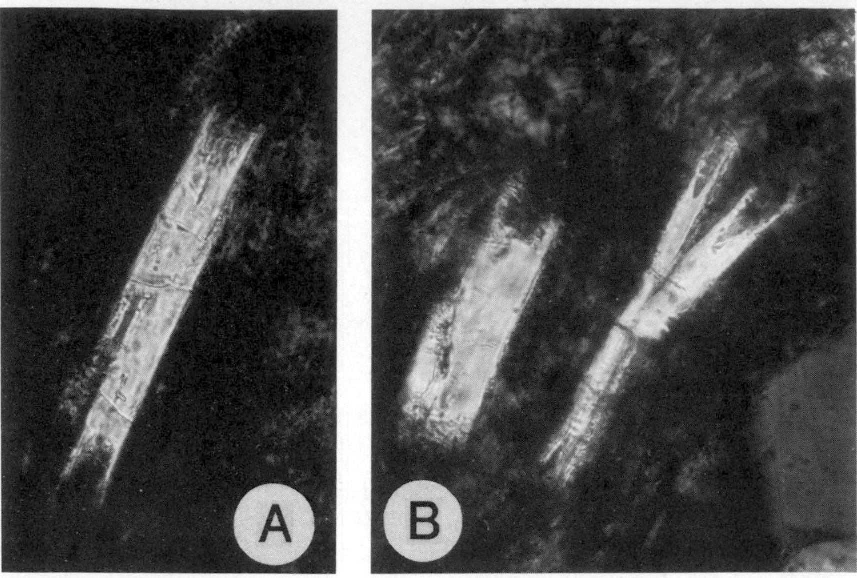

Fig. 4.6 Examples of swallow-tailed plagioclase in trachyte (Remarkable Dyke, Banks Peninsula, NZ). (View lengths measure 0.22 mm (A) and (B).)

grows easily may form a few large crystals, contrasting with a mineral that nucleates easily and forms numerous small crystals. Consider therefore the situation where pyroxene nucleates only occasionally as compared with plagioclase (Fig. 3.32). A pyroxene crystal will envelop smaller plagioclase laths as the pyroxene spreads out from its growth centre. If the entire rock crystallizes in this way, then the large grains will be anhedral with mutual interference boundaries like a metamorphic mosaic (Fig. 4.7). Such a texture is illustrated by MacKenzie *et al.* (1982, p. 37), and they call it 'ophimottled'; it is a variety of poikilitic texture similar to ophitic texture in gabbro. It is also similar to the snowflake texture found in rhyolite to be described later. There is a need for a general term to cover 'ophimottled', 'snowflake', and other similar textures which consist of a mosaic of poikilitic intergrowths: here I coin the name **poikilomosaic** for poikilitic mosaics in volcanic and plutonic rocks. Examples abound in a wide variety of rocks, and I list some of those discovered in the collections at the University of Canterbury, New Zealand: poikilomosaics of ophitic texture in dolerites and gabbros, poikilomosaics of quartz enclosing microphenocrysts of feldspar in rhyolites (Fig. 4.8A), poikilomosaics of feldspar enclosing flow oriented phlogopite and pyroxene in lamproites from North America (Fig. 4.8B), poikilomosaics of feldspar enclosing flow-oriented

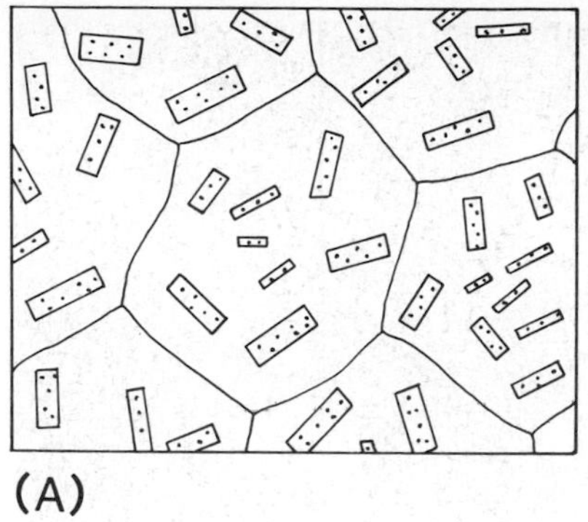

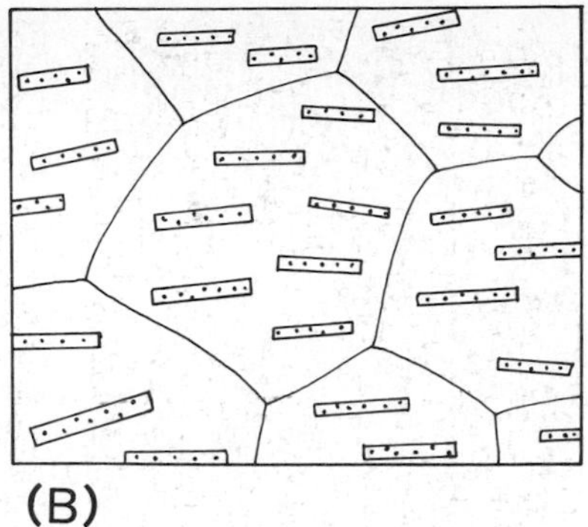

Fig. 4.7 Poikilomosaic texture. (A): The enclosed and host crystals grew simultaneously. (B): The enclosed crystals grew before the host. Discussion in text.

feldspar in trachyte from Lyttelton volcano, and poikilomosaics of feldspar enclosing pyroxene and nepheline in nepheline-rich volcanics from the Cook Islands.

There is a problem in deciding what a poikilitic texture signifies. Does it indicate contrasting nucleation and growth rates, as suggested above, or does it indicate a sequence of crystallization with plagioclase first then pyroxene in the case of the basalt poikilomosaic texture? Of course, these two possibilities are not mutually exclusive. In general, one would expect simultaneous growth to be accompanied by an outwards increase in grain size of the included minerals (McBirney and Noyes, 1979), as in Figs 3.32 and 4.7A. If poikilomosaic texture develops by later growth of the large grains the grain-size increase would be absent and pre-inclusion fabrics (such as trachytic texture) preserved (Fig. 4.7B).

Between plagioclase laths in well-crystallized tholeiitic basalt or dolerite one commonly finds granophyric intergrowth, already discussed in section 3.5.1. This represents eutectic crystallization of the residual differentiate of basalt magma once the crystallization of plagioclase and pyroxene is complete; the often fine-grained intergrowth indicates a degree of undercooling such that crystal growth outpaced diffusion rates; hence, independent development of quartz and alkali-feldspar crystals was not possible.

4.1.4 Rhyolitic crystallization

Rhyolitic magma is typically more viscous than basaltic magma; consequently rhyolitic flows (or domes) have a larger thickness/lateral-extent ratio than basaltic flows. One reason for increased viscosity in rhyolite magma is the greater degree of polymerization in the melt which not only increases viscosity but slows diffusion and inhibits nucleation (polymer

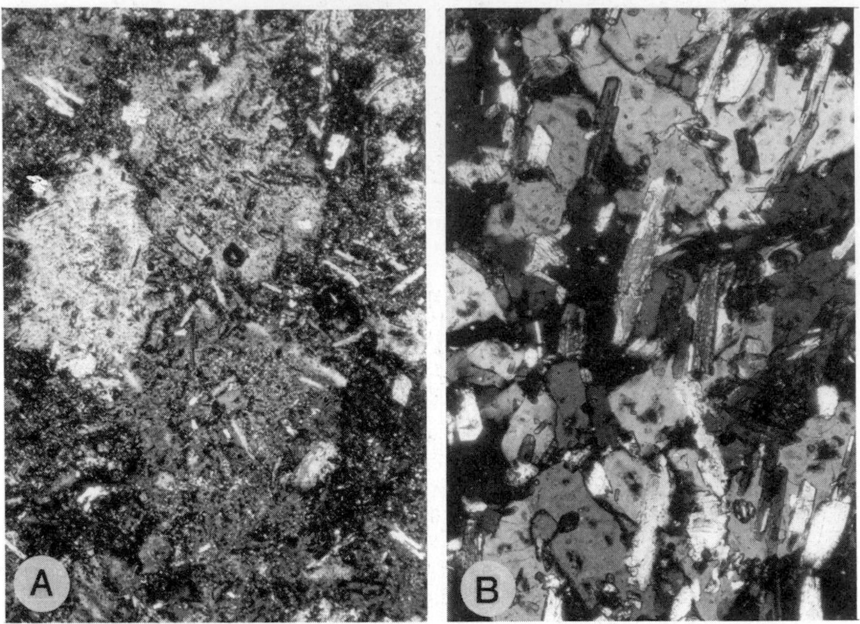

Fig. 4.8 Examples of poikilomosaic texture, as described in text. View lengths measure 0.85 mm (A) and (B).

structures have to be reconstructed). For these reasons, rhyolitic glass (obsidian) is abundant, not just at the chilled margins of bodies, but throughout entire rock masses.

Rhyolites are dominated, therefore, by textures that characterize a large degree of undercooling. We have already seen that crystal form changes from regular habits to skeletal forms as undercooling increases with concomitant decrease in diffusion rates and increases (at first) in nucleation and growth rates.

Phenocrysts in rhyolite usually have regular crystal forms representing the slow and measured growth from slightly undercooled magma before it was emplaced in its final setting. Microphenocrysts are not as common in rhyolite as basalt; when they do occur they may display skeletal forms as evidence of mild undercooling (Swanson *et al.*, 1989).

With a higher degree of undercooling, crystal habit is partly suppressed so that the crystalline material grows rapidly outwards in all directions from a focus of nucleation as an aggregate of acicular branches; these radiating masses are called spherulites (Figs 1.30B and 4.9). Often, traces of flow layering are enclosed by spherulites, showing clearly that they post-date flow. In the case of rhyolite, spherulites are most commonly

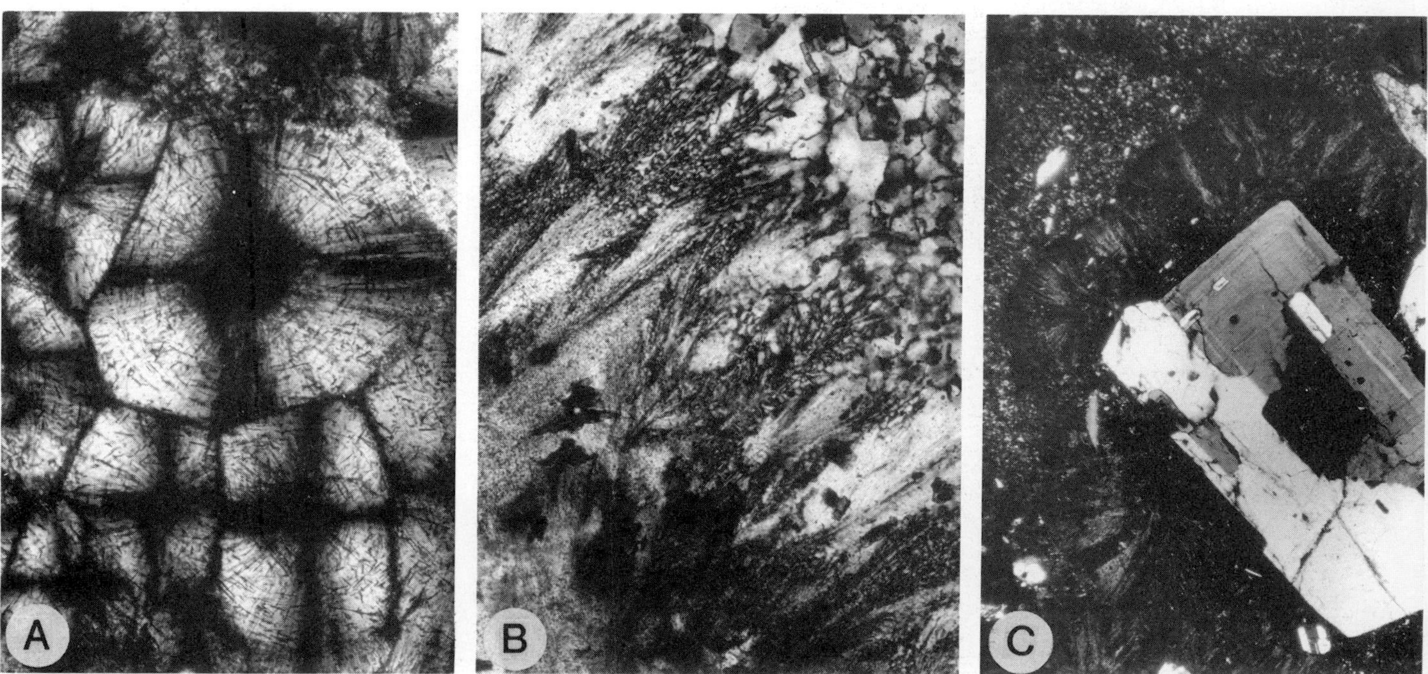

Fig. 4.9 Spherulites in rhyolite. (A): A number of spherulites have merged to form a mosaic. (B): A close-up view of the coarser terminations of the spherulites illustrated in Fig. 1.30B to show the granophyric intergrowth. (C): Spherulites grown around a phenocryst. (View lengths measure 0.81 mm (A) and (B), and 3.13 mm (C).)

densely packed and composed of silica (usually quartz) and alkali feldspar; two phases are evident from the distinct Becke lines in these aggregates. The outer ends of some coarse spherulitic growths consist of an extremely fine-grained pattern of interconnected quartz grains set in a single feldspar host (Fig. 4.9B), a fine-grained example of granophyric intergrowth (section 3.5.1). Such coupled growths are not surprising given the more-or-less cotectic composition of rhyolites (Figs 3.18 and 4.19).

Spherulites indicate diffusion to be the dominant rate-limiting factor during crystallization. Material cannot be moved easily to all possible sites of crystal growth. Consequently, each crystal grew only along its fastest growth direction towards the sites of new material. The most efficient way that such crystallization can proceed is as a radiating cluster of acicular crystals; crystal form is lacking except as expressed in the direction of elongation.

The focus of nucleation for spherulitic growth may be an existing phenocryst (Fig. 4.9C). In other cases, spherulites are centred on cavities (Swanson *et al.*, 1989), and indeed, large cavities called lithophysae, bordered by spherulites, are common in rhyolite. This suggests that an initial nucleation and crystal growth event produced a release of volatiles which in turn made further nucleation and growth easier. This kind of chain reaction could cause multiple points of nucleation to develop at the one focal point. The fact that cavities and lithophysae develop indicates the glass was still hot and soft enough to deform; Swanson *et al.* (1989) suggest a temperature of at least 620°C. In some cases, there is no apparent cavity or phenocryst at the centre of a spherulite, and the reasons for nucleation are obscure. Some spherulites develop from a 'bow-tie' structure (cf. Fig. 5.9), where an initial crystal develops branches that curve and radiate outwards.

If spherulites occupy the entire volume of rock, they merge into a polyhedral mosaic (Fig. 4.9A). A number of generations of spherulites may occur, often shown by a few large, truly spherical bodies embedded in a later generation of smaller spherulites merged into a mosaic.

Spherulites made up of densely packed crystals are perhaps the most spectacular and distinctive of rhyolitic crystallization products. They do form in other felsic rocks too, including trachyte, but are not common in more mafic rocks.

Spherulites may also consist of an open, less-densely packed cluster of radiating crystals, usually as a result of a lesser degree of undercooling than for the densely packed varieties. They are reported in experimental plagioclase crystallization by Lofgren (1980). In rhyolites, open spherulites of feldspar are usually enclosed poikilitically by quartz (sometimes another silica polymorph) forming a snowflake (poikilomosaic) texture (Swanson *et al.*, 1989). Snowflake texture also develops as an array of feldspar

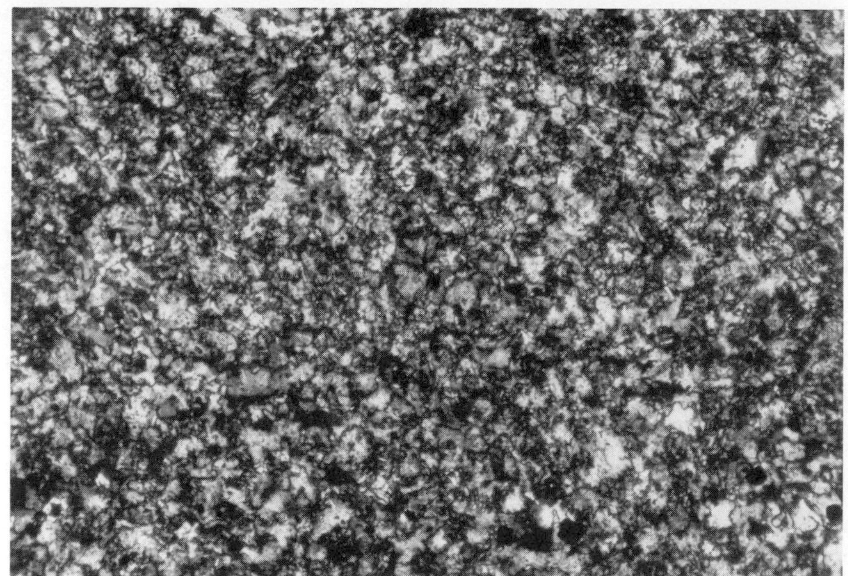

Fig. 4.10 Felsitic texture. (Length of view measures 0.85 mm.)

microphenocrysts enclosed in quartz (Anderson, 1969), and an example was given earlier (Fig. 4.8). As already noted for poikilomosaic textures in general, the problem interpreting snowflake texture may be: did quartz nucleate less frequently than feldspar but grow simultaneously with it, or did quartz grow after feldspar? Swanson *et al.* (1989) point out that snowflake texture in the Obsidian Dome, California, developed in the interior of the dome where cooling was slower and water content higher. Unravelling the various effects of cooling rate, degree of undercooling, and the effects of changing volatile content on nucleation, growth and diffusion rates is not easy.

Very high degrees of undercooling may result in felsitic texture (Fig. 4.10). Crystal form is totally suppressed, and the constituent minerals form a very fine-grained mosaic. The mosaic is not a regular polyhedral one, and grain boundaries anastomose in an irregular way. Felsitic texture in rhyolite is composed of silica and alkali feldspar, and Becke lines show clearly the presence of two minerals; it otherwise resembles the typical texture of chert. Spherulites are frequently embedded in a matrix of felsitic texture, indicating that the spherulites developed first. Felsitic texture represents a combination of very slow diffusion and growth rates; crystal growth was little more than sufficient to reorganize the glass structure into crystalline order, more or less *in situ*. Felsitic textures are likely to coarsen progressively during any subsequent thermal event. A diagnostic feature of glass is perlitic cracking, a nested aggregate of spherical fractures (Fig. 1.30C). Glasses that have devitrified to felsitic texture may still display perlitic cracks.

4.1.5 Quench histories and orders of crystallization

In the above discussion on basaltic and rhyolitic textures, little account has been taken of the timing and possibly variable rates of quenching. Consider the following possibilities: fresh basalt magma is erupted from a fissure directly into water where it is abruptly quenched; further along that fissure the same magma is air-quenched; elsewhere it forms a lava flow that crystallizes progressively as temperature falls by 100°C at which point it is air quenched at the edge of the flow; parts of the same flow enter water where it is more rapidly quenched. Or consider rhyolite magma some of which devolatilized and extruded to form the crust of a dome; other parts devolatilized but cooled slowly at the dome centre; lower intrusive levels retained volatiles and cooled slowly.

In the example of basalt we find not only various rates of quenching but varying degrees of undercooling and crystallization before the most rapid phase of quenching. The relatively rapid cooling of basalt in air may well be considered a quenching event, even though when the flow enters water the rate of quenching suddenly accelerates. In general, basalt glass represents very rapid quenching, skeletal textures much less rapid quenching, and it is clear that different rates of quenching in the same rock are not mutually exclusive.

Swanson *et al.* (1989) provide an interesting account of the textures produced in rhyolite that underwent various degrees of undercooling, quenching and devolatilization. They show that devolatilization produces a sudden increase in undercooling (liquidus–solidus curves rise with loss of H_2O) with concomitant nucleation of microlites.

It is normal in thin-section petrography to determine orders of crystallization, and in principle one should attempt to reconcile thin-section observations with experimental phase data. In basalt one often finds that olivine crystallizes first, followed by plagioclase and pyroxene, often in cotectic relationship, followed by granophyric material in tholeiites. Evidence for such an order may be partial or complete enclosure of olivine by other minerals, and granophyric material in the interstices. The Mg-rich olivines react with saturated or oversaturated basaltic melt if cooled sufficiently slowly. The reaction is in response to the early excess production of olivine and its subsequent reaction at a peritectic point to produce pyroxene, as illustrated in the well-known phase diagram for the pure system forsterite–silica (Fig. 4.11). Evidence for such reaction is often

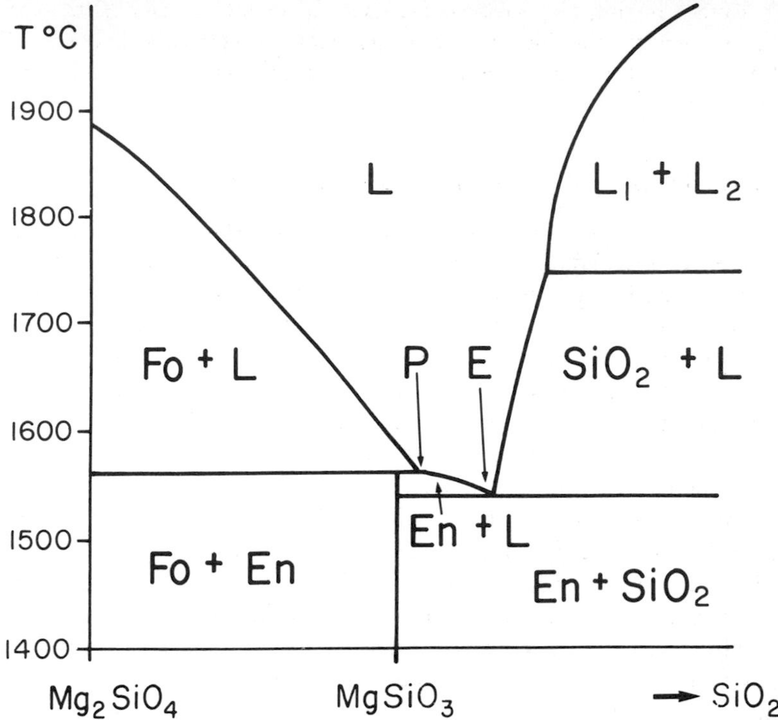

Fig. 4.11 Part of the low-pressure system forsterite (Fo)–silica. Olivine first crystallizes in excess, then reacts with the melt at the peritectic (= P) to form enstatite (En). If olivine fails to react or is extracted from the liquid, crystallization proceeds to the eutectic point (E) when enstatite and silica crystallize together. After Bowen and Anderson (1914) and Greig (1927), with permission of the *American Journal of Science*.

found as corroded olivine crystals; Peck *et al.* (1966), for example, describe the mantling of corroded olivine crystals with its reaction product pyroxene together with plagioclase in tholeiitic basalt in the Alae lava lake, Hawaii.

The purpose of this book is to deal with the principles of textural study, not the welter of possible sequences of crystallization and phase equilibria in all possible volcanic rocks (French and Cameron, 1981, for example, subdivide basalts in terms of four possible sequences of crystallization of the main phases). One must, however, be cautioned not always to expect that crystallization will proceed as experimentally predicted. For example, if one of two cotectic phases (e.g., pyroxene and plagioclase in basalt, as shown in the phase diagram in Fig. 4.12) does not nucleate immediately the cotectic is reached, or is simply more difficult to nucleate than the other, the two phases may not crystallize together as expected. Further-

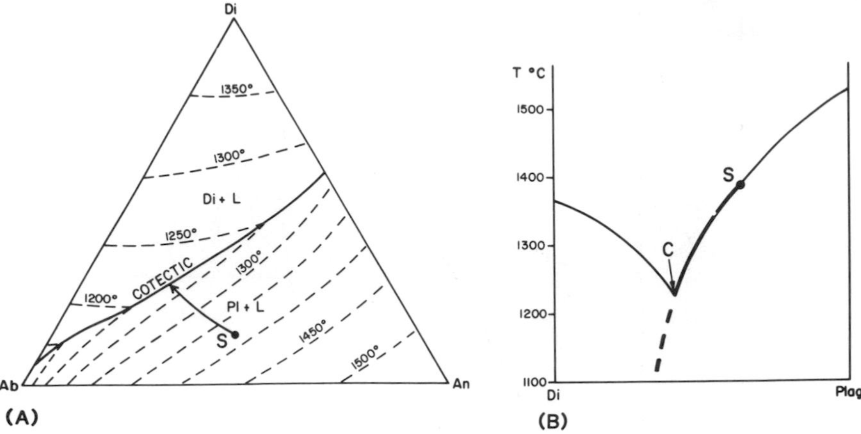

Fig. 4.12 (A): System albite–anorthite–diopside (after Bowen, 1915, with permission of the *American Journal of Science*). Melt *S* first crystallizes plagioclase, and follows a curved path towards the cotectic line where plagioclase and diopside crystallize together. (B): If diopside fails to crystallize, plagioclase crystallization continues along the dashed line, creating considerable undercooling with respect to diopside.

more, Corrigan (1982) shows how the order of crystallization of olivine, pyroxene and plagioclase may change with different rates of cooling (Fig. 4.13). The delay in the appearance of one phase may have unexpected effects on the appearance of other phases, and the degree of undercooling with respect to absent phases will increase (Fig. 4.12B). When a delayed phase does eventually nucleate it is likely to create numerous small crystals in a sudden burst of growth; in turn this affects the crystallization dynamics of the phase that had already been crystallizing. In rhyolite crystallization, Swanson *et al.* (1989) point to the additional complication of metastable phase nucleation during sudden undercooling events.

One must also be cautioned against careless interpretation of inclusion textures. A mineral enclosed by another **may** indicate an order of crystallization; on the other hand it may be an odd cut-effect in two dimensions which gives a false impression (Fig. 4.14), or as discussed for poikilomosaic and other poikilitic textures it may be a matter of differing nucleation and growth rates.

4.1.6 Phenocrysts in volcanic rocks

Aphyric lavas are not common, and this indicates that magma is nearly always crystallizing *en route* to the surface, and it shows that magmas are seldom superheated. This is an important general constraint on possible heat budgets in geological modelling.

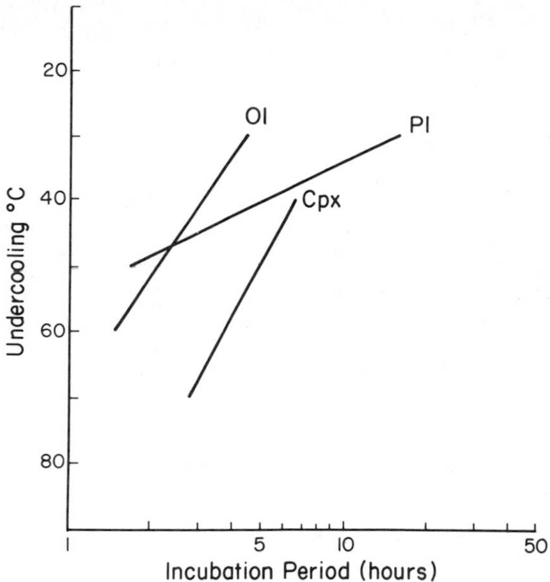

Fig. 4.13 Diagram to show how the order of appearance of olivine, plagioclase, and clinopyroxene changes with rate of cooling and the degree of undercooling. Reproduced with permission from Corrigan (1982).

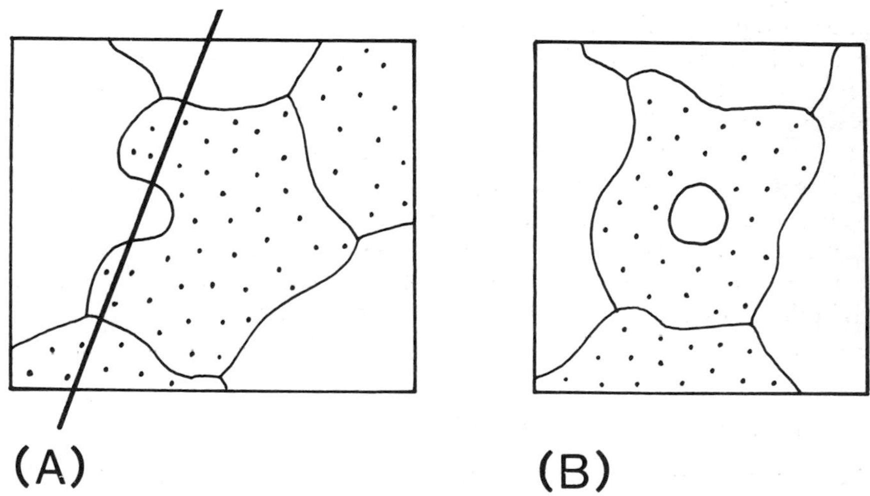

Fig. 4.14 A section cut across an irregular or bulging boundary, along the direction indicated in (A), for example, may give the false impression that one mineral includes another, as in (B).

Microphenocrysts and the matrix of volcanic rocks, discussed in the previous section, record the final stages of cooling and crystallization. Phenocrysts usually predate final emplacement of the rock, and give clues to the earlier evolution of the magma. Two preliminary words of caution:

1. do not confuse early phenocrysts with the large crystals that sometimes form during the late stages of volcanic crystallization in poikilomosaic textures, as discussed above;
2. some early-formed minerals, like apatite and zircon, are present in insufficient quantities to grow large crystals, but evidence for their early growth is often given by inclusion within phenocrysts.

(a) Phenocrysts and crystal/liquid fractionation

Minerals crystallizing in magma usually belong to a solid-solution series and/or a eutectic system; consequently, crystallization will produce a progressive change in magma composition, as discussed in Chapter 3. The presence of phenocrysts therefore indicates that the composition of the magma has already evolved from its original state.

This is the basis of much of the modelling of natural crystal fractionation systems. In the simplest model, one takes a whole-rock analysis (which corresponds with a batch of magma) and subtracts that portion of the analysis represented by the phenocrysts (using modal analyses and a knowledge of phenocryst composition); the result represents the composition of the evolved magma. Somewhere in space and/or time that evolved magma may be segregated from the crystals and crystallize a new and different assemblage of crystals; in turn this leads to a further evolution of the magma composition. Further details of the calculation methods can be found in Cox *et al.* (1979), McBirney (1984) and Wilson (1989).

Wilson (1989) points out some of the difficulties involved in this approach including: uncertainty about the processes of fractionation in the magma chamber (discussed further in section 4.2); the problem that amphibole is possibly an important phenocryst phase at deeper levels though evidence is lost because of the low survival rate of amphibole at low pressures; the problem that high-level fractionation masks high-pressure fractionation involving quite different phases; the difficulty of modelling fractionation involving multiple phases. The common recharge of an evolving chamber with fresh magma introduces further complexities, and one can add to the list of problems the smooth variation of some crystal-size distribution curves (Marsh, 1988); a clearly defined population of phenocrysts may be an illusion rather than fact. Until these problems are better resolved, the calculations of trends of magmatic differentiation using phenocrysts need to be treated with caution. The answers to the questions will ultimately derive from an integrated geological, geochemical, petrological and mineralogical investigation, which is beyond the scope of this book.

What is important here is the documentation of textural features displayed by phenocrysts which may, in the context of such an integrated study, allow one to make a more realistic assessment of the processes involved.

(b) Zoning in phenocrysts: magmatic stratigraphies and magma mixing

Many phenocrysts exhibit evidence of an involved history. As well as the complex compositional zonal patterns displayed by plagioclase, phenocrysts may show evidence of resorption (Fig. 1.31D), reaction with the magma (Fig. 1.31C and D), and fracturing. These features have the potential for use in reconstructing 'magmatic stratigraphies', sequences of events analogous to conventional stratigraphic records.

Pearce *et al.* (1987), for example, reconstruct the history of the Mount St Helens dacite magma erupted on 18 May 1980, as follows:

1. Crystallization of plagioclase, amphibole, orthopyroxene, magnetite, and apatite phenocrysts in a magma chamber at depths >7 km. Plagioclase growth took approximately 2000 days.
2. Massive resorption took place as magma was transported to a higher-level chamber. The lowering of pressure induced the solution of all phases, and plagioclase developed a sieve texture with magma filling the hollows. Some phenocrysts are significantly less resorbed than others and therefore it is thought that magma from different levels in the system got caught up and mixed in with the main batch of rising magma.
3. Only plagioclase continued to crystallize at lower *P*, and the zone patterns in the mantles covering early corroded plagioclase are the same as those in a new generation of smaller phenocrysts. The zoning patterns are complex, and differ in detail from crystal to crystal; the significance of this is uncertain. Growth again took place over approximately 2000 days.
4. The large plagioclase phenocrysts were shattered during eruption as magma trapped in the sieve textured cores expanded. The speed of eruption was so rapid that no quench textures formed, only glass.

Many workers have discovered evidence for magma mixing from phenocryst textures. Thus Nixon and Pearce (1987) describe plagioclases from andesite and dacite which, in addition to complex oscillatory zoning, have as many as twelve major discontinuities in composition coinciding with resorption surfaces. In each case, resorption was followed by the

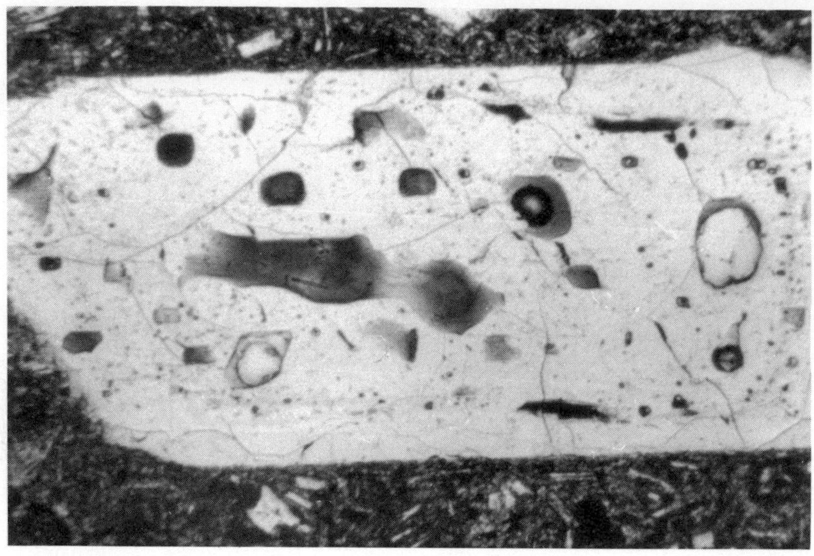

Fig. 4.15 Areas of clear rhyolitic glass enclosed in a plagioclase phenocryst set in andesitic matrix. This texture is the result of magma mixing, as discussed in text. Specimen from Toliman Volcano, Guatemala. Photograph kindly provided by S. D. Halsor. (View length is 1.2 mm.)

crystallization of markedly more calcic (10–33 mol. %) plagioclase (Fig. 3.12A). Patterns cannot be correlated from crystal to crystal in any one rock. The features suggest repeated injection of fresh basalt magma into a reservoir of dacite magma with consequential increases in *T*, resorption of existing crystals and subsequent crystallization of more calcic plagioclase. Turbulent mixing of hybrid magma in an open system is thought to be the reason for the mixing of phenocrysts with different zone patterns.

Somewhat different evidence that andesite may be produced by magma mixing is given by Halsor (1989), who describes andesites with mixed assemblages of plagioclase phenocrysts, the more sodic of which enclose large areas of clear rhyolitic glass (Fig. 4.15). This contrasts with the usually dusty, smaller areas of glass in typical sieve textures. The clear glass was not in equilibrium, either with the whole rock or the groundmass. Halsor believes that fresh hydrous basaltic magma was injected into an existing rhyolite magma; density instabilities were created as volatiles were released during crystallization. The ensuing mixing caused very rapid skeletal growth of plagioclase in the rhyolite, trapping pockets of the magma.

But it is not simply in plagioclase or calc-alkaline magmas that evidence of magma mixing is found. Thus, Helz (1987) shows that the patterns of

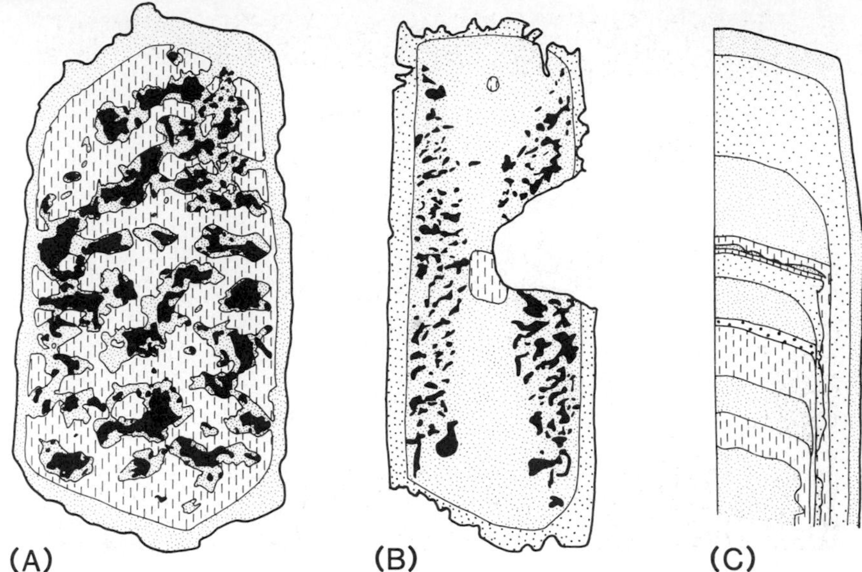

Fig. 4.16 Pyroxene crystals affected by magma mixing. (A): Mg-rich diopside resorbed then overgrown by Fe-rich diopside. (B): Small relic of resorbed Mg-rich diopside (at core) overgrown by Fe-rich diopside which itself is resorbed preferentially in two sectors, then overgrown by further Fe-rich diopside. (C): Portion of crystal with complex zoning and resorption pattern. Stippled is the Fe-rich diopside, lined is the Mg-rich diopside, and the black represents the groundmass. Reproduced from O'Brien *et al.* (1988, Figs 8, 10 and 13) with permission. Copyright: Mineralogical Society of America.

olivine zoning in basalt from the 1959 Kilauea eruption vary in detail from one eruptive phase to the next, indicating magma mixing below the surface. And O'Brien *et al.* (1988) describe complex zoning, reaction and resorption of clinopyroxene phenocrysts, caused by the mixing of phonolite and minette magmas. Magnesium-rich diopside, derived from the minette magma, is often rounded, embayed and coated with Fe-rich diopside overgrowths derived from the phonolite magma (Fig. 4.16A). In some cases the Fe-rich diopside is itself resorbed and overgrown by more Fe-rich diopside (Fig. 4.16B); the Fe-rich diopside has sector and oscillatory zoning, and preferential resorption took place in the [001] sector. Other crystals display numerous repetitions of Mg-rich and Fe-rich diopside, often separated by zones of resorption (Fig. 4.16C), and repeated magma mixing is indicated. During resorption, the composition of existing pyroxene adjacent to the resorption surfaces is changed by reaction with the mixed melt in an effort to regain equilibrium. Final quenching was often accompanied by crystallization of an outer rim of Al-rich hedenbergite.

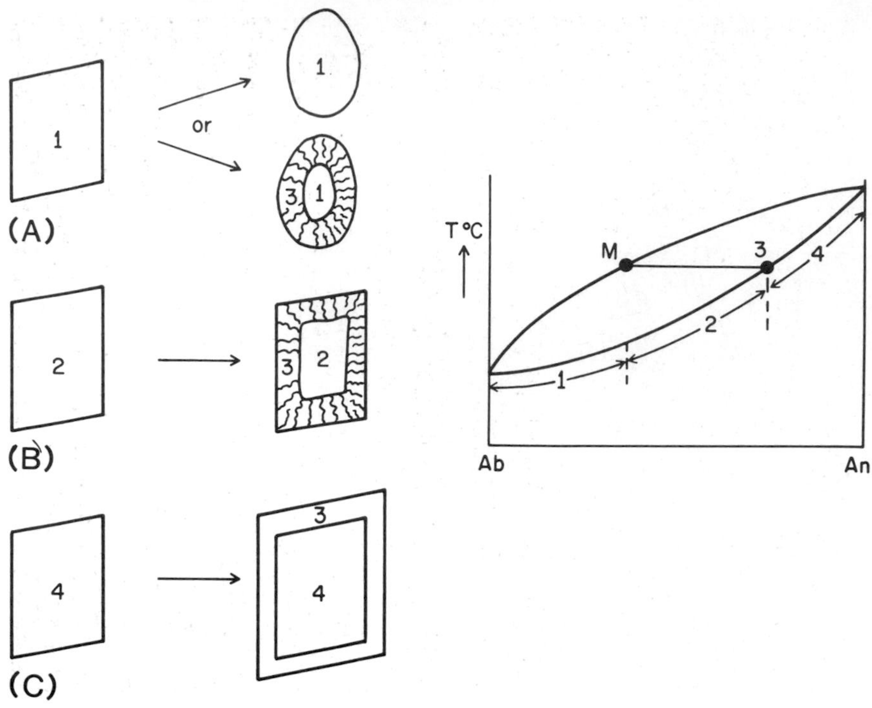

Fig. 4.17 Results of mixing plagioclase crystals of various compositions with magma of composition (*M*) that is in equilibrium with plagioclase of composition (3). (A): Crystals of composition (1), mixed with *M*, may possibly melt, but in any case dissolve to give rounded shapes (top left). If *M* penetrates the dissolving crystal (1), or if partial melting occurs internally, sieve texture is produced as (1) reacts with the melt to form a more calcic rim (no more calcic than 3). (B): Crystals of composition (2) cannot melt completely and cannot simply dissolve in *M*, but they will react during partial solution or melting to give a sieve texture and more calcic rims (no more calcic than 3). The former euhedral outline is preserved. (C): Crystals of composition (4) continue to grow with euhedral overgrowths of (3), although some equilibration of (4) with M may take place by diffusion. After Tsuchiyama (1985).

(c) Sieve textures, solution and reaction

The above interpretations are reinforced by the experiments of Tsuchiyama (1985). He showed that heating above the plagioclase liquidus causes solution of plagioclase phenocrysts and a rounding of shapes (Fig. 4.17A). If the An % of the phenocryst is less than the plagioclase in equilibrium with the melt, the corroded surface becomes very rough and finely indented to form a sieve texture (Figs 4.17A,B, and 4.18) which is filled with magma, and in which the plagioclase reacts to become more calcic. If basic and acid magmas mix, sodic feldspar in the acid magma dissolves,

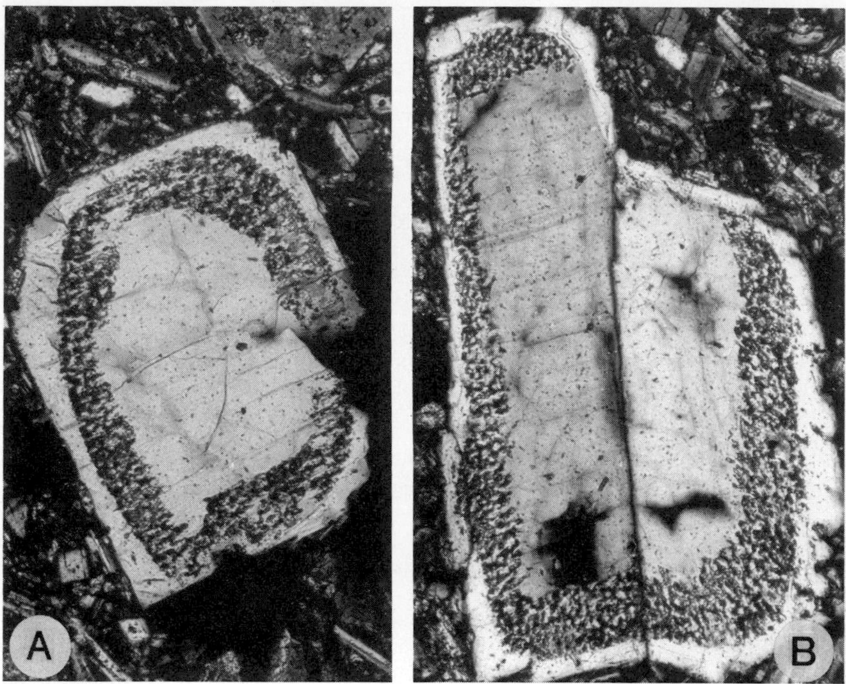

Fig. 4.18 Examples of the solution of plagioclase, the development of sieve texture, and the following euhedral overgrowth. Basaltic trachyandesite, Pukepoto Cone, NZ. View lengths measure 0.85 mm (A) and (B).

develops sieve textures and the corroded parts react to become more calcic; when crystallization eventually resumes, the sieve texture is coated by euhedral growths of the more calcic plagioclase. Calcic plagioclase in the more basic magma simply continues to grow as the magma cools during mixing, although the euhedral crystal growth becomes more sodic (Fig. 4.17C).

It is also possible to achieve solution by decompression as magma rises towards the surface (Pearce *et al.*, 1987), in which case there should be no evidence that corrosion was accompanied by a change in the composition of the magma. In the case of feldspar, dissolution takes place in hydrous magma because the liquidus and solidus curves are depressed as pressure decreases and volatiles are released. In the case of hornblende or mica, the release of volatiles as the magma rises towards the surface, reduces the stability field of the mineral, causing reaction with the melt. In other cases, it may be that a mineral is stable at high pressure but not at low pressure, or that the order of crystallization has changed with pressure. Thus Thompson (1972) showed from experimental crystallization of Snake River basalt that the first phase to crystallize at the surface is olivine, but at depths of 15–30 km it is plagioclase, below 30 km it is pyroxene and below 100 km garnet.

The situation with feldspars in quartz-normative trachytes is peculiarly complex due to reaction relations in the ternary Or–Ab–An system. Plagioclase and alkali feldspar may both crystallize at first, but because of peritectic-type behaviour (similar to that in the low-*P* forsterite–silica system) the plagioclase subsequently may be completely or partially dissolved (Nekvasil, 1990). If water activity increases during further cooling, plagioclase may begin to crystallize again as a late phenocryst phase. Textural evidence includes the preservation of partially resorbed plagioclase within anorthoclase phenocrysts in many trachytes; what must be remembered is that evidence of the primary plagioclase phenocrysts will be completely lacking if resorption is complete.

(d) Embayments in quartz and olivine phenocrysts

It is a very common observation that quartz phenocrysts in rhyolite and olivine in basalt are embayed, giving the immediate impression of corrosion (Fig. 1.30A). Doubtless, in many cases this has happened. In the case of quartz, one can appeal to changing positions of the quartz alkali-feldspar eutectic with changing pressure (Fig. 4.19) to explain the corrosion of quartz as magma rises towards the surface. In the case of olivine one can point to the fact that the low-pressure system forsterite–silica (Fig. 4.11) first crystallizes excess olivine which later reacts with the melt due to the system containing the incongruently melting orthopyroxene.

One needs, however, to ask why embayments form, because in many experiments involving the dissolution of silicates, smooth, rounded forms without embayments are produced. Indeed, as Donaldson and Henderson (1988) point out, solution at a crystal surface should actually be inhibited by an embayment. To get around this problem they show experimentally that gas bubbles provide a mechanism for solution at localized points on a crystal face, because of the continuous turbulent motion of the liquid about them. This enables continuous and rapid solution to take place where a bubble contacts a crystal face, and in this way a bubble can effectively drill a hole through a crystal.

Donaldson and Henderson (1988) also remind us that embayments can result from unstable primary growth. If an embayed crystal has sharp corners and edges, and especially if zones of (fluid) inclusions follow the shape of the embayments, then primary disequilibrium growth rather than corrosion is the cause.

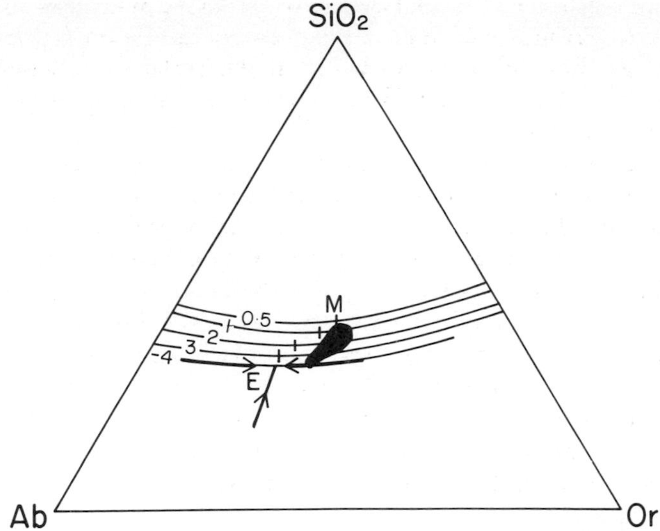

Fig. 4.19 Effect of increasing water pressure on crystallization in the system K-feldspar (Or), albite (Ab) and silica. Note the changing position of the cotectic line and temperature minimum (*M*) as pressures increase from 0.5 to 3 kbar. By 4 kbar the minimum has become a eutectic point. The compositions of granites and rhyolites (shaded) are close to *M* and *E*. After Tuttle and Bowen (1958). Compare Fig. 3.18.

(e) Glomeroporphyritic texture, and cognate xenoliths

Clusters of phenocrysts or microphenocrysts are very common (Figs 4.20 and 3.11A), and they could play an important role in igneous differentiation. For example, a large glomeroporphyritic mass will sink in magma more easily than small individual grains, and light minerals such as feldspar, which might otherwise rise, sink if caught up in a cluster of heavy minerals. In the basalts from Hawaiian lava lakes described by Kirkpatrick (1977), all the phenocrysts occur in glomeroporphyritic clots. He ascribed the texture to heterogeneous nucleation so that plagioclase, for example, nucleates on pre-existing plagioclase, pyroxene or olivine, the frequency being in that order. In general there appears to be no specific crystallographic relationship between host and nucleating phase. The clusters have olivine and pyroxene concentrated at the centre, surrounded by pyroxene and plagioclase.

The alternative mechanism of synneusis is suggested by Helz (1987) for aggregates of subhedral olivine in basalts of the 1959 Kilauean eruptions. As discussed under twinning in Chapter 3, crystals that accidentally collide are likely to stick together if they touch in crystallographic continuity or in twinned or some favourable epitaxial relationship. The example (Fig. 4.21)

Fig. 4.20 Glomeroporphyritic texture in andesite, White Island, NZ. (Length of view measures 3.3 mm.)

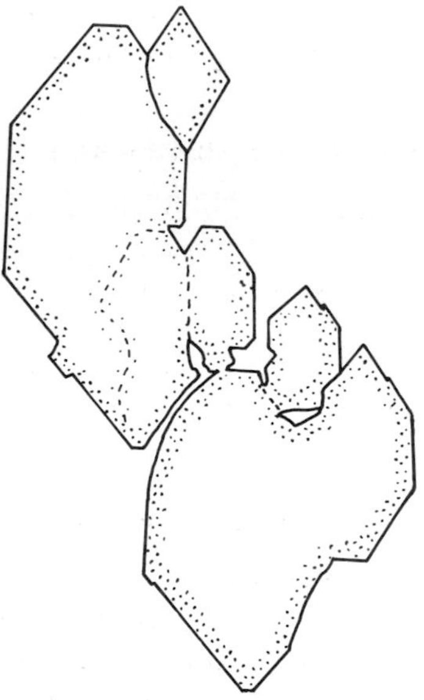

Fig. 4.21 Coherent aggregate of olivine crystals formed by synneusis. Cluster is approximately 6 mm long. Based on photographs in Helz (1987).

taken from Helz (1987), shows the crystallographic continuity of the olivines, quite different from the clusters described by Kirkpatrick (1977).

It may be difficult to distinguish glomeroporphyritic clusters from xenoliths. This is especially the case with cognate xenoliths which represent disrupted cumulates of crystals from the same or closely related magma. Thus the floor, roof or walls of a magma chamber may be lined with crystals which are plucked off in clusters by some later change in movement of magma. The question of exotic xenoliths and the production of xenocrysts is discussed later in section 4.1.7.

(f) Megacrysts

The non-genetic term megacryst is used for exceptionally large crystals in igneous and metamorphic rocks. The mere fact that they are exceptional (larger than phenocrysts in the same volcanic rock) suggests an exotic origin, and indeed most are found to be xenocrysts that are not in chemical and/or textural equilibrium with the rest of the rock. Helz (1987) describes megacrysts of olivine in basalt from Hawaii, believed to have been ripped off a conduit wall deep in the magma system, and Aspen *et al.* (1990) describe megacrysts of alkali feldspar in alkali basalts, believed to be high-pressure syenitic debris from an upper-mantle source. Although neither of the above examples of megacryst can be regarded simply as large phenocrysts, both examples are thought to have been derived from the deeper regions of the magma plumbing system, and related to the same cycle of magma generation as the extruded magma.

4.1.7 Volcanic structures in thin section

Many volcanic structures are not appropriate subjects for this book which is concerned only with what may be studied directly with the microscope. Those structures that are observable in thin section include chilled margins, flow layering, blebs, patches and veins resulting from segregation or immiscibility, vesicles and their infillings and xenoliths. It must be remembered that field-work is necessary to ascertain their spatial distribution and geometry.

(a) Chilled margins and flow differentiation

Heat loss is greatest at the margins of a magma body, and the effects of this are seen in numerous flows which display a general decrease in grain size towards flow boundaries. Rhyolitic flows in particular may be encased in a very thick glassy marginal zone. However, it is in thin dykes and sills that the thermal gradient is sufficiently marked for the progressive changes to be observable in one thin section. Laminar rather than turbulent flow is necessary if the first wall coating of a dyke is to be preserved as a chilled margin. Even so, dykes are seldom emplaced in one simple pulse, and early chilled margins may be invaded and ripped up by later dyke material. Such finer-grained chilled fragments are often observed in thin section. Considering the fact that dykes commonly represent transport of magma over horizontal distances of tens, even hundreds of kilometres, it is not surprising that marginal relationships become complex.

The matrix in a chilled margin is generally finer grained than that in the main rock body. Glass is frequently present or dominant, and it may develop totally different textures (e.g., spherulitic) from the main rock body. Observations of grain-size variation in microphenocrysts across a chilled margin provide valuable insight into the timing and rates of crystal growth.

Phenocrysts are presumed to have grown before dyke intrusion, and one might therefore expect the phenocryst mode to be the same in the chilled margin as the rest of the dyke. This is not always the case, providing evidence of either the intrusion of an already differentiated or mixed magma, or flow differentiation. Komar (1972) showed that grain-dispersive pressures (Bagnold effect) during laminar flow causes a flow differentiation by moving phenocrysts towards the centre of a dyke (Fig. 4.22). The effect increases with grain size so that grain-size sorting also occurs. The effect is most marked in relatively small dykes and sills (<100 m thick according to Barrière, 1976) where there is a marked velocity gradient across the dyke due to the 'drag-effect' of the walls. The geochemical consequences of flow differentiation can be significant: thus Brouxel (1991) analyses the fractionation of magma caused by flow differentiation where clinopyroxene is concentrated at dyke centres.

(b) Flow layering

Millimetre-scale flow-layering is exhibited by a wide range of dyke and flow rocks, but especially the more felsic ones. Layering is often in the form of folds and irregular swirling patterns (Figs 4.23 and 4.24). Possible origins include: differentiation (particularly of volatiles) due to differential shear; the streaking out of mixed magmas; rheomorphism and/or welding of pyroclastic deposits; superimposed textural variations.

The idea that variations in volatile content, hence texture, occur as a result of differential laminar flow is an old idea put in its latest form by Nelson (1981). He proposes that differential laminar flow creates local variations in temperature because of shear heating; in turn these create variations in the solubility of volatiles and diffusion and nucleation rates. An example given is red and black layered obsidian, caused by an increase in oxygen fugacity (in the oxidized red layers) due to a local increase in temperature.

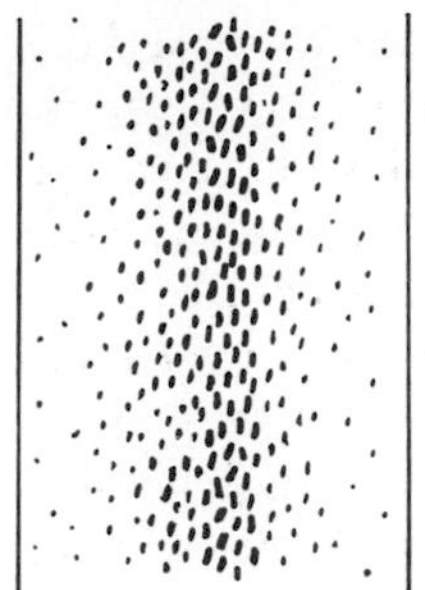

Fig. 4.22 Schematic illustration of concentration of phenocrysts towards the centre of a dyke due to the Bagnold effect.

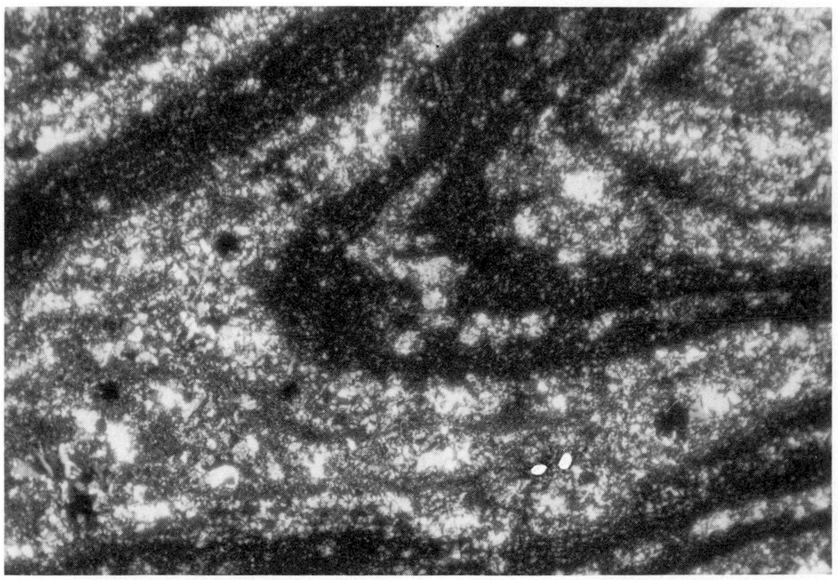

Fig. 4.23 Fold patterns in flow layering in rhyolite. (Length of view measures 3.3 mm.)

The simultaneous eruption of two or more different composition magmas is common. The magmas may be as profoundly different as basalt and rhyolite (Sigurdsson and Sparks, 1981; McGarvie *et al.*, 1990), or silicate and carbonatite (Clarke and Le Bas, 1990), or they may be closely similar as described by Sampson (1987) for two mixed rhyolitic magmas. The mixing usually seems to take place in the vent, and the streaking out into layers has been duplicated experimentally by Kouchi and Sunagawa (1985).

Fig. 4.24 Fold patterns in flow layering in andesite, Ruapehu, NZ.

Mixed populations of phenocrysts and reaction between phenocrysts and hybrid magmas are the attendant consequences of magma mixing, but McGarvie *et al.* (1990) note that re-equilibration is often restricted to thin rim zones because of the short time between mixing, eruption and solidification. Kouchi and Sunagawa (1985) show that basalt and dacite magma mix easily to produce homogeneous andesitic magma, but if mixing, eruption and solidification take place in too short a time (McGarvie *et al.*, 1990), various stages of partial blending are produced. On the other hand, some flow layering may represent immiscible liquids streaked out together (Clarke and Le Bas, 1990), and homogenization is then impossible.

Glassy fragments in pyroclastic deposits, if erupted sufficiently quickly and in large enough volume, may retain enough heat to be flattened and welded by the accumulating mass of material on top. The result is a streaky 'layered' rock, and the deposit may move down a slope and develop the character of a flow (a phenomenon called rheomorphism). Rheomorphic flows are common on basaltic volcanoes where thick accumulations of spatter and scoria around a fissure move off to form a flow characterized by drawn-out basaltic clasts (eutaxitic texture). Each clast usually fines or varies in texture towards its margins, and hence forms a distinctive layer in thin section. Eutaxitic textures are better known from the rhyolitic end of the spectrum, where huge volumes of pumice are erupted to form ignimbrites, characterized (in part) by welding and flattening of pumice clasts and shards (Fig. 1.4). The flattened cuspate forms of the shards are distinctive of an ignimbrite, but the general streaked-out appearance is not unlike other forms of flow layering.

Rhyolitic magmas may develop considerable textural diversity during crystallization (section 4.1.4). Thus cracks (possibly originating as drawn-out vesicles) may be the sites of nucleation of layers of spherulites. In this way, post-flow crystallization may follow flow anisotropies to produce a secondary layering that mimics flow layering. Not only this, but vapour-phase crystallization may produce layering so that Hausback (1987), for example, describes white and brown layered rhyolite in which the white layers are lithophysae filled with vapour-phase minerals, and the brown layers are more-or-less original magmatic material.

(c) Blebs, patches and veins, resulting from segregation or immiscibility

A wide variety of basic and intermediate igneous rocks exhibit felsic patches or veins interpreted to be the products of segregation of the residual liquid differentiate (due to compaction, perhaps – section 8.1.2). There is little doubt, for example, that tholeiitic basalts develop a residuum of rhyolitic liquid which may crystallize as granophyric intergrowth in the groundmass, or which may segregate into dilational veins. Sometimes the residual fluids are injected into vesicles (Anderson *et al.*, 1984).

On the other hand, some of these felsic patches have been interpreted as immiscible liquid fractions. Spherical or ellipsoidal bodies (called ocelli) up to a few millimetres across are interpreted as globules of one immiscible liquid in the other; the ocelli may coalesce to form layers which because of their low density become diapiric in some sills (Fig. 4.25). Note, however, that some ocelli in plutonic rocks are interpreted to result from magma mixing rather than unmixing (Bussell, 1985). Immiscible liquid fractions may be streaked out by flow to produce a layering, as discussed above.

The only unequivocal textural evidence for liquid immiscibility in volcanic rocks is the presence of tiny globules of glass in another glass (Fig. 4.26), and numerous examples of this, particularly in tholeiitic basalt, have been documented by Philpotts (1982).

(d) Vesicles and vesicle infillings

Volatiles are released from magma because of decompression, especially on reaching the surface (an event which may be rather like taking the cork

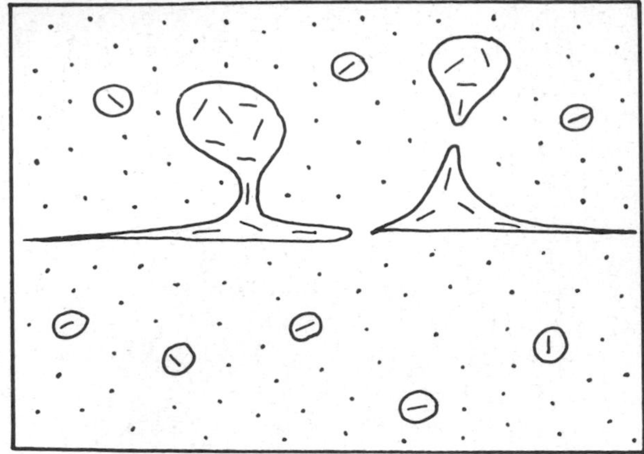

Fig. 4.25 Immiscibility phenomena described by Philpotts (1972). Shown are feldspar-rich ocelli, a layer formed by coalescence of ocelli, and diapirs that originated in that layer. Diapirs are approximately 1 cm across.

out of a bottle of champagne), and as a result of cooling and crystallization. It is clear from any volcanic eruption that some volatiles are lost to the system completely; others are trapped within the igneous rock as vesicles, gas blisters or lithophysae.

Vesicles that do not escape may grow by decompression (as they rise in a body of magma, or as magma rises), by the addition of more gas by diffusion through the host magma, and by coalescence. They may be spherical, but are commonly deformed into ellipsoidal shapes (Fig. 4.27), or kneaded out altogether by shear. Severely deformed vesicles may impart a platiness to lava (Fig. 4.27). Related to vesicles are diktytaxitic voids, which form angular spaces between microphenocrysts, probably as a result of the release of gases during crystallization. Some vesicles are pipe-like in shape, the pipes running perpendicular to cooling surfaces. Pipe vesicles probably form by exsolution of gas which is trapped as bubbles at the solidification front; as the front progressively advances through the lava, bubbles grow in pipe-form normal to it (Philpotts and Lewis, 1987). A similar process creates the familiar gas tubes in ice cubes. Most vesicles are <1 cm in mean diameter, but sometimes the gases accumulate as gas blisters (of the order of 1 m in length), or lithophysae. Walker (1989) describes how gas blisters develop in basalt flows, either at the upper boundary between a central flowing mass of lava and its solidified carapace, or simply in the central parts of highly vesicular *pahoehoe* flows. Lithophysae, or irregular cavities in rhyolitic rocks, form by the exsolution of gases, especially during spherulitic crystallization (section 4.1.4).

Walker (1989) discusses the size and shape of vesicles in a Hawaiian

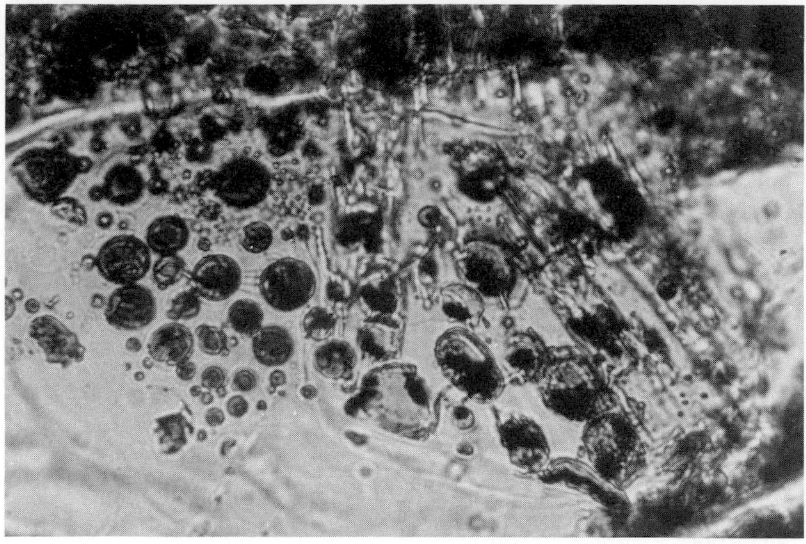

Fig. 4.26 Immiscible globules of glass in basaltic glass, Kilauea, Hawaii. Globules on the left are preserved as dark-brown glass, those on the right have crystallized pyroxene and an opaque phase. (Length of view is 0.17 mm. Photograph courtesy of A. R. Philpotts.)

pahoehoe flow. The volume of vesicles is the same in chilled flow contacts as in the main body of lava, and therefore he believes them to have formed primarily in the vent. However, sizes and shapes are modified after eruption; thus, in more distal parts of the flow, vesicles are spherical (the shape must be secondary because flow would have deformed them), and the size of vesicles increases towards the centre of the flow. He ascribes this mainly to coalescence of vesicles with the aid of diktytaxitic voids (Fig. 4.28) which puncture vesicles and provide channels for gases. Obviously the magma must have been plastic enough for these size and shape changes to take place, but Walker shows that the magma was not fluid enough to allow gases to escape altogether.

Vesicles may be completely or partially filled with secondary material; if the filling is complete, or nearly so, the resultant spherical or ellipsoidal mass is called an **amygdale**. Infillings fall into four main categories:

1. Volcanic glass.
2. Crystals of essentially igneous character
3. Crystals of secondary character deposited from fluids during the immediate post-eruptive cooling.
4. Crystals grown during post-eruptive burial.

Fig. 4.27 Platy structure in lava caused by the streaking out of vesicles. Lyttelton volcano, NZ.

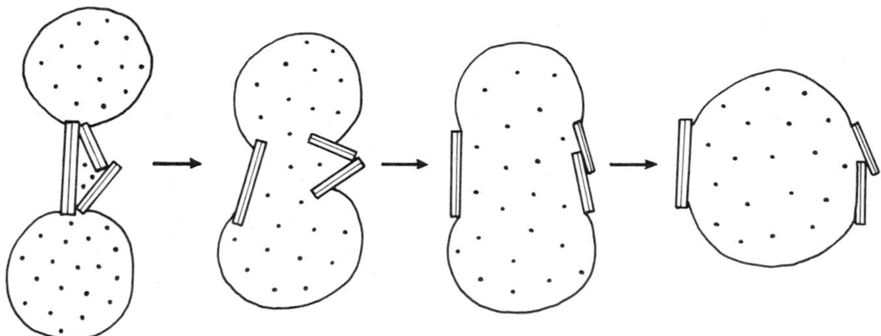

Fig. 4.28 Schematic illustration of how a diktytaxitic void (between the three plagioclase microphenocrysts) may puncture the walls of adjoining vesicles, enabling them to coalesce. The crystals may project into the vesicles during the process.

Cooper (1979) describes vesicles in lamprophyre filled with glass, and proposed that glass was drawn into the vesicles as the vapour phase was reduced in volume during cooling. Anderson *et al.* (1984) also describe vesicles filled with black glass, and suggest that volatiles released during crystallization created higher fluid pressures in the host magma than in the vesicles. These high pressures, together with compaction (section 8.1.2), caused residual liquids to be segregated into the vesicles. An alternative explanation to vesicle infilling for some ellipsoidal masses is that of liquid immiscibility (discussed above), but this cannot apply to the ocelli described by Cooper (1979), since the glass in the groundmass is identical to that filling the vesicles.

Puncturing and coalescence of vesicles by diktytaxitic voids (Walker, 1989) is likely to leave microphenocrysts projecting into some vesicles, but, in addition, a number of workers have described essentially igneous assemblages in vesicles, the minerals having grown from volatile-rich residual fluids. Birch (1980), for example, describes olivine, clinopyroxene, apatite, titanomagnetite, mica, nepheline and feldspar lining vesicles in a leucitite, and he suggests crystallization temperatures of 700–900°C. The pyroxenes tend to be hollow, suggesting rapid crystallization. The vesicles are associated with irregular patches, veins and cavity fillings of pegmatitic material of similar mineralogy.

Secondary minerals in vesicles usually have some close relationship to the composition of the host rock. Thus in basaltic rocks the most common infillings are zeolitic, whereas in felsic and acidic rocks, some form of silica is most frequently found. Some crystals have euhedral terminations or forms projecting into the central space in the vesicle (Figs 4.29A and B), others are compact, fine-grained, perhaps radiating anhedral or subhedral aggregates (Fig. 4.29C). Some of the zeolite varieties may be identified by their distinctive crystal habits; thus, the distinctive pseudo-cubic habit of chabazite (Fig. 4.29A), for example, is totally different from the acicular habit of natrolite. In rhyolitic rocks, the silica in vesicles may be in the form of opaline, chalcedonic or euhedrally terminated quartz crystals. It is not uncommon to find all three forms of silica present as successive, concentric vesicle linings. Layered chalcedony (agate) is sometimes deposited at the base of a vesicle in flat successive planes, rather like a sediment, in which case it provides a palaeohorizontal orientation (and facing direction if the vesicle is not completely filled). Many trachytes contain the silica polymorph tridymite in vesicles. Although tridymite is well known for its ability to grow metastably out of its high-temperature stability field, in trachyte it probably does represent a genuine high-temperature crystallization from late, silica-rich fluids. For example, thick trachyte dykes of the Miocene Lyttelton volcano, New Zealand, have tridymite infillings near their margins, but quartz infillings near the centre, and the quartz probably formed as a replacement of tridymite during the slower cooling of dyke centres. Other minerals commonly found in vesicles include celadonite, chlorite, carbonates, members of the epidote family and various other alteration products from within the rock, precipitated in the vesicles. Calcite is particularly common in the vesicles of

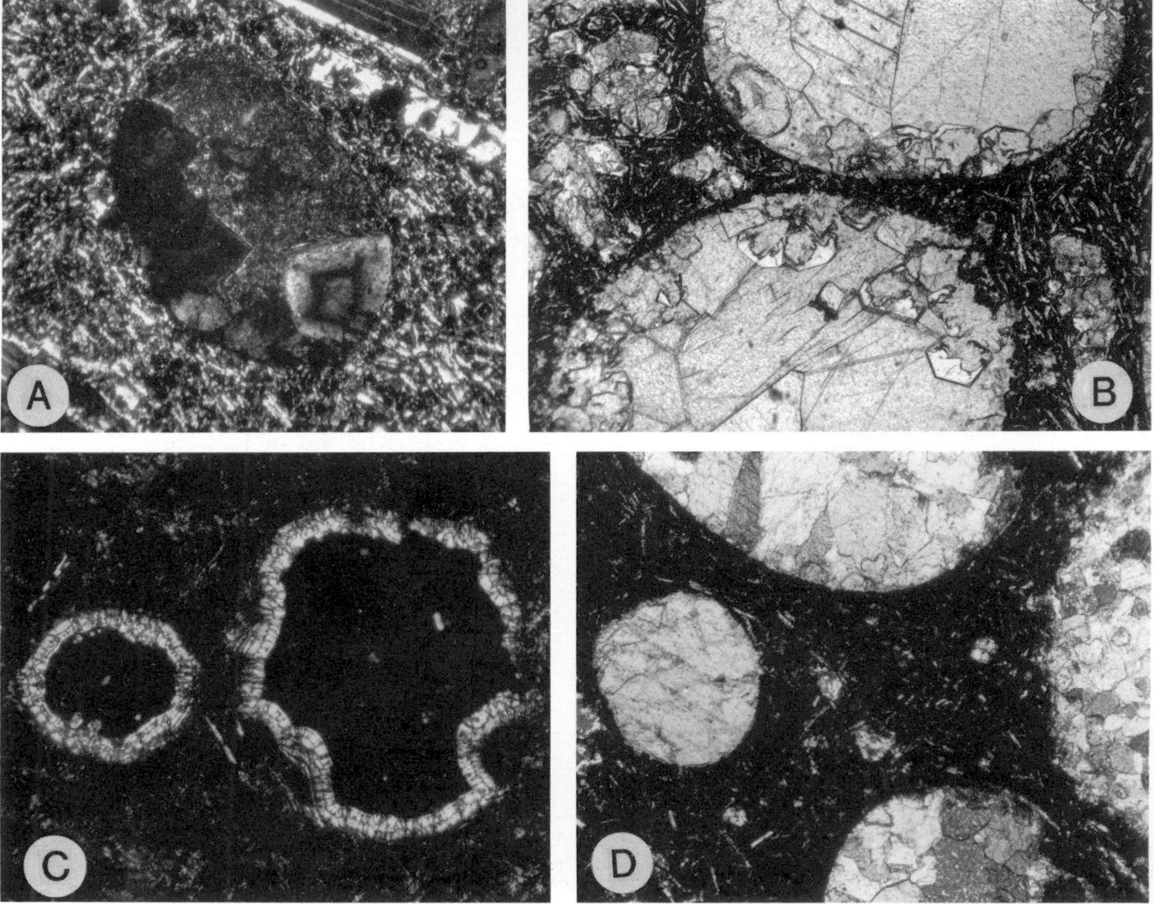

Fig. 4.29 Vesicle infillings. (A): Euhedral chabazite projecting into a single partially filled vesicle in Lyttelton volcano hawaiite. (B): Euhedral apophyllite and calcite lining vesicle, later filled by calcite. (C): Radiating growths of chlorite lining vesicles. (D): Complete filling of vesicles by calcite in spilite. (View lengths measure 2.44 mm (A), 2.97 mm (B), 0.55 mm (C), 2.76 mm (D).)

altered submarine basalts (Fig. 4.29D), and often the infilling consists of a single crystal.

Secondary minerals in vesicles that form during burial or metamorphism may be difficult to distinguish from those formed from late igneous fluids; in large rhyolitic fields such as the Taupo Volcanic Zone of New Zealand, hydrothermal activity may be continuous over very long periods of time, and there may be no sharp boundary between metamorphic and late igneous processes. However, the effects of burial of lava flows and their subsequent alteration (and vesicle infilling) may be discerned in distinctive mineral zonations. Thus, Walker (1960) describes the following sequence of zeolites related to burial of a thick sequence of Tertiary basaltic lavas in Iceland: from top to bottom, a zeolite-free zone, then zones characterized by chabazite–thomsonite, analcime, and finally at the base mesolite–scolecite.

(e) Xenoliths and xenocrysts

Xenoliths are 'foreign' rock fragments somehow incorporated into the magma. There are three principal types:

1. Restite material, the refractory residue brought up from the site of mantle or crustal melting.
2. Totally exotic material, ripped off the walls of conduits or the walls or roof of a magma chamber.
3. Cognate material, representing perhaps an accumulation of crystals already formed in a magma chamber, later ripped up during fresh magma movement.

There are inevitable difficulties in deciding which category a xenolith belongs to, and indeed it is not always possible to draw sharp lines between the three categories. For example, olivine-rich material in basalt could represent restite peridotite, an exotic peridotite from the conduit wall, a cognate accumulation of crystals somewhere in the conduit system or, for that matter, a glomeroporphyritic cluster of olivine (which is not really a xenolith at all). Mantle peridotite fragments are likely to exhibit metamorphic textures, distinguishing them from cognate or porphyritic material, but further distinction may be difficult. In any case, a sharp boundary between glomeroporphyritic clusters and cognate xenoliths cannot always be drawn; a decision on the matter is often arbitrary! Deciding whether a single crystal is a xenocryst, a phenocryst or a phenocryst caught up in reactions due to magma mixing or changing pressures and temperatures can be even more difficult.

Some xenoliths are obviously exotic (quartzite in basalt, for example), but some cognate material may appear quite exotic too. For example, basalt injected into rhyolite magma may mingle as a mass of quenched pillows which resemble xenoliths, and may be veined by the heated and fluid rhyolite (Blake *et al.*, 1965). The basalt pillows form simultaneously with the rhyolite, so they can hardly be termed xenolithic, but if these bodies are carried along in a flow of rhyolite they eventually become cognate xenoliths. A similar situation exists in many plutonic complexes, and the phenomenon of net veining and basic enclaves in granitic rocks is discussed in section 4.2.5.

Truly exotic xenoliths and restite material provide clues about the history of geological events (xenoliths pre-date the host rock), as well as important information on the nature of the mantle and lower crust (locally and globally). The information plays a vital role in creating realistic models of magma genesis.

Xenoliths that are not in equilibrium with the magma may melt, dissolve or react, and these processes may lead to their eventual assimilation. Therefore the absence of xenoliths does not necessarily indicate that no contamination of magma occurred. It may in fact mean that numerous xenoliths were totally assimilated. On the other hand, the presence of numerous xenoliths does not necessarily imply the magma has been severely contaminated; they may have been incorporated without much reaction just before magma solidification. Assimilation changes the chemistry and mineral content of the igneous rock. This may occasionally be obvious, especially where an unusual xenolith composition causes a strong localized effect, but, in general, trace-element and isotope studies are required to determine the extent of any contamination.

It is worth examining, in turn, the three ways that xenoliths can be assimilated. First, the effects of melting. Take, for example, a quartz–feldspar sandstone incorporated in some intermediate magma (Fig. 4.30) at 800°C. At a water pressure of 2 kbar, quartz, plagioclase and K-feldspar individually melt at temperatures in excess of 800°C, but minimum-melting-point mixtures of quartz with either one or both feldspars, or feldspar with feldspar all form at temperatures <800°C. This means that melting could occur everywhere quartz and feldspar or two different types of feldspar are in contact (Fig. 4.30). The important point to note is that melting occurs through the entire xenolith at appropriate sites (in contrast, solution can only occur at the margins of a xenolith). Evidence of melting is sometimes preserved as a film of glass along grain boundaries of xenoliths in volcanic rocks (Maaløe and Printzlau, 1979; Aranda-Gómez and Ortega-Gutiérrez, 1987), and if reactions take place in association with the melting (Tsuchiyama, 1986), a spongy (sieve) texture also develops similar to that already described for plagioclase (Figs 4.17 and 4.18). Melting is a powerful mechanism for disaggregating a xenolith, hence creating xenocrysts and promoting more rapid assimilation; if xenoliths do break up,

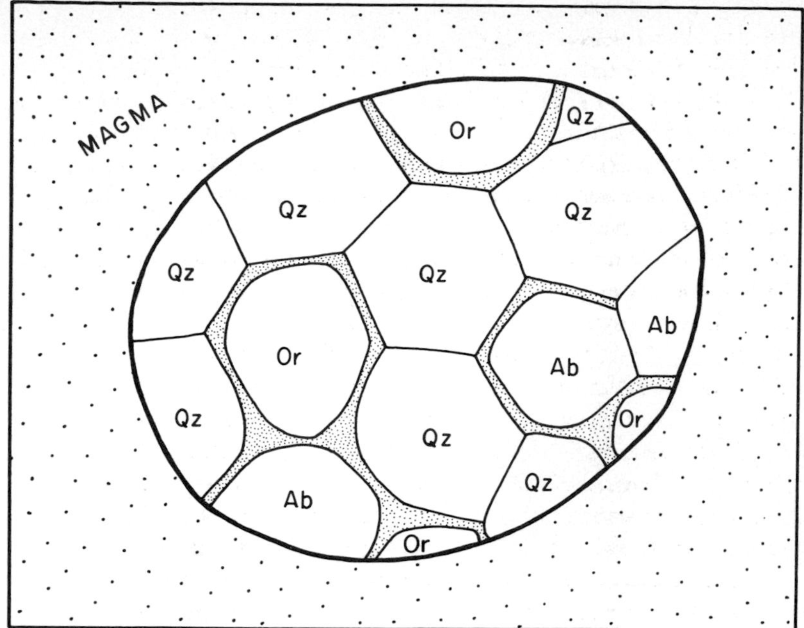

Fig. 4.30 Melting phenomena in a xenolith containing quartz, orthoclase and albite. Melt along grain boundaries is stippled. Explanation in text.

evidence of melting is likely to be lost. The principal energy limit on melting is the need for latent heat, and since most natural magmas are not superheated this means that melting has to be balanced by an equivalent amount of crystallization in the melt which yields latent heat of crystallization. Readers who wish to pursue the quantitative aspects of this should refer to Tsuchiyama (1986).

Solution differs from melting in that it takes place only where the xenolith contacts magma (at its margins). A relatively sodic feldspar, for example, caught up in magma crystallizing a more calcic plagioclase, may not be sufficiently sodic to melt, but it may still partially dissolve and react (Fig. 4.17). The process will not in itself lead to a disaggregation of a xenolith. Xenocrysts which undergo such solution become rounded, just like some phenocrysts during magma mixing (Tsuchiyama, 1985).

Reaction, like solution, sometimes takes place only where xenolith contacts magma. For example, the reaction of quartz with basic magma produces pyroxenes, usually as a fine-grained aggregate encrusting the margins of xenolith or xenocryst (Fig. 4.31). However, reaction may involve more pervasive diffusion, especially if water is involved, and the effects are similar in many ways to those of metamorphism. Referring

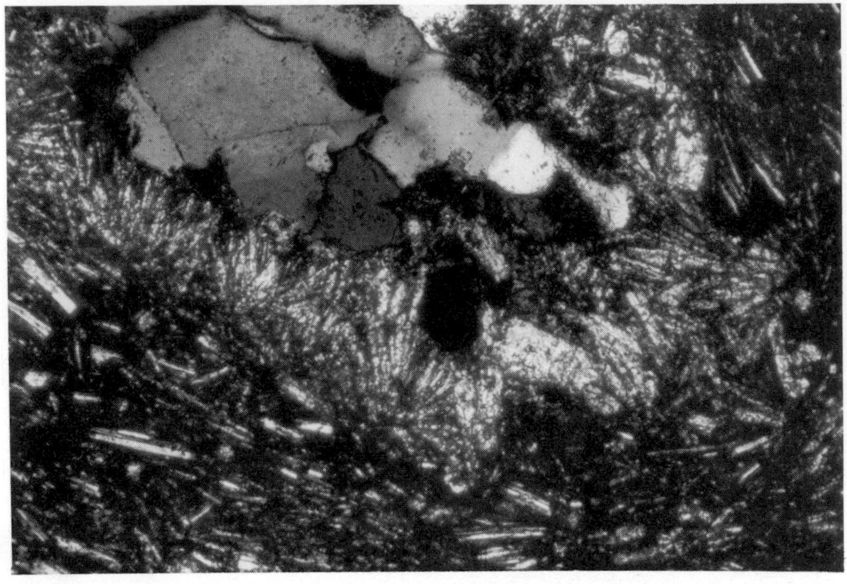

Fig. 4.31 Quartzite xenolith in basalt with a reaction rim of pyroxene. (Length of view measures 0.85 mm.)

back to Fig. 4.17, one can see that plagioclase in a xenolith, more calcic than that crystallizing from the magma, cannot melt, and neither can it dissolve because the liquid is already saturated in those components. Instead it may slowly equilibrate and change composition by diffusion, adding An to the melt. This does not cause the melt to crystallize a more calcic plagioclase, but it does reduce the amount of differentiation possible subsequently.

Some of the more important reactions affecting rock-forming minerals during assimilation were assessed by Bowen (1922). As a result he proposed the now famous (Bowen's) reaction series. There are two which he merged together at low temperatures (Fig. 4.32). Thus a high-temperature pyroxene may change by reaction to a lower-temperature amphibole, and so on. There are, of course, other common reactions such as those that occur when pelitic rocks are caught up in dry basaltic magma, lose water and other mobile elements to the magma, and develop anhydrous Al-rich minerals such as cordierite and sillimanite. Metamorphic and metasomatic changes when carbonate rocks are incorporated in magma can be much more spectacular, leading to the production of skarn minerals such as Ca-rich garnet, diopside, tremolite, anorthite and wollastonite.

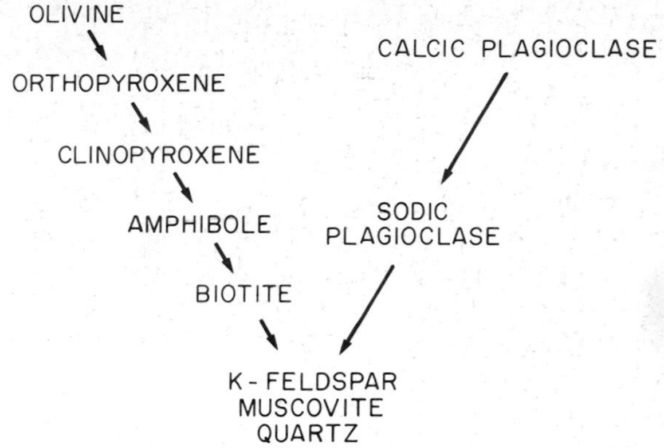

Fig. 4.32 Bowen's reaction series: the series from olivine to biotite is discontinuous, the other from An-rich to Ab-rich plagioclase is continuous. Bowen proposed the series merely as simplifications and generalizations of admittedly more complex natural systems.

4.2 PLUTONIC ROCKS

4.2.1 The contrasting characteristics of plutonic and volcanic rocks

Plutonic igneous rocks are the products of slow or very slow cooling and crystallization of magma. The rate of cooling depends on a number of factors including the volume of magma involved and the ambient temperature, this being related to the regional geothermal gradient and depth of emplacement. In sharp contrast to volcanic crystallization, complete within hours or days, crystallization within a large magma chamber may take hundreds of thousands of years, and the general consequences of this are worth examining before looking at plutonic textures in detail.

Perhaps the most striking textural difference between plutonic and volcanic rocks is their grain-size ranges. The very wide range (from microlite to phenocryst) that characterizes most volcanics is lacking in plutonic rocks. Although phenocrysts do occur in plutonics, the total range of grain size is very much more even. This is in part a consequence of the lack of quenching in the plutonic environment. A striking contrast in grain shapes also exists: volcanic crystals are generally euhedral or subhedral (because they grow freely suspended in magma) whereas plutonic crystals are more commonly subhedral or anhedral (because the crystals ultimately must fill the entire volume of rock necessitating mutual interference boundaries rather than crystal faces between the grains). Furthermore, because cooling is slow, primary crystallization of magma in a plutonic environment can be achieved at low degrees of undercooling. This means two things: firstly, the quench textures typical of strongly undercooled volcanics are rare; secondly, compositional zoning in crystals is much less marked.

These comments seem to suggest that plutonic crystallization is a good deal simpler than volcanic crystallization. But that is true only in regard to the points noted above. In other ways plutonic crystallization is much more complex. For example, one can state generally that a volcanic rock represents a crystal-laden magma that solidified *in situ* at the position of emplacement. Apart from loss of volatiles, whole-rock chemistry is a good measure of the composition of that particular batch of magma. On the other hand, a plutonic rock may represent a sediment of crystals from which all host magma has been removed; or it may represent a kind of 'crystal garden' growing at the magma chamber wall from which the residual fluids have been fractionated. Not only must the possibility of mechanical separation of liquid from crystal be considered, but so must the possibility of long-range diffusion in the magma, and the consequences of that. Plutonic rocks are much less likely to represent the composition of natural magmas than are volcanics.

There are also marked contrasts in the responses of volcanic and plutonic rocks to secondary processes. Although volcanic rocks display considerable diversity in grain size and shape, related perhaps to a cooling history that is complex in detail, the textures nevertheless relate primarily to crystallization over a relatively short period of time (at least in so far as the groundmass and final solidification is concerned). On the other hand, plutonic rocks may cool so slowly that they undergo continuous textural readjustment. For example, Marsh (1988) notes that the even grain size of many plutonics is unexpected in terms of steady nucleation and growth. The general absence of small crystals (representing the latest to nucleate) suggests they have been resorbed in complex processes of post-solidification textural re-equilibration (discussed further for layered rocks in section 4.2.2). Textures related to exsolution (e.g., perthite) and grain-boundary reactions (various symplectites such as myrmekite) are typical of the plutonic environment of slowly changing conditions, but rare or absent in volcanics. The type of adjustment depends to a large degree on the presence or absence of fluids; for example, corona textures indicate a limited fluid supply, and uralitization a somewhat more generous supply of fluids (these textures are discussed in Chapter 3).

Subject to progressive metamorphism, volcanic rocks respond more quickly and pervasively than plutonic rocks. The reasons are: volcanics contain inherently unstable materials such as glass which are readily replaced during metamorphism; the finer grain-size of volcanic rocks presents a larger grain-boundary surface area, thus promoting reaction; volcanics often comprise a stratified complex of heterogeneous material more easily deformed than typical massive plutonic rocks.

Returning to primary igneous processes, viscosity plays an important role in determining the behaviour of magma within the crust. Basaltic magma is very fluid, Newtonian or close to Newtonian in character and it has a low viscosity in sharp contrast to that of the crust; consequently, basalt magma conduits and chambers tend to be fracture controlled, and magma can rise to the surface in large volumes (Ramberg, 1981). The viscosity of silica-rich magma is higher than basaltic magma but still very much lower than that of the crust, so the rise of many granitoids is also controlled by fractures, particularly ring fractures leading to plutons of the bell-jar type (Pitcher *et al.*, 1985). However, relatively few granitic magmas reach the surface as lava flows, and the reasons for this probably lie in the way that water interacts with melting and crystallization in the granite system. The large-scale extrusion of silicic magma only takes place when the release of volatiles causes explosive fragmentation of magma (leading to the formation of an ignimbrite). Some silicic plutons are diapiric, that is they shoulder crustal rocks aside rather than occupying fracture-controlled chambers; according to Ramberg (1981) this shows that the granitoid was almost totally crystalline with similar viscosity to the crust.

There is a marked contrast between most basic and granitic plutonic complexes; the former are almost always conspicuously layered, the latter only rarely so. It is therefore logical and convenient to discuss plutonic textures mainly by reference to these two contrasting types of plutonic complex. A variety of rock types, not just gabbroic and granitoid, will enter the discussion where appropriate.

4.2.2 Layered intrusives

Studies on the origin of layering involve such a wide range of geological disciplines that a comprehensive account is beyond the scope of this book. Some introduction to the subject is appropriate here, but the reader is referred to Wager and Brown (1968), Parsons (1987) and the references given in the next paragraph for more detailed information.

Layering is of two principal types: phase layering, involving variations in mineral composition, and less-obvious layering represented by a slow systematic change in bulk composition from layer to layer (cryptic layering). Composition may also change laterally within any one layer, in which case cryptic and phase layering are discordant. Regardless of scale, the various layerings demonstrate that differentiation took place, and that crystals and magma were capable of being segregated. The question is how, and answers have to be sought in various processes of magmatic differentiation currently the subject of much debate. For example, is it crystal settling or liquid fractionation (McBirney *et al.*, 1985) that separates liquids and crystals? What are the relative roles of crystal settling and *in situ* crystallization at the borders of a chamber (Langmuir, 1989)? Do magma chambers become naturally stratified into layers characterized by double diffusion of heat and mass, and if so does crystallization take place along the boundaries between layers (Robins *et al.*, 1987)? What is the role of surge-like density currents, laden with crystals, in the evolution of a magma chamber (Irvine, 1987)? To what extent are the less dense minerals such as quartz and feldspar carried along with denser minerals in convection and density currents (Irvine, 1987)? What effect does the injection of new batches of primitive magma have on the system (Tait and Kerr, 1987)? How easily do different magmas mix, and what role does the shape of the magma chamber play (Baker and McBirney, 1985; Turner and Campbell, 1986)?

In the following sections we shall look at those aspects that impinge most directly on what can be done with the microscope.

(a) Phase layering

Most layered complexes consist of a main succession of layers, built up from the base of the chamber, and thinner successions of layers formed around the sides and at the roof. Layer thickness is usually in the range of millimetres to a few metres. The main succession of layers typically displays wide lithological variation. A regularly repeated change in mineralogy is called rhythmic layering. Layers may be characterized by density grading (Fig. 4.33C), with heavy minerals such as olivine concentrated at the base, lighter minerals at the top, or grain-size sorting and grading (Fig. 4.33D), or some combination of these. Other layers are simply uniform in grain size and mineralogy. Sedimentary-type structures such as cross-stratification, scouring and slumping are common. Inequidimensional mineral grains often display a preferred orientation with long-axes more-or-less parallel to the layering. Two principal mechanisms have been proposed to explain these structures: either crystal accumulation by a sedimentary-like process, or accumulation during *in situ* crystal growth. The mechanisms are not mutually exclusive, and we shall show later that both can be effective in appropriate circumstances. Whichever is the dominant process, structures may be significantly modified by secondary processes of compaction and laminar flow (section 8.1).

Side-wall layering may also be rhythmic and complex in structure, but because of the steep attitude of the layering it is necessary to appeal to some form of *in situ* crystallization rather than gravity-controlled crystal settling. Colloform layering, reminiscent of some organic growths, is a special structure that forms as magma progressively congeals against the side of the intrusion; domal layers face towards the centre of the intrusion, separated by sharp cusps that point towards the walls (Fig. 4.33A). In addition, side-wall layering commonly features crystals, attached to the

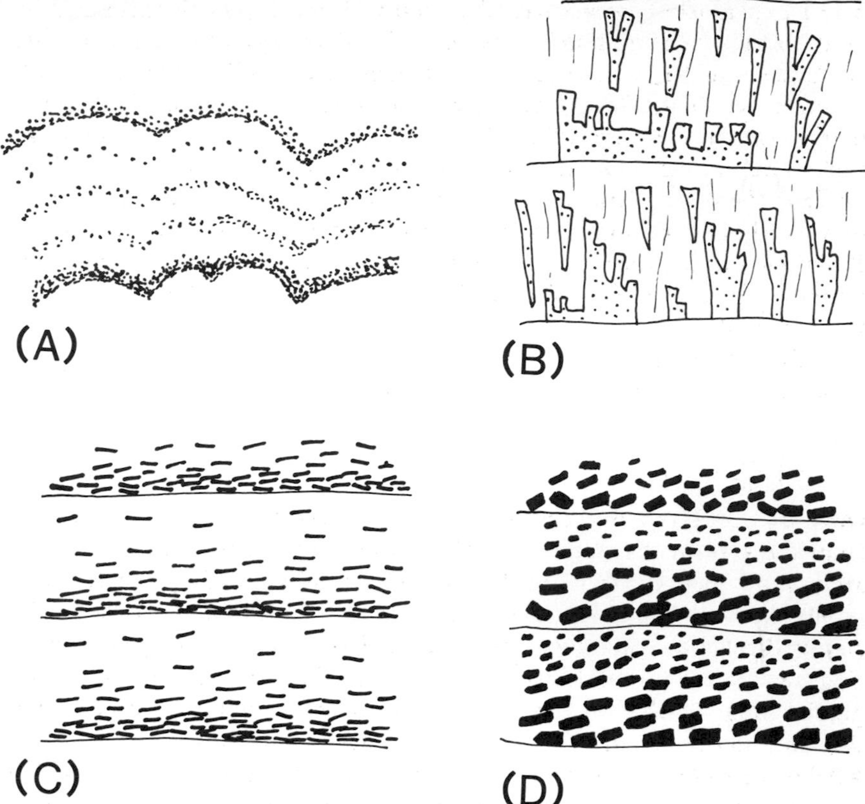

Fig. 4.33 Schematic views of features associated with layering in plutonic rocks. (A): Colloform layering with individual mounds of the order of 1 m across. (B): Crescumulate texture with crystals elongate at a high angle to layering. (C): Density grading, with heavier minerals typically concentrated at the base of layers. (D): Grain-size grading with coarser grains typically at the base of layers.

wall, that grew with long axes perpendicular to the layering (crescumulate texture – Fig. 4.33B, discussed later).

Crescumulate texture is also found in roof layering (e.g., Sørensen and Larsen, 1987, Fig. 10). However, in the roof zone, besides *in situ* growth mechanisms, it is possible to consider the idea that crystals could have floated there, or at least have been carried to the roof by currents and left there.

(b) Cumulate textures

The rocks in a layered complex are called cumulates, the term being appropriate for the products of crystal settling or *in situ* crystal growth.

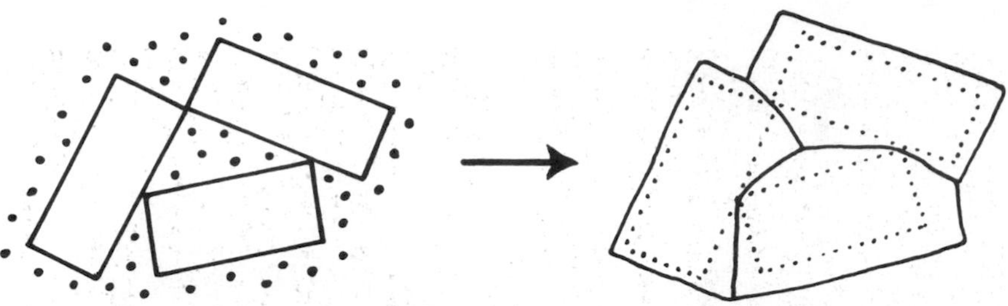

Fig. 4.34 Progressive growth of crystals, and the development of mutual growth interference boundaries to form adcumulate texture. Discussed further in text.

The crystals accumulating are called cumulus crystals, and the liquid in which they are immersed is called the intercumulus liquid. The main textural types are adcumulate, orthocumulate and crescumulate.

Adcumulate texture (Fig. 4.34) represents an accumulation of crystals, often of one phase only, from which the intercumulus liquid was removed, possibly as a result of crystal settling, and/or compaction, and/or diffusion of material to and from the adjacent magma. Efficient diffusion is supported if there is a lack of mineral zoning. The cumulate phases occupy the entire space of the rock, and so the crystals cannot be euhedral, and crystal shapes are defined by the jointly merged boundaries of adjacent crystals. Nevertheless, if crystal habit is markedly inequidimensional (e.g., tabular plagioclase), that habit is usually still discernible (Fig. 4.34). In adcumulates where one mineral encloses another poikilitically the rock is said to have a heteradcumulate texture. It is clear from these textural features that the final stages of an adcumulate represent *in situ* crystallization. Whether the primary accumulation of crystals was due to crystal settling or *in situ* crystallization has to be decided by some other means.

Orthocumulates supposedly represent cumulus crystals around which a differentiating intercumulus liquid crystallizes; movement of material to and from the main body of magma was restricted. The intercumulus material may poikilitically enclose cumulus phases, and one interpretation of ophitic texture (Fig. 1.17) is that it represents cumulus plagioclase enclosed by intercumulus pyroxene. In fact, poikilitic textures do not necessarily indicate such a simple order of crystallization. As discussed in Chapter 3, the simultaneous crystallization of several small crystals of one phase and a single large crystal of another may be a better explanation, supported by the observation that plagioclase increases in grain size towards the outside of the texture (Fig. 3.32 – and also Cox *et al.*, 1979, Fig. 12.4; McBirney and Noyes, 1979, Fig. 9). McBirney and Noyes also pointed out that in many ophitic textures plagioclases are well separated from each other, and this gives no support to the idea that the plagioclase

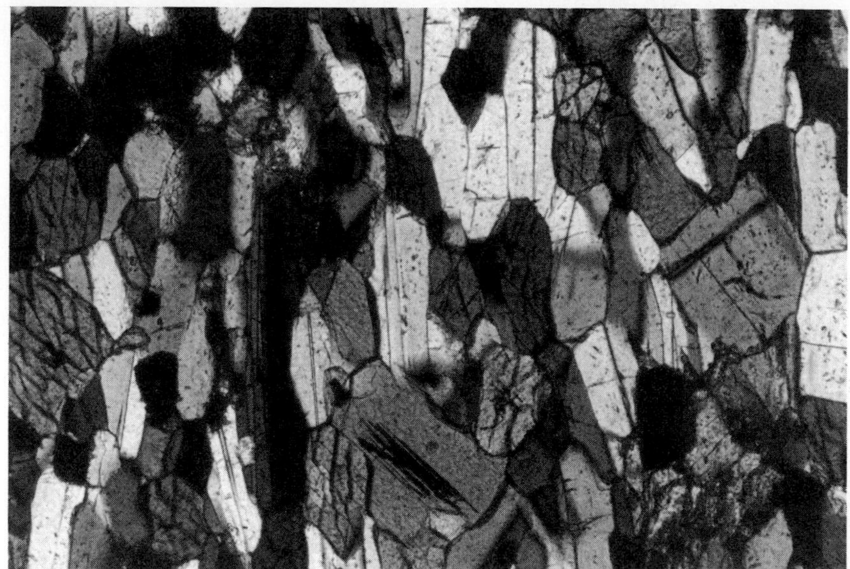

Fig. 4.35 Igneous lamination in hornblende gabbro, Bluff, NZ. Crystals visible are calcic plagioclase and hornblende. (Length of view measures 0.85 mm.)

grains were an accumulation of crystals in contact with each other (as conceived in crystal settling theory) before pyroxene grew. The evidence of much ophitic texture points to *in situ* crystallization.

Adcumulates and orthocumulates often display some degree of post-cumulus maturation in an attempt to attain textural equilibrium. This is discussed below under 'secondary textures'.

Crescumulates display a prominent growth fabric with elongate crystals of olivine, pyroxene, plagioclase or apatite, often many centimetres long, at high angles to layering (Fig. 4.33B). There seems to be no doubt that this texture represents *in situ* crystal growth. In examples where crystals are skeletal or dendritic, very rapid growth is indicated. Non-skeletal forms may represent slower crystallization, or may result from late-stage overgrowth which obscures an earlier skeletal habit; in the absence of compositional zoning it is difficult to decide between those options.

Foliation (often called igneous lamination) is the general term used to describe the preferred orientation formed by inequidimensional grains with their long axes more-or-less parallel to the layering (Fig. 4.35). It is commonly displayed by plagioclase, but also other minerals such as olivine. It could be produced by sedimentation, laminar flow of a crystal mush or by the close packing of crystals during compaction and the removal of pore fluids (section 8.1); laminar flow is suggested if long-axes form a lineation within the layering, as described, for example, by Sørensen and Larsen (1987) for arfvedsonite prisms. McBirney and Noyes (1979) also suggest that grains might develop long axes parallel to layering by growth *in situ* when growth rates are very slow. This seems unlikely, however, because crystals growing parallel to layering would be growing towards each other and competing for the same material; crystals are not likely to grow at right angles to the compositional gradient.

(c) Density grading

The classical explanation for density grading (Fig. 4.33C) in the main layered series of the Skaergaard Complex is that magma cools and crystallizes near the roof, descends along the side walls, then sweeps across the floor of the chamber depositing the crystals in order of density (the heaviest first) as the current slows. An accelerating current will keep the crystals in suspension and scour the underlying incompletely solidified cumulus layers. Certainly the evidence of scouring and cross-stratification supports the idea of current activity and an erodable mush of crystals, but the fundamental objection to the idea is that similarly graded layers occur on the side walls and roof. Crystal settling on nearly vertical walls is not feasible, and neither is it feasible to have heavy minerals floating preferentially to the roof. Neither does the fact that heat loss occurs at the roof necessitate crystallization there. Irvine (1970) showed that basaltic magma, rising to the roof by convection, cools on an adiabatic path to a lesser degree than the associated lowering of the liquidus curves (due to a P decrease in the anhydrous system); even though heat is lost at the roof, crystallization takes place as the colder current returns to the floor and pressure increases. A further problem with the classical explanation for density grading is the presence of *inverted* density grading, which is the norm in the main layered series of the Klokken Complex, for example (Parsons and Becker, 1987). Crystal settling may sometimes be the answer, but not always.

An alternative explanation, provided by Wager (1959) was that density grading simply reflects the easier nucleation of the mafic minerals (olivine has a much simpler atomic structure than plagioclase, for example). If the oxides and olivine crystallize first and settle out to the base of a magma layer at the bottom of the chamber, then the residual melt is enriched in feldspar components. This eventually crystallizes to form the feldspar-rich top of a layer. This theme is taken up by Komor and Elthon (1990) for *in situ* crystallization in a stagnant bottom layer of magma. They believe that latent heat of crystallization, released as feldspar crystallizes, heats up the magma sufficiently to stop all further crystallization. Cooling then sets in

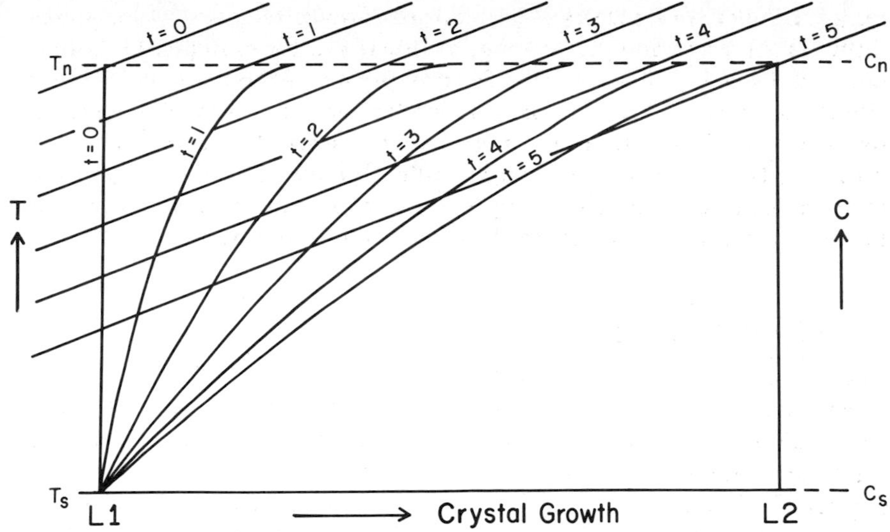

Fig. 4.36 Oscillatory crystallization mechanism to explain layering in terms of the differing rates of diffusion of heat and chemical components adjacent to a layer of growing crystals. Nucleation requires a certain combination of temperature, T_n, and supersaturation, C_n (both at the top of the vertical axis). A layer (*L*1) starts to grow at time $t = 0$ when the magma is supersaturated (C_n) and when $T = T_n$. Crystal growth causes a sudden drop in concentration to saturation (C_s), and as growth proceeds, different concentration gradients are established (curved lines) for various times $t = 1$ to 5 (the gradients become less steep as growth slows). The temperature gradients for times $t = 1$ to 5 are given as straight lines. At time $t = 5$ the curved and straight lines intersect again at C_n and T_n, causing nucleation of a new layer (*L*2). Reprinted with permission from McBirney (1984), *Igneous Petrology*, © 1992 – 2nd edition, Boston: Jones and Bartlett Publishers.

until conditions are right for a new cycle starting with mafic nucleation and crystallization.

McBirney and Noyes (1979) tried to answer this problem with an oscillatory crystallization mechanism (Fig. 4.36) involving:

1. Supersaturation of magma before nucleation of a mineral.
2. Initial rapid growth causing the magma adjacent to the crystal to be depleted in those components needed for growth. A steep compositional gradient is therefore established.
3. Growth slows, and diffusion gradually decreases the initial steep compositional gradient.
4. All the time, temperature is slowly decreasing.
5. Eventually temperature and compositional gradients are such that a fresh burst of nucleation is initiated, and a new cycle of growth begins.

Various phases crystallizing from a magma require different degrees of supersaturation for nucleation, and therefore a number of complementary cycles will produce a phase layering. The mechanism may possibly explain centimetres thick layering, but McBirney (1984) states that diffusion is too slow for it to explain the metres thick layering found in many layered complexes.

The concept that liquid in a magma chamber becomes stratified into cells of differing temperature and composition has also been used to explain layering. Each layer of liquid loses heat to the one above until eventually heat is lost through the roof; the bottom liquid layer may start crystallizing first. Thus, Sørensen and Larsen (1987) propose that layering in the Ilimaussaq alkaline intrusion represents the successive crystallization of each liquid (and stagnant) layer. Density grading within each layer (up to 15 m thick) is due to the heaviest minerals (arfvedsonite and eudialyte) nucleating first and settling out to the bottom.

Inverted density grading in the neighbouring Klokken Complex (Parsons and Becker, 1987) was produced during crystallization of alkali feldspar and hedenbergite from a eutectic melt. Early stages of crystallization were dominated by feldspar (it makes up 80% of the eutectic mixture), and the crystallization led to a build-up of vapour pressure. This caused a lowering of the liquidus, and because hedenbergite nucleates at a lower undercooling than feldspar, feldspar nucleation was suppressed. Hence, pyroxene came to dominate the late stages of crystallization (at the tops of layers) producing inverted density grading. Periodically, pressure release through a ring-fracture system overlying the intrusion, initiated another cycle of crystallization.

These ideas involve a degree of crystal settling and crystallization almost but not quite *in situ*. The slumping of layers metres thick in the Ilimaussaq Complex, and the disturbance of some layers in the Klokken Complex rule out the idea that crystals grew firmly attached to the floor of the chamber. However, the classical idea of crystal settling from crystal-laden currents has not been abandoned for the Skaergaard Complex. Thus Irvine (1987) describes density-graded beds which he ascribes to surge-like density currents which periodically swept across the floor. Part of the evidence is that large blocks from the roof zone were carried along in the currents. At the same time, he advocates *in situ* crystallization for much of the less-obviously layered material at the Skaergaard chamber floor.

McBirney and Noyes (1979) pointed to the difficulty of crystals sinking through a magma that is possibly non-Newtonian and has a yield strength. There is a critical crystal size that must be exceeded before a crystal can overcome the yield strength, and either sink or float. Very small crystals (less than a millimetre or so) are unlikely to sink or float, but will remain in suspension. We are still uncertain as to the composition (especially in relation to volatile content) and viscosity of natural magmas below the surface, but it is probable that the density of plagioclase feldspar is similar to the Skaergaard magma, and even large crystals would have shown little tendency to sink or float. Plagioclase could, however, be swept along in suspension. Its presence in the Skaergaard density-graded layers can therefore be explained as follows:

1. plagioclase was carried in suspension to the base of the chamber by dense crystal-laden currents;
2. showing little tendency to rise or sink, plagioclase would become concentrated at the top of a layer simply because the heavier minerals sank to the base;
3. the layer would solidify as crystals matured by diffusion and crystallization of the intercumulus liquid.

(d) Grain-size grading

Hydraulic sorting should produce grain-size sorting within layers deposited by crystal settling (Fig. 4.33D). Crystals of any one phase should fine upwards, and if density-grading is present, both heavy and light phases should fine upwards. McBirney and Noyes (1979) pointed to a lack of such sorting in many of the Skaergaard layers. In practice, there are severe problems in measuring grain sizes in a completely crystalline rock. Firstly, the rock cannot be disaggregated (as is the standard practice with sediments), there is the problem of how to measure inequidimensional grains, and a major problem is that it may not be possible to determine what is 'cumulus' rather than 'post-cumulus' growth. Nevertheless, the predicted grain-size grading is well developed and perfectly obvious in some layers from the Duke Island Complex (Irvine, 1987). For Skaergaard, Irvine (1987) points out that grain-size sorting cannot occur if the grains were initially uniform in size!

Robins *et al.* (1987) describe an upwards increase in grain size of olivine in cumulates which grew more-or-less *in situ* within a chamber where the liquid was stratified and periodically recharged from outside.

Much more documentation and analysis of grain-size variation in layered complexes is desirable and would undoubtedly reveal further important clues to the formation of cumulates.

(e) Cryptic layering

Igneous differentiation within a magma chamber takes place by the progressive separation of crystals from liquid, and possibly by liquid stratification within the chamber if compositional and temperature gradients exist. Progressive changes may be interrupted and reversed by recharge of the magma chamber. In the Skaergaard Complex, mineral composition (and therefore rock composition) changes progressively from the base to

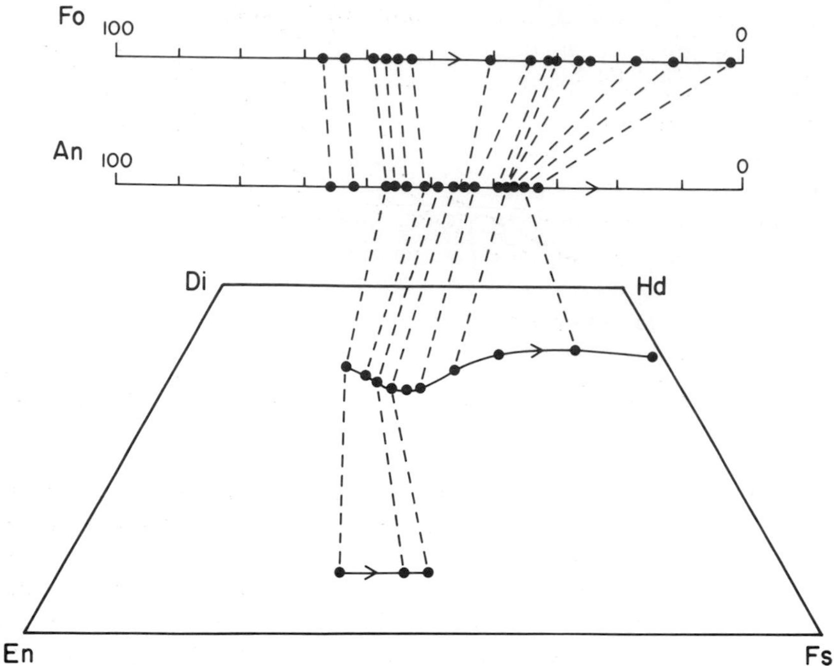

Fig. 4.37 Progressive upwards changes in pyroxene, plagioclase, and olivine compositions in the Skaergaard layered series based on the data of Figs 14 and 15 in Wager and Brown (1968). The dashed lines connect coexisting inverted pigeonite, augite, plagioclase and olivine.

the top of the main layered series: plagioclase from An_{66} to An_{33}, olivine from Fo_{67} to Fo_2 and the pyroxenes as shown in Fig. 4.37. The compositional changes can be related to crystallization within solid-solution series, as described for plagioclase (Fig. 3.10).

The term 'cryptic layering' is used to describe these progressive changes in mineral composition. 'Layering' is really a misnomer for the phenomenon ('zoning' may have been more appropriate): conceptually one has to imagine contours connecting rocks with phases of the same composition. The details of cryptic layering within the many layered complexes described from around the world is beyond the scope of this book; one must remember that the changes are not evident in any one rock specimen.

However, with regard to the origin of layering and textures, it is relevant to note that cryptic layering is not always parallel to phase layering. Even for Skaergaard, cryptic variations parallel to phase layering are now well established (McBirney, 1989). Huppert *et al.* (1987) summarize the literature on discordant cryptic and phase layering, and they take it as strong evidence for liquid stratification in the magma chamber.

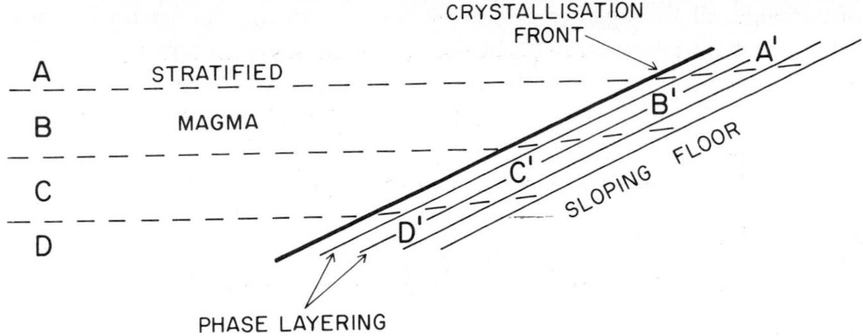

Fig. 4.38 Possible relationship of cryptic layering (A′ to D′) to phase layering and layers in stratified magma (A to D). Explanation in text. Based on model of Wilson and Larsen (1985).

Fig. 4.38 shows the sloping floor of a chamber, and the horizontal stratification of the liquid. According to this model, the cryptic layering simply reflects the changing composition of liquid layers. The crystallization front is parallel to the floor of the chamber, and local phase layering, due perhaps to localized current activity, will be parallel to this. The cryptic layering dips towards the centre of the intrusion even though the liquid stratification is horizontal; this is the result of post-cumulus compaction and the thinning of liquid layers as buoyant residual liquids at the crystal front migrate upwards.

On the other hand, Parsons and Becker (1987) and McBirney (1989) suggest that post-cumulus deuteric or late-stage magmatic redistribution of material is responsible for discordant cryptic layering in the Klokken and Skaergaard Complexes.

(f) Secondary textures in gabbroic rocks

Owing to the slow cooling of a plutonic complex, crystals may be in contact with magma for an extended period of time. We have already seen that this results in a relatively even grain size compared with volcanic rocks, and compositional zoning is also less prominent than in volcanic minerals.

(i) Towards textural equilibrium during cooling

Crystals have high surface energies, and freely suspended in a liquid they tend to form euhedral crystals the faces of which minimize those surface energies. Slow crystallization, as in a pluton, might be expected to produce crystals with simple euhedral form, reflecting the lattice structure. However, it is not geometrically feasible that such free-floating euhedral forms could all fit together to fill the volume of an even-grained, completely crystallized rock. This problem is solved in volcanics when the magma is quenched, and euhedral phenocrysts are 'frozen' into a fine-grained matrix. In plutonic rocks, the euhedral shapes of early-formed crystals may be preserved, especially if the residual fluids crystallize different minerals in the interstices, but accommodation with neighbouring grains, and the development of mutual interference boundaries is usually necessary (Fig. 4.34).

Such accommodations typically produce rather irregular angular boundaries. A stable arrangement of interlocking grains would involve a minimization of the grain-boundary surface area to produce a mosaic similar to that found in some metamorphic rocks. If the rock cools slowly enough (high temperatures are maintained for a long time), and particularly if late residual fluids are still present, the grain boundaries may adjust towards textural equilibrium by processes involving diffusion, solution and reprecipitation. In principle, textural equilibrium means the lowest achievable grain boundary energy, and it involves:

1. Grain coarsening, possibly at the expense of the smallest grains which go into solution because of their high surface area to volume ratio (this may explain the anomalous absence of small grains in plutonics – representing the latest to nucleate – noted by Marsh, 1988);
2. The development of equant grain shapes (regular mosaics) with dihedral angles appropriate to the relative surface tensions of neighbouring grains.

Take, for example, two phases, *A* and *B*, possibly with differing surface tensions. If the surface tension of *B* is less than *A*, then *B* will tend to 'wet' the grain boundaries of *A* (Fig. 4.39A). Melts have a lower surface tension than crystals, and usually fill spaces with dihedral angles in the range 40°–60°. If the surface tensions of *A* and *B* are the same, then the dihedral angle will be 120° (Fig. 4.39B). Note that this requires the grain boundaries to be curved. If *B* has a higher surface tension than *A*, than the dihedral angle will be >120° (Fig. 4.39C).

Hunter (1987) has explored this subject with regard to the maturation of cumulate textures. He illustrates the situation for an ideal monomineralic adcumulate (Fig. 4.40A) and two kinds of orthocumulate, one where the included mineral has a much higher surface tension than the other (a to c in Fig. 4.40B), the other where both phases have very similar surface tensions (d to f).

Hunter (1987) shows that many cumulates have matured in this way, and the following dihedral angles are observed: 120° for monomineralic domains and rocks; >100° for plag + plag + mafic or oxide minerals; 40°–60° for two mafic or oxide minerals + plag; <30° for cumulus olivine

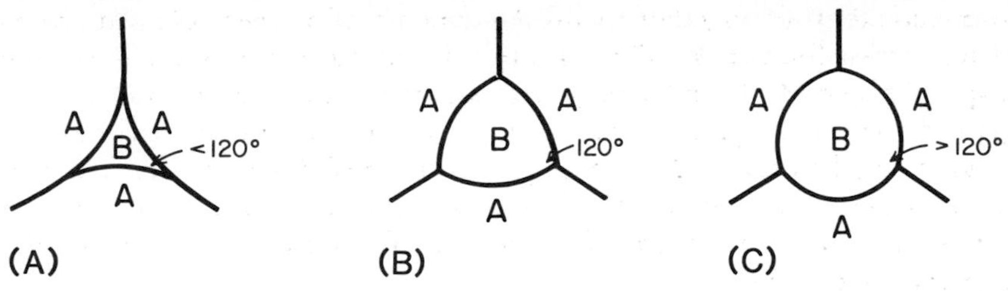

Fig. 4.39 Dihedral angles between minerals A and B with surface tensions in (A), $B < A$, in (B), $B = A$, and in (C), $B > A$.

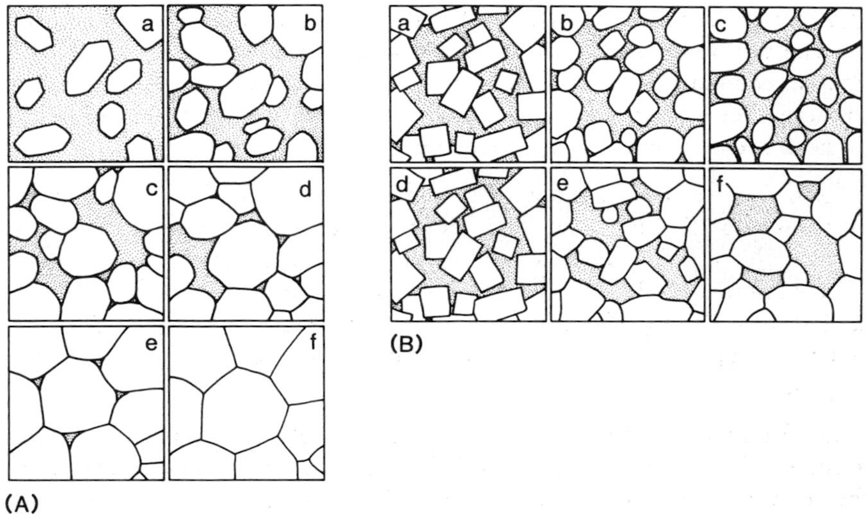

Fig. 4.40 (A): Stages in crystallization and maturation of an ideal monomineralic adcumulate involving coalescence of grains and porosity elimination. (B): Stages in crystallization and textural maturation of orthocumulates. (a) to (c) where the two phases have dihedral angles of 0°, (d) to (f) where dihedral angles are 120°. Reproduced with permission from Hunter (1987).

or orthopyroxene + post-cumulus pyroxenes. The degree of maturation depends on the rate of cooling. In general, he found that the most primitive and most mafic cumulates (that is those with the highest initial temperature) show the greatest degree of maturation.

(ii) Other subsolidus textural changes

A plutonic rock may pass very slowly through conditions appropriate to metamorphic change. Subsolidus changes in plutonic rocks that involve unmixing of minerals, and grain-boundary or more pervasive alterations were discussed in Chapter 3.

In gabbros, unmixing of pyroxene is optically visible, and exsolution lamellae provide interesting clues about the nature of crystallization and cooling. For example, orthopyroxenes may be the product of the sub-solidus inversion of monoclinic pigeonite, the evidence being lamellae close to the relic {001} plane of the pigeonite. The pigeonite may initially have exsolved augite on planes close (but not parallel) to both {001} and {100}. After inversion to orthopyroxene it exsolves augite strictly on {100}. Some orthopyroxene crystallizes as the primary mineral and it will have lamellae only on {100}. Similarly, augite may give clues as to whether it was exsolving pigeonite or orthopyroxene. More details are given in section 3.5.3, and it is worth noting that most textbooks give incorrect information on the orientation of lamellae in clinopyroxene – clinopyroxene pairs. Unlike pyroxene, plagioclase usually unmixes on a submicroscopic scale. An unusual texture concerning pyroxene is reported by McBirney (1984) from the extremely Fe-rich gabbros of the Skaergaard Complex; a pyroxene with a wollastonite structure crystallizes first, and when this inverts to hedenbergite it produces an incoherent mosaic.

Gabbroic rocks are particularly dry and peculiarly resistant to metamorphic change. Nevertheless, prolonged exposure to small quantities of fluid will effect various grain-boundary reactions producing the corona textures and symplectites discussed in sections 3.5.2 and 3.5.4. The reactions may be restricted to the boundaries between particular pairs of minerals, especially olivine and plagioclase, and the most common corona textures indicate amphibolite- or granulite-facies metamorphism (Fig. 3.25). If fluids invade the rock under greenschist-facies conditions, reactions such as serpentinization of olivine, chloritization or uralitization of pyroxene, and saussuritization of plagioclase are typical. The complex mix of fine-grained alteration products in a thoroughly altered gabbro spells confusion for many a student.

4.2.3 Granitic textures

(a) The effects of water on crystallization

Most granitoids are thought to originate by crustal melting – I-types from the melting of infracrustal unweathered igneous parents, and S-types from supracrustal weathered and sedimentary parents (Part One gives details of granitoid types). In the presence of water, melting of crustal rocks can take place at remarkably low temperatures, and it is very tempting to explain granitic rocks in this way. There are, however, two snags: firstly, there is the question of where the water comes from; secondly, the nega-

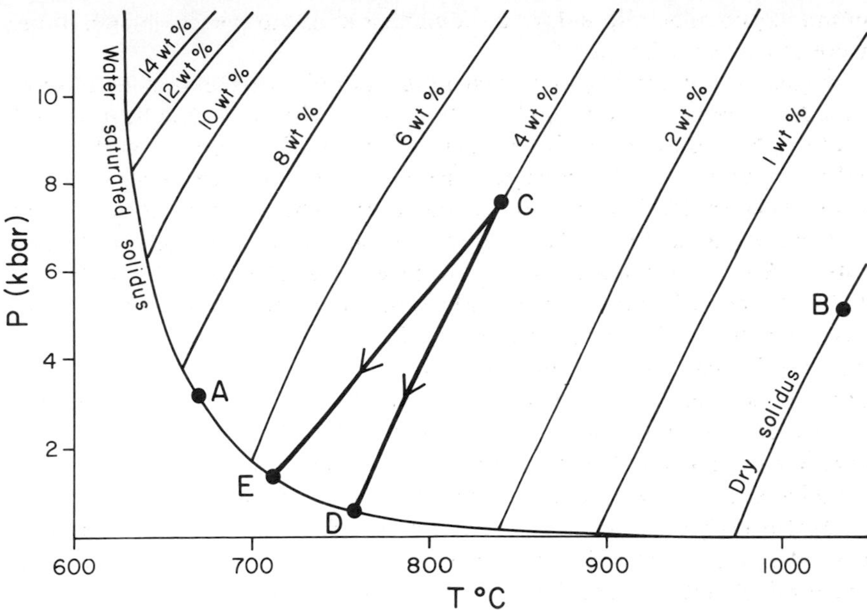

Fig. 4.41 Solidi for granitic melts containing various amounts of water, from water saturated to dry. Discussion of points A to E is in the text. Curves plotted using the data in Nekvasil (1988).

tive slope of water-saturated melting curves means that such melts cannot rise in the crust (Fig. 4.41). Granite magmas must therefore be under-saturated in water at their source, although they may possibly become saturated as they approach the surface.

An important source of water is in the breakdown of mica. For example, one possible reaction in the granulite facies is:

Biotite + silica = K-feldspar + orthopyroxene + almandine + H_2O

and in the amphibolite facies:

muscovite + silica = K-feldspar + sillimanite + H_2O

The first appearance of migmatites usually coincides with the break-down of muscovite, but this reaction is unlikely to produce enough water for complete melting at the minimum temperature. We still know little about the general movement and availability of fluids in the lower crust and mantle, and the actual amount of water present in natural granitic magmas is often uncertain; at least some water is indicated by the presence of the hydrous phases mica and amphibole. In contrast, basaltic magmas are relatively dry, which helps to explain why they reach the surface so often.

The effect that water has on melting and crystallization is illustrated by Fig. 4.41. The water-saturated curve has a negative slope, and it is clear that a low-temperature melt at point *A* is prevented from rising in the crust. In contrast, a dry melt at point *B* requires a higher initial temperature but it is able to rise to the surface provided that it does not lose too much heat along the way. Superheating is rare (as shown by the almost universal presence of phenocrysts in volcanics), so magma with 4.0 wt.% water at point *C* will probably lose heat as it rises, and either follow the 4.0% curve to *D* or diverge from it (to *E*, for example), crystallizing as it rises; crystallization may slow or halt the rise of the magma. At best, magma with 4.0% water cannot rise beyond point *D* without an increase in temperature. Unless all the water can be accommodated in minerals like hornblende and mica, crystallization leads to the phenomenon of second boiling (Burnham, 1979) which may result in a dramatic pressure increase and the catastrophic loss of volatiles along fractures; the attendant fragmentation of magma may cause the eruption of ignimbrite. If volatiles are lost without catastrophic effects, the liquidus/solidus curves are raised, magma becomes moderately undercooled and granophyric rather than ordinary granitic textures develop.

(b) Hypersolvus and subsolvus granites

Another important influence on granitic textures is the effect water pressure has on the alkali-feldspar system (Fig. 3.19). At low pressures (<5 kbar), the solvus lies below the solidus curves so that a single K–Na feldspar of intermediate composition can crystallize from melt (hypersolvus crystallization). At higher water pressures, the solvus and solidus curves intersect so that intermediate Or – Ab composition melts crystallize two feldspars, plagioclase and K-feldspar (subsolvus crystallization). Hypersolvus granites therefore indicate either low pressures or very dry conditions.

(c) Primary textures

Cumulate terminology is not generally used for granitic rocks. The reasons are historical rather than logical since cumulate terminology is now applied to layers in basic complexes that crystallized *in situ* in much the same way that granitic bodies crystallize. Although granitic rocks are not usually as conspicuously layered as gabbro (perhaps the crystallizing phases are not disparate enough to separate on a fine scale, or perhaps magma viscosity is too high), the following process of crystallization, advocated by Sawka *et al.* (1990) for some California plutons, is precisely analogous to that suggested for *in situ* crystallization of some basic layered complexes. Side-wall crystallization created a buoyant SiO_2-rich residual liquid (at the crystal–magma interface) which migrated to the upper parts of one particular pluton to create a zoning with marginal quartz monzonite and a granite core; the idea that sharply defined zones in a granitic pluton can form simultaneously contrasts with the traditional interpretation of sequential intrusion. Many of these rocks are cumulates in the modern sense, and not only do some Sierra Nevada plutons have well-developed crescumulate textures at their margins (Vernon, 1986), but poikilitic K-feldspar-bearing rocks are similar in essence to the orthocumulates of basic layered complexes.

Putting aside secondary textures and flow structures, there are three common textural arrangements in granitic rocks:

1. An equigranular mix of euhedral to anhedral grains called granitic texture.
2. Textures dominated by K-feldspar phenocrysts or megacrysts (often poikilitic).
3. Granophyric textures.

Granitic texture (Fig. 4.42A) represents the final merging together at the end of crystallization of a mix of more-or-less even-shaped and sized grains, some slightly tabular and euhedral, others anhedral. The most euhedral minerals are usually hornblende (if present), biotite, and plagioclase; K-feldspar (if present) varies from euhedral to anhedral; quartz is characteristically anhedral, filling the interstices, although bipyramidal quartz may be found included in feldspar. Plagioclases are strongly zoned, sometimes from very calcic plagioclase but more typically from andesine to sodic oligoclase or albite.

Unusually large K-feldspar phenocrysts, or megacrysts, are common in granitic rocks, and they often poikilitically enclose all the other minerals (Fig. 4.42B). Poikilitic K-feldspar is frequently misinterpreted. The texture does not necessarily imply an order of crystallization, and in most cases it is at least partly due to simultaneous crystallization of the enclosed minerals and K-feldspar. If the K-feldspar has a lower nucleation rate than the enclosed minerals, then the texture can be explained by the latter minerals nucleating on the surface of the faster-growing K-feldspar, or simply being overtaken by it. Only where inclusions are consistently concentrated at the centres of K-feldspar can the outer parts of the feldspar be safely interpreted as late (Flood and Vernon, 1988). The vague idea that megacrysts of K-feldspar develop during a K-metasomatism (this idea goes back to the time when granitization was in vogue) is still commonly floated without compelling evidence. The matter has been addressed by Vernon (1986) who finds that the megacrysts have all the chemical and textural hallmarks of magmatic crystallization (e.g., euhedral shape, fine

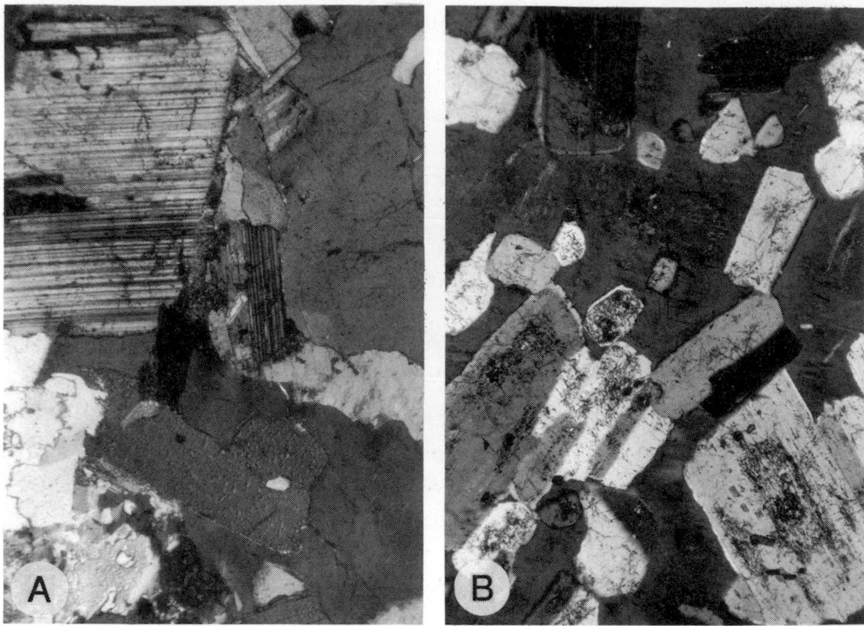

Fig. 4.42 (A): Granitic texture in granodiorite with subhedral plagioclase and biotite, anhedral quartz. Kosciusko Batholith, Australia. (B): Poikilitic K-feldspar (it occupies the entire view and encloses numerous subhedral plagioclases) in granite (Separation Point Batholith, NZ). (Lengths of view measure 3.3 mm (A) and (B).)

zoning patterns, and simple Carlsbad twins). Their supposed growth in the contact rocks around a granitoid can be reinterpreted in terms of porphyroclasts in the mylonitized marginal zone of a pluton, and many enclaves that contain large K-feldspars do so because they are comagmatic hybrids or mixed rocks caught up in the main body of the pluton. Finally, a few cautionary words about recognizing poikilitic K-feldspar. Some examples in thin section may appear as discrete, separate areas interstitial to the other phases. The fact that these areas form one large crystal (perhaps larger than the area of the thin section) may not be at all obvious until one appreciates their optical or twin continuity. Crystals this rich in inclusions may not be obvious in hand specimen either.

Granophyric texture, discussed in section 3.5.1, is common in hypersolvus granites subjacent to volcanic complexes. The loss of vapour through fractures is responsible for the raising of the liquidus/solidus curves with consequential increase in undercooling and rate of growth. The texture (Fig. 3.20) represents the rapid coupled growth of quartz and a hypersolvus alkali feldspar; this contrasts with the independent quartz and feldspar crystal growths typical of plutons with a low degree of undercooling.

(d) Secondary textures in granitic rocks

There are three main categories of secondary texture in granitic rocks:

1. Textures that evolved during cooling or later metamorphism without a major change of mineralogy.
2. Textures due to a change of mineralogy as a result of deuteric or later hydrothermal activity.
3. Textures that represent strain during deformation. In contrast to basic plutonics, strain effects are common in granitic rocks because they contain relatively ductile quartz and mica.

The first category includes perthite in K-feldspar, swapped-rim development at K-feldspar boundaries, myrmekite and the initiation and coarsening of cross-hatch twinning in microcline. All are discussed in detail in Chapter 3. Granitic rocks contain two feldspars and quartz, all of which have a low relief and grey interference colours in thin section and therefore the complex of primary and secondary textures can be very difficult for the beginning student to sort out. Secondary features can effectively obscure the primary features, and in fact granitic rocks are much more difficult to deal with than basic plutonics. My advice for the student is to develop a very systematic approach on two fronts. Firstly, list all the features that commonly develop (including plagioclase twinning and the deformational features of quartz, discussed below), then determine, one by one, whether those features are present. Secondly, it is vital to develop a sound microscope technique. Always look at a thin section with low magnification first, and always look carefully for slight differences in refractive index (learn to close the lower diaphragm very firmly when using plane polarized light). In particular, perthite and twinning are frequently confused but easily distinguished by the small but observable differences in RI.

Into the second category come such alterations as chloritization of biotite and hornblende, sericitization, saussuritization, and kaolinitization of feldspar. They reflect a ready availability of water under conditions equivalent to the greenschist or sub-greenschist facies. Further details about these alteration processes are given in Chapter 3; they are found in a wide variety of rocks, not just granitoids. It is perhaps worthwhile remembering here that sericitization and saussuritization preferentially affect the more An-rich zones in plagioclase; sodic rims and K-feldspar may remain relatively unaffected. Kaolinitization affects K-feldspar preferentially, and the products are usually so fine grained that thoroughly altered material takes on a grey, almost opaque, amorphous appearance in thin section. Of course, quartz usually stands out in thoroughly altered granitic rocks as the only remaining fresh material.

These common alterations involve the addition of water and a redistribution and/or removal of K, Na and Ca, and the changes may occur during cooling or a later metamorphic or hydrothermal event. Some granitic rocks undergo much greater chemical change, brought about by the action of residual magmatic fluids enriched in elements not found in the common rock-forming minerals. The elements include B, F, Li, Sn, etc., and the minerals formed include tourmaline, topaz, fluorite, Li-rich mica and tin minerals. The resulting rocks, 'greisen' and tourmalinized varieties, are found near granitoid contacts and in the vicinity of veins and pegmatites where the flow of rare elements was concentrated.

Greisen is the end product of Li and F action on granitic rocks, and it consists of Li-rich mica ± fluorite ± topaz + quartz. In transitions from granitoid to greisen, pseudomorphs of feldspar are found, but the extreme end products represent a complete textural and mineralogical reconstitution. Stout prisms of the borosilicate tourmaline in granitic rocks may be primary. However, granitic rocks develop acicular and radiating clusters under the action of late-stage B-rich fluids, capable in extreme cases of completely transforming a granitoid to a pure tourmaline–quartz rock (Fig. 4.43).

The third category of secondary texture involves strain effects. Secondary twinning in plagioclase and cross-hatch twinning in microcline represent some slight degree of strain, but otherwise feldspars (in common with pyroxene and amphibole) are not readily deformed plastically, especially under conditions of the greenschist or sub-greenschist facies. In contrast, micas are easily bent, and quartz behaves plastically over a wide range of conditions by various dislocation/glide mechanisms. Owing to the plasticity of quartz and mica relative to feldspar and pyroxene, granitoids deform very much more easily than gabbros, for example. The topic is discussed further in Part Three.

Many granitoids are deformed during or immediately after emplacement. Strain may develop as a granitoid rises diapirically, the effects being concentrated at the margins of the pluton (e.g., the Ardara pluton, Donegal – Pitcher and Berger, 1972), or it may simply be that the granitoid was emplaced in a tectonically active environment so that it naturally became involved in the regional deformation (e.g., the Main Granite, Donegal – Pitcher and Berger, 1972). Other strain effects may be imposed during completely independent post-emplacement tectonic events (e.g., the Tertiary–Recent mylonitization of Palaeozoic and Cretaceous granitoids along the Alpine Fault in New Zealand – Sibson *et al.*, 1979).

If strain is minor, then the only visible effects may be undulatory extinction in interstitial quartz, and minor bending of mica. Quartz rib-

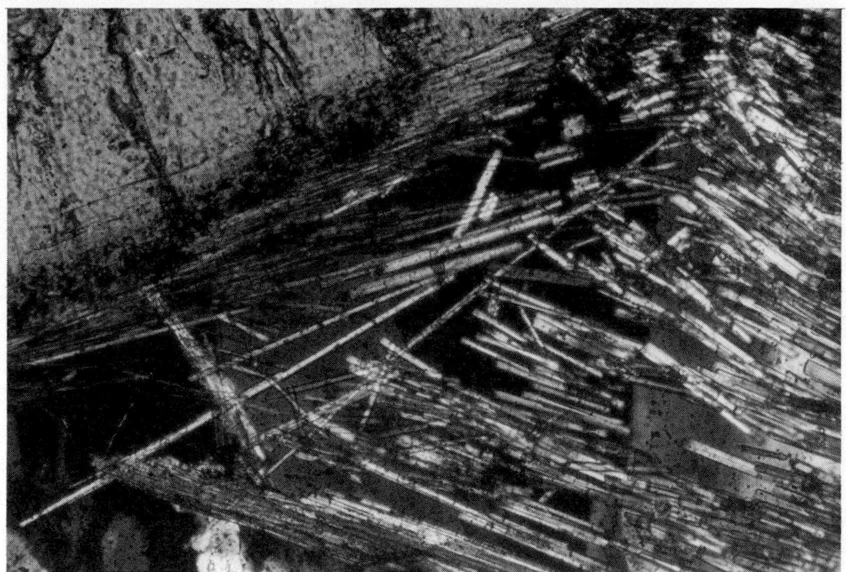

Fig. 4.43 Tourmaline–quartz rock, Cornwall, UK. In one corner is part of a stout crystal of tourmaline; the rest of the view is made of slender prisms of tourmaline set in quartz. (Length of view measures 3.3 mm.)

bons and fine-grained recrystallization of quartz characterize more-intense deformation, and elongate quartz zones become wrapped around resistant feldspars to form augen structure. At the same time, micas are severely bent and possibly smeared out along foliation planes (Figs 2.6 and 2.7). If the strain within mineral grains is obvious, severe and widespread, the rock is best termed a granite mylonite; if the strain effects have been annealed by recrystallization (Chapter 5), the rock may be best termed a gneissic granitoid, presuming the original igneous character of the rock is still preserved (usually as primary feldspar textures). If all (or most) primary textures are lost during recrystallization, the rock simply becomes a gneiss (an orthogneiss if the general geological evidence points to an igneous origin).

4.2.4 Primary and secondary granitic structures

(a) Primary foliation, schlieren and segregation layers

Laminar flow of magma orientates inequidimensional feldspars and micas at a small angle to the flow plane, and with long-axes more-or-less parallel to the flow direction (Fig. 4.44); the differential rates of flow (drag against

Fig. 4.44 Vertical alignment of K-feldspar megacrysts in granite due to primary magmatic flow. Hammer head is 125 mm long. Meybelle Bay Granite, NZ.

a magma chamber wall, for example) produce an imbrication of tablet-shaped crystals (Figs 8.3 and 8.4), as described by Blumenfeld and Bouchez (1988). Such alignments of crystals define a visible foliation in thin section and hand specimen, which will be discussed further in section 8.1. It must be remembered that foliation is a penetrative structure independent of phase layering; nevertheless the two structures often occur together.

Phase layering similar to that in basic layered complexes does occur in granitic rocks (Moore and Lockwood, 1973; Parsons, 1987 pp. 422–3), but it is much less common. Perhaps because of this, and perhaps because granitoids are more often deformed, one finds a somewhat different approach in the literature to granitic structures. For example, the word 'schlieren' (originally meaning a flaw in glass) is used most commonly in granitic rocks for any abnormality, be it a diffuse streak of xenolithic material or a well-defined phase layering. The mechanisms already discussed for layering in basic complexes need to be considered but, in addition, other ideas are commonly advocated for granitoids. One is that deformation of a crystal mush will segregate liquid from crystals, partly by the inevitable effects of dilatancy in any fracture or shear zone, and partly by the Bagnold effect (Fig. 4.22) which moves crystals away from zones of high differential shear (Komar, 1972). By this mechanism, a crystal mush shearing past a chamber wall or along a feeder dyke, or affected by tectonic shearing, will segregate into schlieren of relatively phenocryst-free and phenocryst-rich material. A case in point is the remarkable layering that pervades much of the Main Donegal Granite (Pitcher and Berger, 1972). The 100 mm-scale layers consist of alternating finer-grained dark trondhjemite and coarser light granite, the main difference being the respective absence and presence of microcline. Layering is steep and fairly constant in its NE–SW strike, regardless of position relative to the boundaries of the pluton. Pitcher and Berger (1972) invoke tectonic shearing of a crystal mush and segregation of K-feldspar-rich fluids into low-pressure areas within the shear zones. According to Hutton (1982), emplacement of the pluton and its shearing took place in a major sinistral fault zone, at the site of what is essentially a transtensional basin; doubtless the presence of hot granite would have led to increased local strain.

There is really no sharp boundary between what is a primary and a secondary foliation. The Main Donegal Granite layering illustrates this well because the deformation that sheared the primary crystal mush continued to shear the completely solidified pluton, imposing a secondary foliation on the layering; the two structures are generally more-or-less parallel but cross-cut in locations where the primary layering is at an angle to the main trend.

(b) Secondary foliation

The textural responses to strain that ultimately produce ribbon quartz and smeared-out micas define a secondary foliation, akin to that found in metamorphic rocks. The quartz and mica form a plastic medium in which rigid feldspars rotate into a secondary preferred orientation (Fig. 4.45). Secondary foliations may simply enhance an existing primary foliation, possibly during diapiric uprise of a pluton. It is usually easy to see in thin section that post-solidification strain has occurred; it is not always easy to judge that a primary fabric existed before that strain.

An example of foliation of supposed secondary origin due to inflation during diapirism is found in the Ardara pluton of Donegal (Holder, 1979). The pluton has a tonalitic outer zone, sharply bounded inwards by two

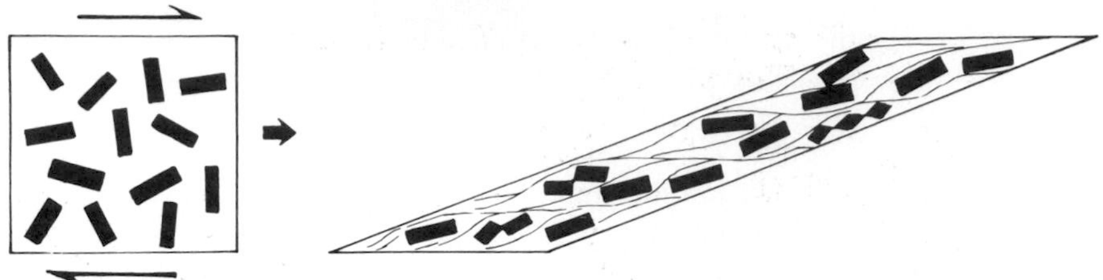

Fig. 4.45 Illustration to show randomly oriented K-feldspar megacrysts in granite oriented by strain (in this case simple shear) to define a secondary foliation. Plastic deformation and cataclasis of matrix quartz and mica also produce layers that are wrapped around the more rigid, sometimes broken, feldspars.

varieties of granodiorite. The already solidified outer zones of the pluton, together with the country rocks, had the new foliation imposed on them as the pluton rose; in the tonalite it is defined by an alignment of plagioclase and biotite. Enclaves are flattened in the plane of the foliation, and they were used by Holder (1979) to show that strain decreases systematically inwards from the margins of the pluton (Fig. 4.46); strain contours are strictly parallel to the margins of the pluton, and cut across lithological boundaries within it. This type of deformation is known as balloon tectonics because the strain is likened to that produced in the plastic skin of an inflated balloon.

An alternative interpretation of the Ardara pluton has been given by Paterson *et al.* (1989 and 1991). They believe that the preferred orientation of subhedral feldspar, and the general lack of evidence for high-temperature plastic deformation argues against balloon tectonics. Instead they suggest the foliation is due to some combination of magmatic flow and post-emplacement regional deformation. In their model, the enclaves represent deformed magma globules rather than xenoliths. The supposed balloon tectonics of the Papoose Flat pluton, California, are reinterpreted entirely in terms of post-emplacement regional tectonism (Paterson *et al.*, 1991). In addition, Schmeling *et al.* (1988) show that diapirs may develop strain simply due to ascent in the crust; inflation from within is not necessary. The concept of balloon tectonics is therefore under considerable attack, and may no longer be sustainable.

In mylonitized granitic rocks, the quartz and micas may be sufficiently strained and smeared out to produce a fine layering (Figs. 2.7A and D). Sometimes the strain is not evenly spread, but is concentrated in discrete shear zones. Shear zones in granitoids typically appear as dark layers in the field, and at first sight may be mistaken for phase layering of some kind. In fact the dark colour is due to their finer grain size, and on closer examination the tectonic features are usually clear, especially in thin

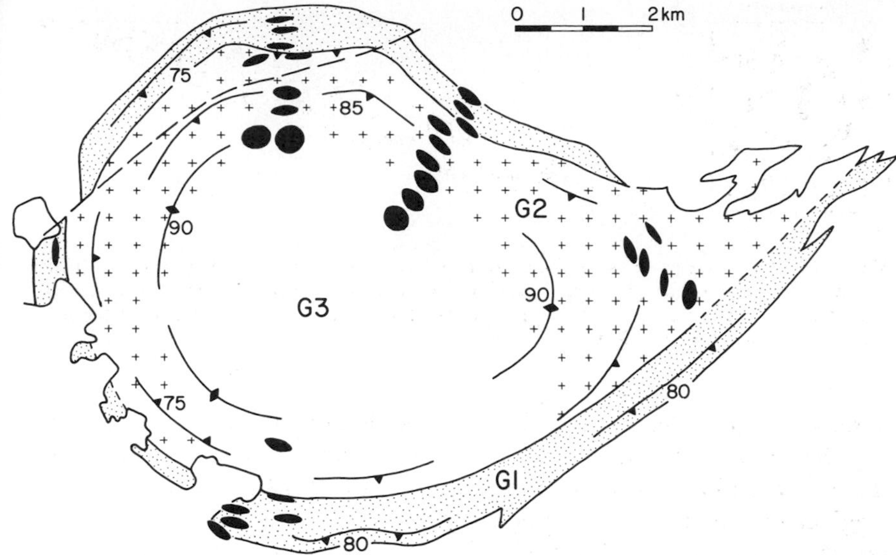

Fig. 4.46 Sketch map of Ardara pluton (Eire), subdivided into G1, G2 and G3, Ellipses represent the finite strain recorded by enclaves. Numbers give the dip of foliation. Reproduced with permission from Holder (1979).

section. In the field, the shear zones have distinctive boundary relationships with the less-deformed granitoid (Fig. 7.14).

(c) Orbicular structure

It is not always possible to draw a sharp line between what is a texture and what is a structure. Orbicular structures are ovoid bodies composed of concentric layers of radiating crystals, especially feldspar, and tangentially arranged crystals such as mica (Fig. 4.47); the structure usually grows around a xenolith. In many ways orbs resemble spherulitic texture, although orbs are generally many centimetres in diameter, too large for a standard-sized thin section. Size alone is not sufficient to class a feature as a structure rather than texture, because crescumulate refers to an arrangement of very large crystals in what is designated a texture. The criterion that finally determines an orb as a structure is the concentric shell-like layering rather than the radiating texture.

Crescumulate texture, spherulitic texture and orbs all reflect an initial difficulty in nucleating crystals followed by rapid outward growth without the development of equilibrium crystal shapes. In each case, some foreign substance, respectively the magma chamber wall, a phenocryst and a xenolith, acted as the centre of nucleation.

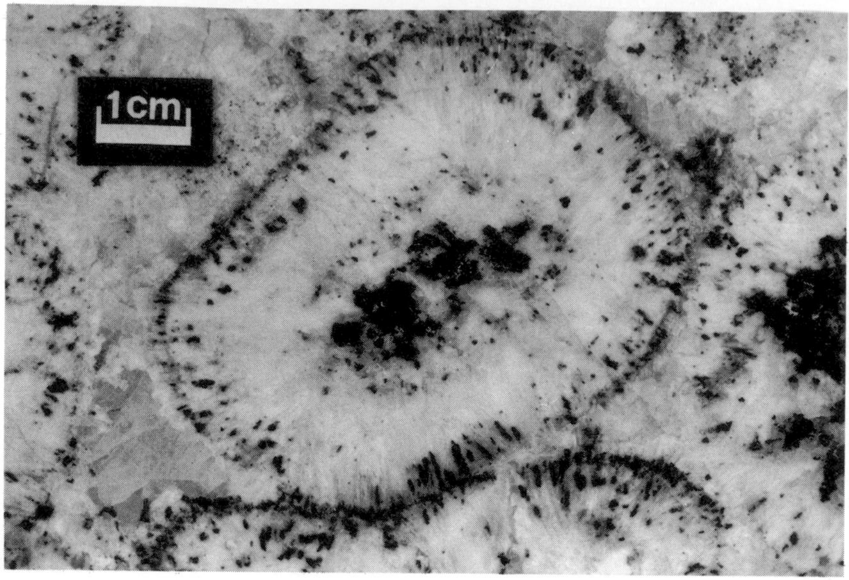

Fig. 4.47 Orbicular structure in granite from Karamea, New Zealand.

In some Californian plutons, crescumulate texture and orbs are so similar and in such close association that Moore and Lockwood (1973) believe that the only difference between them is the site of nucleation. These authors and Vernon (1985) both point to the following features: orbs are often mechanically sorted and concentrated, indicating the presence of currents in a fluid; orbs are often broken with more normal granitic material separating the fragments (some Californian orbs grow around broken crescumulate layering); orbs are most common near the margins of plutons, or in conduits; and the fluids that crystallize orbs lack pre-existing crystals, which suggests superheating; a water-rich fluid is advocated. Vernon favours an increase in water to superheat the magma and destroy any existing phenocrysts, and this might occur at the margins of a chamber by absorption from adjacent country rock. Other ways of superheating include injection of basalt into the granitic magma (although there is no evidence for this in connection with orb formation), or a rapid ascent of magma in a conduit. The non-equilibrium orb texture and radiating growth habit suggest that the superheated magma was rapidly undercooled, possibly as the result of sudden loss of volatiles released along a fracture system. The crystallization of orbs seems to require a rapid succession of events: superheating, probably due to rapid ascent in a conduit and/or absorption of water from the country rock, followed by release of volatiles, probably along fractures activated as the magma moved towards the surface. For these reasons, they are restricted to the marginal facies of plutons or minor intrusives.

The concentric layering of an orb is probably explained by an oscillatory supersaturation, nucleation, growth mechanism, such as advocated by McBirney and Noyes (1979) and discussed in section 4.2.2.

Orbs occur in rocks other than granitoids. Thus Symes *et al.* (1987) describe plagioclase, pyroxene and hornblende orbs in diorite and gabbro. They too are associated with crescumulate textured layering, they occur at the margins of the intrusion, form around xenoliths, and were frequently broken during and after formation.

4.2.5 Mixed rocks: magma mixing and xenoliths

In the discussion of volcanic rocks we have already seen abundant evidence for magma mixing. One can visualize a magma chamber or conduit system being periodically recharged with fresh magma, causing minor compositional changes in the case of one eruptive sequence in Hawaii (Helz, 1987), but more substantial changes where magma has already fractionated in the chamber (Nixon and Pearce, 1987). Some flow layering in volcanics is a direct sampling of two magmas erupted together. In a very real sense, volcanic rocks provide snapshot views of the process of magma mixing in subjacent chambers.

In the plutonic arena, the record of magma mixing is much more likely to have been erased by the passing of time. The great granitic batholiths of the world are generated by crustal melting in regions where there is an influx of heat from the underlying mantle, and naturally enough the situation is ripe for mixing basaltic and granitic magmas. However, the two magmas mix quite easily (Kouchi and Sunagawa, 1985) and, given time, the granite-derived phenocrysts are likely to undergo complete resorption, and the basalt-derived ones complete reaction, so that a record may not be preserved. Nevertheless, textural and structural features of magma mixing do survive, and some are briefly noted below. The question as to whether zoned or complexly layered plutons ultimately derive from magma mixing is beyond the scope of this book.

(a) Rapakivi and other textures indicating magma mixing

Rounding of feldspar phenocrysts in volcanic rocks, sieve textures and overgrowths have been discussed in section 4.1.6. They often represent magma mixing and the consequential resorption and/or reaction of feldspar as it undergoes re-equilibration with the hybrid liquid. Such textures are sometimes preserved in plutonic rocks, but in addition there is the rapakivi texture, characteristic of some granitic rocks, and consisting of crystals of K-feldspar, typically rounded, then overgrown coherently by plagioclase.

Much has been written on the subject, but the explanation that seems to be most appropriate involves magma mixing (Hibbard, 1981). In this scenario, K-feldspar, derived from the cooler more felsic magma, is rounded by solution as temperature rises due to mixing with more basic magma. The mixed magma is richer in plagioclase components, and the hybridized magma may have become undercooled with respect to those components, explaining the skeletal or dendritic base of some plagioclase overgrowths.

Other ideas on rapakivi are less promising. Thus the rounding of the K-feldspar and the occurrence of repeated overgrowths in some examples, cannot be explained by exsolution, and although swapped-rim perthite may superficially look like rapakivi, swapped rims have an incoherent relationship with host K-feldspar. Neither can the simple idea that plagioclase occasionally nucleates on K-feldspar explain the widespread rapakivi texture in some granitoids, nor does it explain the rounding of K-feldspar or the cessation of K-feldspar growth. And although syenites and trachytes may display successive resorption and growth of various Na-rich and K-rich feldspars due to complex reaction relationships in the Or–Ab–An system, Nekvasil (1990) states that in the water- and silica-rich granite system, resorption is precluded except at very low pressures.

In general, if magmas mix, phenocrysts from the more basic magma, that continue to grow, will do so rapidly in skeletal or dendritic form because the basic magma is effectively quenched by felsic magma. An example of this is given by Sutcliffe *et al.* (1990) who describe skeletal hornblende derived from the more mafic magma in a magma-mixing situation. At the same time, phenocrysts from the felsic magma will become rounded by solution as the felsic magma is heated during the mixing event. Such rounded hornblendes in hybrid tonalite are also described by Sutcliffe *et al.* (1990).

(b) Net-vein complexes

Direct evidence of magma mixing is found in net-vein complexes. An Archaean example is described by Sutcliffe *et al.* (1990), although the best-known examples come from intrusive complexes subjacent to Tertiary–Recent North Atlantic volcanoes in Iceland, Ireland, and Scotland (e.g., Blake *et al.*, 1965). Mafic magma is cooled by felsic magma, and so it solidifies rapidly, and may even be chilled against the felsic material. On the other hand, felsic material is heated by mafic magma to become increasingly fluid. The result is a cracking of solid mafic material and the

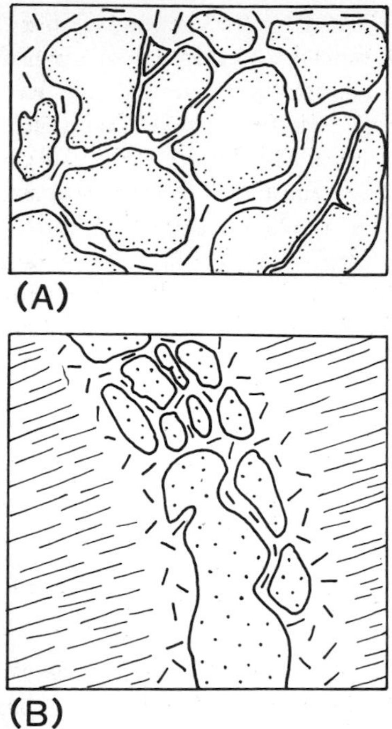

Fig. 4.48 (A): Schematic view of a net-vein complex with basic pillows, some with chilled margins, set in and veined by acidic material, as described by Blake *et al.* (1965). Pillows are *c.* 1 m across. (B): Foliated tonalite remobilized by a basic dyke (*c.* 20 cm across) which formed pillows in the remobilized tonalite. Foliation was destroyed during remobilization. Based on photograph in Sutcliffe *et al.* (1990).

infilling of the fractures by a complex of felsic veins (Fig. 4.48A). In some cases, as mafic material is intruded into increasingly fluid felsic material it forms pillows. This happened in the Archaean example of Sutcliffe *et al.* (1990), and one consequence of the host tonalite flowing between pillows was the loss of its pre-existing foliation (Fig. 4.48B). Occasionally, the fluid felsic material forms globules within the more-viscous mafic material. Thus Bussell (1985) describes subspherical ocelli made of various mixtures of quartz, K-feldspar, plagioclase and hornblende, in which the plagioclase and hornblende grew radially inwards from the margin of the ocelli.

(c) Xenoliths

Many dark xenoliths in granitic plutons consist of mafic igneous material, and Hibbard (1981) cites this as supporting evidence for magma mixing as a cause of rapakivi texture (although such xenoliths are rather more common than rapakivi texture). Vernon *et al.* (1988) describe mafic igneous enclaves from a net-vein complex of tonalite and gabbroic diorite in Australia; the net-veining provides direct evidence of the magma mixing, and it can be seen that mafic magma became quenched and incorporated as globules in the more fluid felsic magma. The globules have lobate and serrated boundaries with wispy protrusions, suggesting magma mingling rather than thorough mixing. The enclaves contain zoned plagioclase, hornblende, and a little biotite, and in areas where the enclaves are strongly elongated (possibly in conduits) the minerals have a strong preferred orientation in the direction of flow. The fact that the crystals could rotate into such a preferred orientation indicates the enclaves were >50% liquid at the time of elongation. There is no evidence of solid-state plastic deformation, and obviously these xenoliths cannot be used as tectonic strain markers.

The textural evidence for magma mixing is also strong. Thus in a tonalite from California (Dorais *et al.*, 1990), enclaves contain fine-grained strongly zoned plagioclase with cores (An_{85-65}) representing the basaltic material and rims (An_{55-45}) the hybridized material. The rims sometimes enclose hornblende which also occurs in the tonalite, and both plagioclase and hornblende may be poikilitically enclosed by quartz or K-feldspar. The enclaves are thought to represent the quenching, hybridization and mixing of basalt intruded into the base of a tonalitic magma chamber (Fig. 4.49). An alternative explanation is that the enclaves represent restite material, but Dorais *et al.* (1990) argue that the strong zoning disproves that; restite plagioclase should be unzoned if it represents the end product of protracted metamorphism which led to melting in the lower parts of the crust.

Restite material is amply represented in S-type granitic rocks by the dark pelitic xenoliths representing the more refractory relics of partially melted sediments. Other pelitic xenoliths may represent stoped material prised off the chamber walls and roof as magma was intruded. Extraction of granite melts from pelitic material increases the Al, Mg, and Fe in the residue, and water is lost to the melt, so composition and mineralogy (sillimanite and cordierite are to be expected) may have changed from the original metasediment. Melting of stoped pelitic material is unlikely, and neither is restite material likely to melt once it is incorporated within a rising magma. However, minerals will react with or dissolve in the melt if they are not in equilibrium with it. Thus xenolithic plagioclase will change in composition towards that of the host granitoid plagioclase, hornblende may change to biotite, and at a late stage, anhydrous Al-rich minerals like sillimanite may change to sheet-silicates. If the degree of assimilation or re-equilibration is high, the xenoliths gradually lose their identity, become diffuse, and merge into the background granitoid. All degrees of

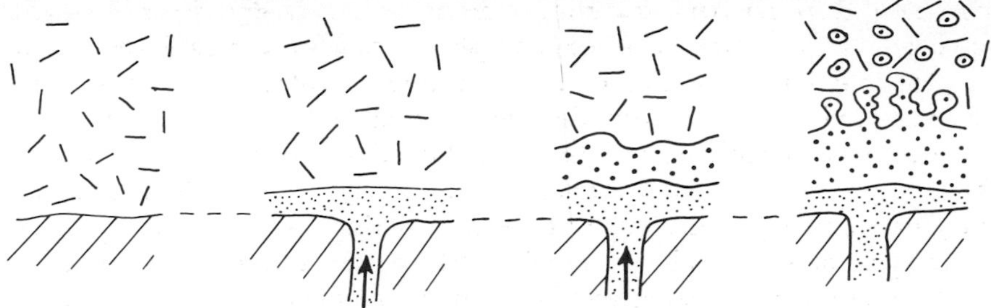

Fig. 4.49 Formation of basic enclaves, as described by Dorais *et al.* (1990). From left to right: base of tonalite magma chember; basalt intruded at base of tonalite; hybrid zone develops; hybrid zone disaggregates to produce magma globules (ultimately basic enclaves).

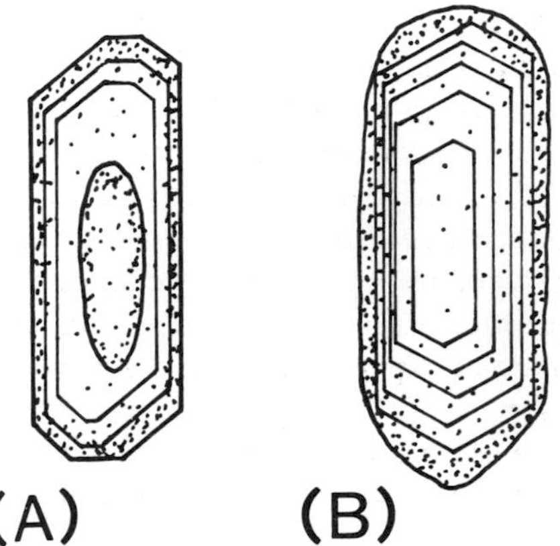

Fig. 4.50 Schematic representation of some zircon morphology types in granitoids. (A): A rounded core is overgrown by a euhedral multiply zoned crystal. (B): A rounded crystal displays multiple internal euhedral zones. The euhedral zones indicate igneous or metamorphic crystallization. Rounded crystals may either be relic sedimentary grains or a consequence of magmatic solution.

assimilation are commonly visible within any one granitoid, and it is usually difficult to determine how much xenolithic material was originally present.

The incorporation, possible disaggregation and/or assimilation of xenoliths imply that xenocrysts could also have been incorporated or left as a refractory or non-reacting residue. Rounded zircons (Fig. 4.50) have been interpreted as possible sedimentary relics, but the rounding could also result from solution in magma (cf. Figs 4.17 and 4.18). Despite difficulties in interpretation, zircons are important; potentially they

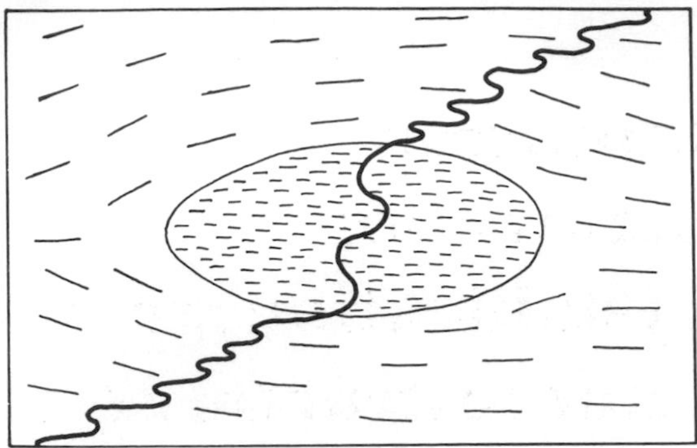

Fig. 4.51 Schematic diagram to show straining of an igneous rock and a relatively competent enclave within it. Buckling of a cross-cutting vein reveals the difference in finite strain of host and enclave.

provide information about the source rock, especially using U–Pb dating to date the source rock or elucidate stages in the rock's evolution.

Granitic rocks deform plastically during tectonism, and xenolithic material can provide a useful measure of tectonic strain. Examples from the Donegal Granites (Hutton, 1982; Pitcher and Berger, 1972) illustrate the effects of the Caledonian Orogeny on granitic plutons in NW Ireland, but it must also be remembered that the example of enclave strain (Fig. 4.46) in the Ardara pluton, Donegal, may be the result of magmatic flow and/or diapiric ascent rather than tectonism. Features such as deformed veins that extend from the xenolith through the host rock can be examined to determine strain contrasts between xenoliths and host granitoid (Fig. 4.51). Textural evidence of subsolidus plastic or brittle deformation usually confirms the tectonic character of strain in pelitic xenoliths (cf. the strain due to magmatic flow in igneous enclaves that were deformed as magma globules).

5

Metamorphic crystallization, recrystallization, textures and microstructures

The origins of crystals in metamorphic rocks can be very diverse indeed, and all of the following need to be considered:

1. Relics from the sedimentary or igneous precursor.
2. The growth of new phases during prograde metamorphism.
3. The products of retrograde metamorphic reactions (some of which have been described in section 3.5.4).
4. Products of unmixing during cooling, as already described in section 3.5.3.
5. The modification of an existing phase by recrystallization.
6. The products of post-metamorphic alteration (e.g., weathering to clay minerals).

A regional metamorphic event typically lasts of the order of 10^6–10^7 years, but metamorphism may be superimposed on metamorphism in complex polymetamorphic sequences that span immense periods of the order of 10^9 years. Not only may the relics of the premetamorphic rock survive, but so may relics of the various stages of complex polymetamorphisms. Some or all of the relics or metamorphic products may undergo recrystallization, and at some (or every) stage of metamorphism the rock and its constituent crystals may be severely strained. One of the more interesting and rewarding aspects of thin-section petrology is the successful unravelling of complex metamorphic histories by way of textural studies. If it is essential for the igneous petrologist to make use of the microscope before embarking on geochemistry, for the metamorphic petrologist it is doubly so.

Below we examine, in turn, prograde and retrograde crystallization, processes of recrystallization and deformation, the formation of metamorphic layering, structures that indicate the sense of shear during deformation and finally the ways in which complex metamorphic histories can be deciphered.

5.1 METAMORPHIC CRYSTALLIZATION

5.1.1 Prograde metamorphic crystallization

Crystallization of a new phase in a metamorphic rock usually signifies reaction involving a pre-existing unstable mineral assemblage. However, the mere presence of an unstable assemblage does not guarantee reaction; the fact that high-pressure and/or high-temperature rocks occur naturally at the surface of the Earth proves that. And neither do all polymorphic transformations occur spontaneously, as attested by the survival of diamonds, for example, well out of their stability field. Except in the special case of polymorphic transformations, dealt with later, the following steps are required for reaction to occur: detachment of material usually from the surface of unstable reacting minerals; diffusion of reaction products to sites of new mineral growth; nucleation of the new mineral; growth of the new mineral. Whichever step is the slowest is rate-limiting; for example, if diffusion were very slow or impossible then it would limit the reaction, and even prevent it happening at all, despite the mineral assemblage being unstable and despite potentially easy nucleation and growth. Prograde reactions involving the loss of volatiles are endothermic except possibly at extreme pressures, and this means that prograde metamorphism cannot occur unless the heat flow is in excess of that required to balance the heat loss during reaction. In other words, heat-flow rates can also be a limiting factor in reactions. It is not always easy to decide what the rate-limiting factor is.

In metamorphic as in igneous rocks, perfect equilibrium crystallization never occurs. We already know that crystallization from magma requires a degree of undercooling, and disequilibrium is reflected in crystal shape and compositional zoning. In metamorphic rocks, crystallization is more protracted than in most igneous rocks, but overstepping of the ideal boundary conditions is still necessary before reaction takes place.

Examples of common textures will be used below to illustrate further the processes involved in metamorphic crystallization, and to show how decisions on rates and rate-limiting factors might be made.

(a) Garnet- and mica-bearing pelitic schists

Figure 5.1 is a typical thin-section view of a greenschist-facies garnet schist. General characteristics include:

1. Schist chemical composition is similar to that of the mudstone precursor. Apart from the obvious volatile loss, most workers have

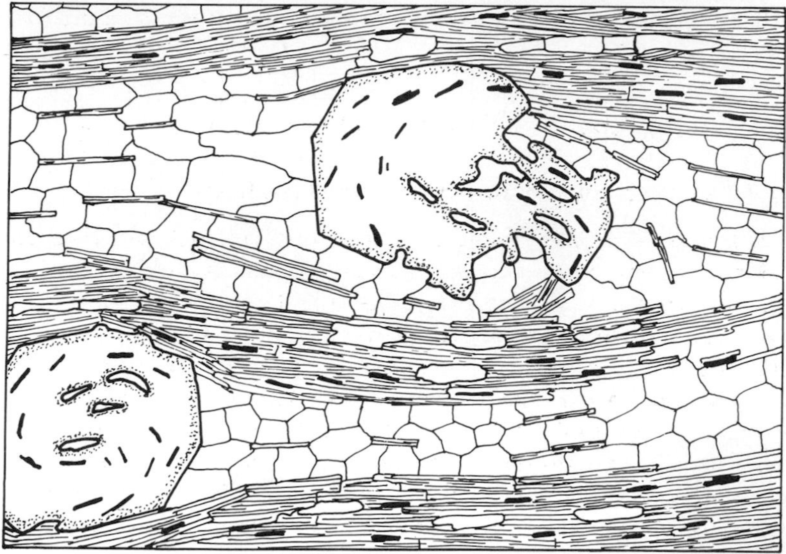

Fig. 5.1 Illustration of some common textural features in pelitic garnet schists, as discussed in text. Based on specimens of Pikikiruna Schist, NZ. The large garnet is approximately 1.6 mm long.

regarded metamorphism as more-or-less isochemical, although this view has been challenged by Bell and Cuff (1989).

2. Individual grains of garnet, mica, quartz and feldspar occupy spaces previously filled by the finer-grained mixture of clays, chlorite, micas and quartz; in other words, the increase in grain size and change of mineralogy represents chemical segregation and diffusion at least on the scale of the spacing of the new minerals.
3. Garnets and micas are euhedral or subhedral, contrasting with the mosaics of anhedral quartz and feldspar.
4. Sheet-silicate preferred orientation is obvious; there may also be less-obvious quartz and feldspar preferred orientations.
5. Garnets contain S-shaped trails of quartz and opaque inclusions, relics which provide clues to the textural and structural development of the schist.
6. Garnets tend to occur in relatively quartz-rich layers. Do these layers represent primary compositional variations, or are they the result of metamorphic segregation?
7. Most garnets in greenschist- and amphibolite-facies schists display concentric compositional zoning (Fig. 5.2); it is usually invisible in thin section and requires electron microprobe study (Atherton and Edmonds, 1966).

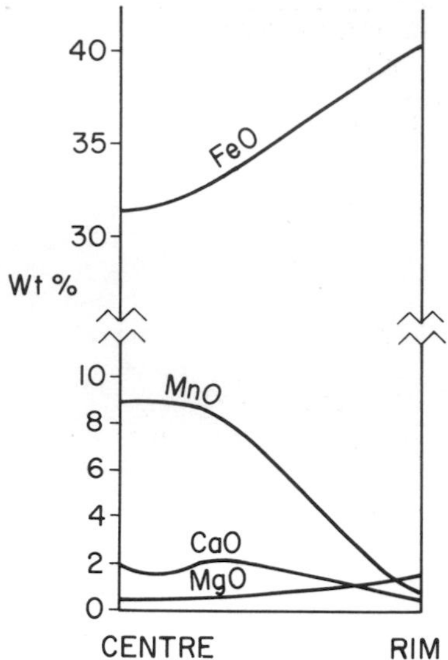

Fig. 5.2 Concentric zoning of garnet, based on electron microprobe data of Carlson (1989). The lines represent typical examples of the variation in four oxides from crystal centre to crystal rim for almandine-rich garnet in pelitic schist.

8. A typical crystal-size distribution curve for porphyroblasts in a regionally metamorphosed rock is bell shaped (Fig. 5.3). Grains smaller than a certain minimum finite size do not exist, and the population density maximum is usually on the fine-grained side of mean grain size (Kretz, 1966; Shelley, 1977; Dougan, 1983).

These features all require explanation. The question of metamorphic layering is pursued in section 5.3, inclusion trails are discussed in section 5.5, and mineral preferred-orientations in Part Three; the other matters are discussed here. Ideally, we should be able to determine for each mineral the reactions responsible, the rates of reaction, diffusion, nucleation and growth for each mineral, and the rate-limiting factors. Complete answers to all these questions are not yet available, but some of the current views and approaches are outlined below.

Numerous prograde metamorphic reactions release fluids, and fluids provide a quick means of transporting reaction products along grain boundaries. Bell and Cuff (1989) point out that expanded sheet-silicate interlayer space, saturated with water, may allow equally fast rates of diffusion. Whether or not sheet-silicate-assisted diffusion or fluid flow was the mechanism, it is generally acknowledged that growth of dispersed phases (such as the garnet) in low–moderate grade rocks depends on

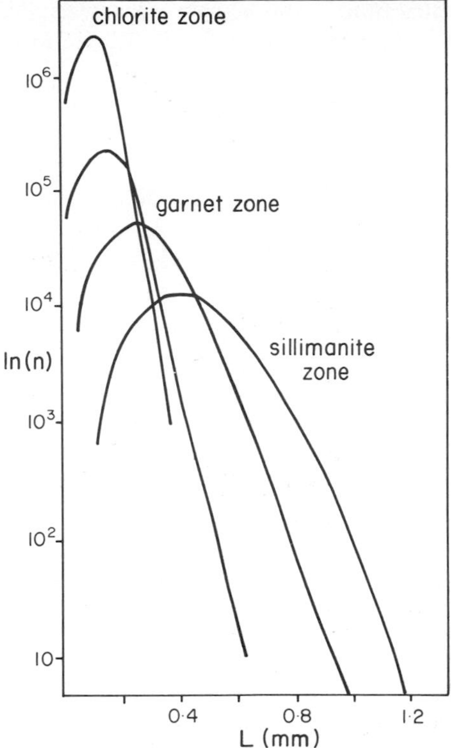

Fig. 5.3 Typical bell-shaped metamorphic CSDs. The four curves show the variation in CSD from chlorite through sillimanite zones in pelitic schists from Maine, USA. L = crystal length, n = population density. Discussion in text. Reproduced with permission from Cashman and Ferry (1988).

efficient diffusion assisted by water. The alternative of solid-state diffusion, either along grain boundaries or within the body of a silicate crystal, is very slow indeed except at very high temperatures. Indeed, survival of compositional zoning in garnet through the amphibolite facies shows that such diffusion has a negligible effect. Only at very high grades ($T > 640°C$) does solid-state diffusion become efficient enough to re-equilibrate and homogenize garnets (Yardley, 1977), although Spear *et al.* (1990) point out that homogenization is size dependent, smaller grains being more easily affected.

Concentric compositional zoning in garnet is possibly in part the result of Rayleigh fractionation (Hollister, 1966), but more generally it can be ascribed to the differential partitioning of elements between coexisting minerals in continuous reactions such as:

$$\text{muscovite} + \text{chlorite} = \text{biotite} + \text{garnet} + \text{quartz} + H_2O$$

The minerals chlorite, biotite and garnet all contain Mg and Fe, but in different proportions. It is found that the Fe/Mg ratio in garnet is always higher than in coexisting biotite, which in turn has a higher ratio than chlorite. Of course, actual amounts of Fe and Mg depend on bulk composition of the original mudstone. In the continuous reaction above, chlorite preferentially loses Fe as the biotite and garnet are produced, and it becomes increasingly Mg rich with increasing grade. As more chlorite is consumed, the biotite and garnet are forced to accept increasing amounts of Mg, but the garnet is always more Fe rich than the biotite. Such continuous reactions explain why the Barrovian biotite zone does not signal the elimination of chlorite which usually survives right through the greenschist facies, and of course they are the basis of the geothermometry and geobarometry employed by workers such as Spear *et al.* (1990).

Two principal factors explain the euhedral or subhedral habit of garnet and mica. Firstly, growth rates must be relatively slow compared with diffusion rates, otherwise dendritic or some other non-equilibrium growth form would result. Secondly, if two crystals share a common boundary, it is energetically most efficient for the crystal with the highest surface energy to have well-developed flat crystal faces. In general, denser minerals have a greater tendency to be euhedral, so that dispersed grains of oxides or sulphides are more commonly euhedral than Fe–Mg minerals, which in turn tend to be more euhedral than dispersed grains of quartz and feldspar. Minerals listed by Spry (1969) in order of their estimated surface energies include (from high to low energy): spinel, muscovite, pyrite, rutile, apatite, hematite, garnet, pyroxene, hornblende, quartz, plagioclase, olivine, calcite, ilmenite and orthoclase. One can see exceptions to the more dense the more euhedral relationship: for example, muscovite, not a very dense mineral, is nevertheless strongly anisotropic with a dense packing of atoms in the {001} plane, so that the tendency to develop planar {001} faces is very strong indeed. Another factor in crystal shape is the fact that the lowest energy faces of an individual crystal prevail as growth proceeds. For example, a mica crystal has a higher concentration of atoms (hence lower surface energy) along {001} than other faces. Growth proceeds most quickly at sites of highest surface energy, so growth on {001} itself is relatively slow (Fig. 5.4), and the eventual result is that the slowest-growing face dominates.

Quartz displays no tendency to develop euhedral crystals in most metamorphic rocks, and mosaics of anhedral quartz may simply represent the mutual interference of quartz against quartz. Figure 2.22 illustrates the typical controlling effect that inequidimensional euhedral mica exerts over quartz mosaic shape. Thus the stable boundary configuration where two quartz grains meet {001} of mica is a 'T' junction, and where two micas are more closely spaced than the average quartz grain size, elongate quartz

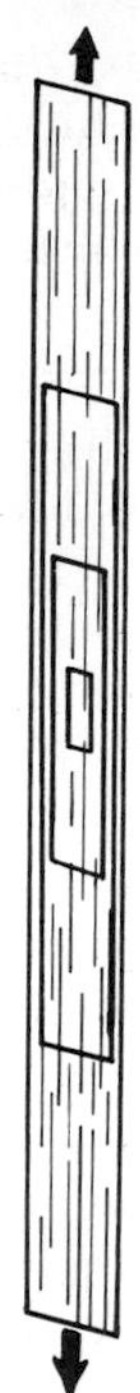

Fig. 5.4 Stages of growth of a mica crystal to illustrate why crystal shape is dominated by slow-growing (in this case {001}) rather than fast-growing faces. Arrows indicate the fastest growth directions.

grains develop. Feldspar forms similar mosaics, either by itself or in combination with quartz, but it may also be euhedral or subhedral when present as a dispersed phase (Fig. 7.24 and Shelley 1989a). Many mosaics, perhaps the majority, represent secondary recrystallization rather than primary growth, and this is discussed later. Interestingly, mica aggregates do not form polygonal mosaics like quartz and feldspar, and {001} continues to dominate (Figs 5.1, 7.8 and 7.13B); this again reflects the very strong anisotropy of mica and its control over crystal form even where grains mutually interfere.

Reaction rates when fluids are present have been studied experimentally by Walther and Wood (1984, 1986). Most importantly they found that rates increase rapidly and more-or-less exponentially with temperature. At $T > 300°C$, dissolution rates for nearly all minerals in H_2O-CO_2 fluids are similar and exhibit similar increases with T. Reactions do not occur without overstepping, so that reaction-rates also depend on the rate of temperature increase. Walther and Wood (1984, 1986) calculated that the time necessary for a 0.1 cm radius crystal to react or to grow completely at moderate metamorphic temperatures during a regional metamorphic event is of the order of 10^3–10^4 years; minimal overstepping is required (<1°C). They assumed that diffusion was not a rate-limiting factor, but Carlson (1989), in a detailed study of garnet porphyroblast growth, concluded otherwise; I discuss this below. The figures of Walther and Wood suggest that prograde reactions can be driven to completion very quickly once the appropriate conditions are reached; in terms of the total duration of a metamorphic cycle (10^6–10^7 years), crystallization may be relatively short lived.

Nucleation is not usually a rate-limiting factor during metamorphism because the great diversity of structure within pre-existing grain boundaries provides ample opportunity for heterogeneous nucleation of new phases. A potential site for nucleation may simply be some completely incidental grain boundary orientation, it may be the site of a lattice dislocation, or an inclusion, but some minerals are heterogeneously nucleated by epitaxy, as, for example, garnet or sillimanite on pre-existing sheet-silicates (Powell, 1966; Kerrick, 1987). Evidence of epitaxis may be lost if the host mineral is subsequently consumed by reaction. If garnets do nucleate on sheet-silicates and/or dislocations, it is hard to imagine any time during regional metamorphism when garnet nucleation would not be possible.

(b) Interpretation of bell-shaped crystal-size distribution (CSD) curves

We have already seen in section 4.1.2 that nucleation and growth rates can be studied, possibly quantified, using crystal-size distribution (CSD) curves. Volcanic rock CSDs show ever-increasing numbers of finer grains (Fig. 4.2), and this can be interpreted in terms of continuous nucleation and growth until crystallization ceased. Typical bell-shaped CSD curves for regional metamorphic porphyroblasts are quite different, and Kretz (1966) interpreted them in terms of fluctuating nucleation rates that increase at first, reach a maximum corresponding with the peak in the CSD curve, then decrease. The idea is taken further by Carlson (1989) who documents size, spacing and compositional zoning of garnet porphyroblasts in several schists. He considers the following two scenarios (Fig. 5.5): 1. There was no restriction on the supply of fluids and reaction products to the nucleation and growth sites, and the rate-limiting step was the actual rate of reaction, or 2. diffusive transfer was limited so that a zone develops around the site of nucleation and growth where fluids are depleted in reaction products. In the first case, there is no restriction on further nucleation close to the existing crystal; in the second case nucleation is suppressed. Carlson discusses the relationships expected between crystal radius and spacing for both scenarios (Fig. 5.6). He

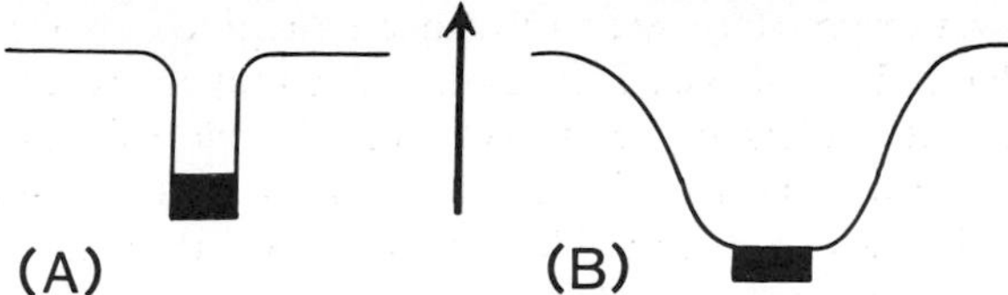

Fig. 5.5 Lines represent typical concentration gradients (concentrations increase in direction of arrow) around growing porphyroblasts (black). (A): Interface reaction or heat-flow controlled growth. (B): Diffusion-controlled growth. Based on Carlson (1989).

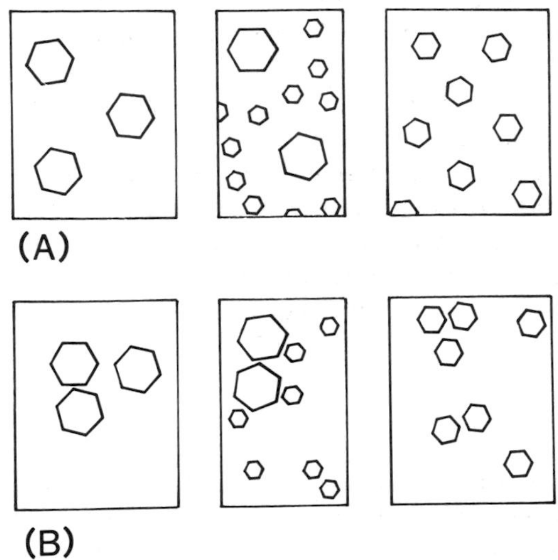

Fig. 5.6 Schematic relationships between crystal radius and spacing. (A): Ordered arrangements if diffusion rates control growth. (B): Non-regular arrangements if interface reaction or heat-flow rates control growth.

believes that his data are best explained by nucleation suppression, in other words with diffusion as the rate-limiting factor. The matrix around the new crystal becomes depleted in reactants, whatever the rate-limiting factor is.

(i) Ostwald ripening

An alternative interpretation of bell-shaped CSD curves is provided by Cashman and Ferry (1988). Building on the work of Marsh (1988) and Cashman and Marsh (1988), they find that high-temperature hornfelses have log-linear CSD curves, similar to volcanic rocks. They propose that the familiar bell-shaped CSDs for regional metamorphic porphyroblasts

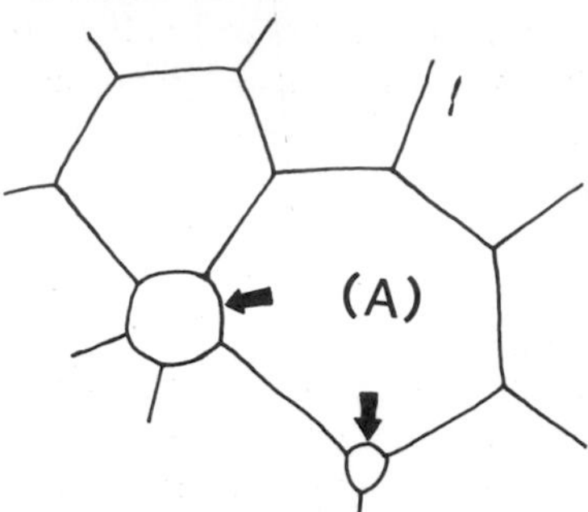

Fig. 5.7 Process of Ostwald ripening. The curved grain boundaries of large grains, such as (A), migrate towards their centres of curvature (arrows) consuming the smallest grains in the process. In contrast to this behaviour of static recrystallization, grain boundaries migrate away from their centres of curvature during dynamic recrystallization (e.g., Fig. 5.25).

are the result of Ostwald ripening, which is the process of annealing (or textural maturing) of a dispersed phase. Ostwald ripening represents an approach towards equilibrium which in any specified rock volume would ideally involve only one crystal of each mineral. Perfect textural equilibrium is never achieved, but it can be approached by grain coarsening and the development of an even grain size (a similar process has already been discussed in the context of layered igneous rocks – section 4.2.2). The idea can be illustrated with the aid of Fig. 5.7. At any one time, only one crystal size is truly in equilibrium with its immediate surroundings. Consequently, the matrix is supersaturated with respect to the larger grains, so these grow larger, and it is undersaturated with respect to the smaller grains which grow smaller and may eventually disappear. Note that smaller grains have boundaries that bulge outwards, and that grain boundaries migrate towards their centres of curvature. As Ostwald ripening proceeds, the size of the crystal in local textural equilibrium increases. Cashman and Ferry (1988) calculate that up to 60% of the garnet mass may be transferred in this way, and they believe that partially resorbed crystals of garnet will retain euhedral shapes because of their high surface energy (the same reason that they grow with euhedral shape). The finer resorbed material nucleated late and has the same composition as the outer zones of larger crystals. Therefore its addition to the larger crystals need not significantly change compositional profiles.

To support their idea, Cashman and Ferry (1988) publish bell-shaped CSD curves for garnets from chlorite through the sillimanite zone (Fig. 5.3). The curves match those obtained during laboratory annealing experiments, and demonstrate a gradual grain-size increase with grade, a gradual increase in the range of grain size, and a population-density peak that migrates gradually towards coarser grain size. Cashman and Ferry (1988) maintain that the essentially linear slope of the right-hand side of the curves is still a good measure of growth and nucleation rates, despite the substantial transfer of mass to the larger grains. Using the growth rate figures of Walther and Wood (1984, 1986) they calculate minimum growth times for garnet that range from <100–40 000 years (they do not change systematically with metamorphic grade), and nucleation rates up to $1.78 \times 10^{-4}\,cm^{-3}\,sec^{-1}$ in the chlorite zone, and $4.22 \times 10^{-6}\,cm^{-3}\,sec^{-1}$ in the sillimanite zone. They conclude that relative nucleation rates are: high-temperature contact metamorphism > low-grade regional metamorphism > high-grade regional metamorphism. In contrast to Carlson (1989), they maintain that consistent development of garnet zoning patterns within any one rock is evidence that diffusion was not a rate-limiting factor or, in other words, the appropriate components for garnet growth were always available to any growth site.

(ii) Minimum grain sizes

One of the crucial points to explain in bell-shaped CSDs is the minimum grain size. Most workers agree that it is real and not an artefact of microscope resolution, although Hunter (1987) notes how fine-grained material with relatively low surface energies becomes concentrated at mosaic grain corners, effectively disappearing from most thin-section views. Garnet, however, has a high surface energy and should not behave in this way. At any one time, Ostwald ripening may remove the very smallest of grains, but at the same time it will reduce the next grain size up to the very smallest possible size by resorption. Thus a complete, continuous spectrum of grain size might be expected at all times. Why therefore a minimum grain size? In the Carlson (1989) type explanation of CSDs, the minimum grain size implies that growth continued after nucleation ceased, and Carlson suggests that this happens when depleted zones (Fig. 5.5) overlap to such an extent that they cover the entire rock volume; this stops nucleation during the dying stages of growth. The heat balance during reaction could also provide an answer (Ridley, 1985). Fig. 5.8 shows temperature increasing during prograde metamorphism and an overstepping of the equilibrium temperature before nucleation starts. As nucleation rates increase, so does the total amount of growth, even if the growth rate is constant; the resulting heat consumption (endothermic reactions) may lower T. Nucleation eventually ceases, but growth on existing nuclei continues, possibly until all reaction is complete.

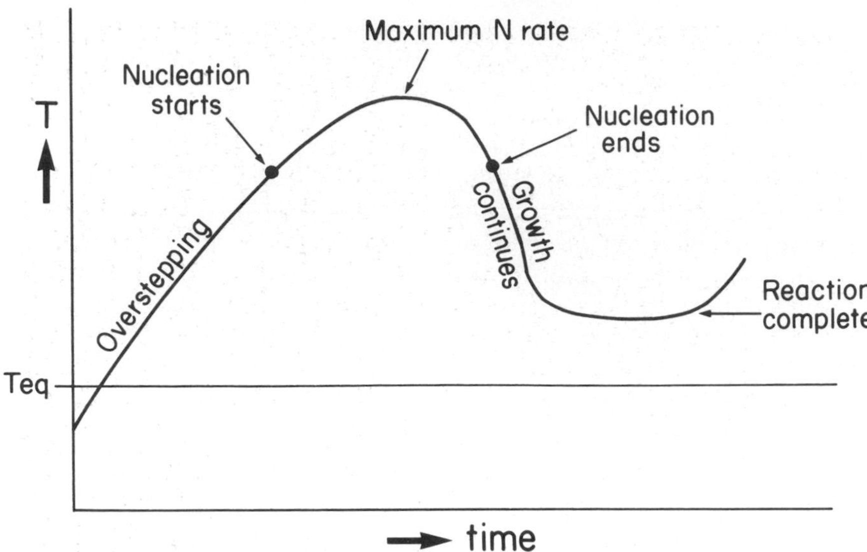

Fig. 5.8 Possible temperature–time history of a strongly endothermic prograde discontinuous reaction. See text for explanation. Reproduced with permission from Ridley (1985).

(c) Very low-grade metamorphic crystallization

Below 300°C the dissolution of silicate minerals becomes slow and variable in rate. Thus Walther and Wood (1986) show that between 25°C and 300°C, dissolution rates vary by up to five orders of magnitude, depending on the pH of the fluid (solubility increases with either very high or very low pH) and the entropy of the mineral per unit volume (the higher the ratio the greater the solubility). The slower and more variable reaction rates are reflected in the incipient and irregular nature of much very low-grade metamorphism. Thus premetamorphic relics are common, often predominant, and new mineral growths tend to be fine grained and severely localized, either in the fine matrix of a metasediment (Fig. 2.4B), as a replacement of particularly unstable precursors such as feldspar and glass (Fig. 2.3A) and in fracture systems (Fig. 2.4A).

Development of very low-grade minerals is irregular, not only on the thin-section scale but on the regional scale. The pioneering work of Coombs (1954) on very low-grade metamorphism in New Zealand caused workers around the world to search for similar effects elsewhere, but often the characteristic Ca–Al-silicate mineral assemblages are absent. The principal reason seems to be that the smallest quantity of CO_2 in the fluids suppresses the formation of laumontite, prehnite and pumpellyite (Liou *et al.*, 1987); carbonates develop instead. In New Zealand, Houghton

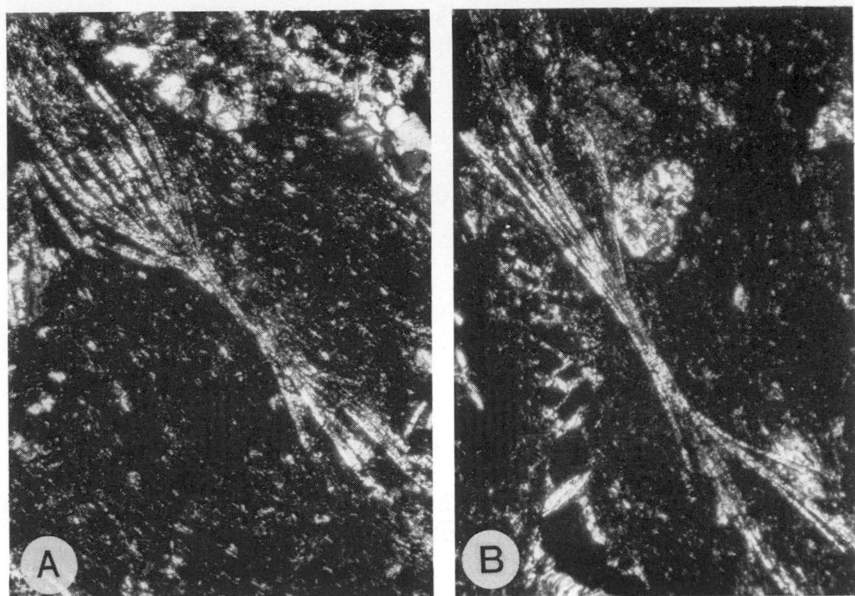

Fig. 5.9 Two examples of lawsonite with bow-tie texture in very low-grade deformed carbonaceous sediments, Nelson, NZ. (View lengths measure 0.79 mm (A) and (B).)

(1982) describes the different mineral assemblages found in shear zones or fractures and the host rock, and he suggests that the reason was a difference between the hydrostatic water pressure (in the fractures) and lithostatic water pressure (in the host rock). Again in New Zealand, Boles and Coombs (1977) note that heulandite develops preferentially in glass where there is a high silica activity, whereas laumontite forms as a cement at the same stratigraphic level and at temperatures as low as 50°C. Furthermore, with progressive metamorphism heulandite alters to prehnite in siltstones but to laumontite in sandstones, the different behaviour being related to differences in the activity of Ca ions.

Little is known about nucleation and growth rates for many minerals in very low-grade metamorphism, but the fine grain-sizes suggest a high nucleation/growth rate ratio. Only rarely do large crystals grow, and an example is laumontite replacing glass (Fig. 2.3A) in tuffs from the classic section of Coombs (1954). The coarse laumontite, often of the order of 3 mm grain size, suggests a relatively high growth/nucleation rate ratio; the lack of a spherulitic-type texture suggests that diffusion was not a rate-limiting factor. However, curved and highly elongate crystals, and bow-tie textures (similar to spherulitic), are found in some very low-grade rocks (Fig. 5.9), and these suggest suppressed nucleation, diffusion as the rate-limiting factor and/or very fast growth rates; crystals grew rapidly towards the source of material, more rapidly than material could be moved towards the sites of growth. The figured example of curved lawsonite crystals is from a black, fine-grained carbonaceous metasediment, presumably lacking permeability, thus inhibiting diffusion. Similar textures occasionally occur in higher-grade rocks, the best known example being the garbenschiefer with large centimetre-scale bow-tie amphiboles, resembling sheaves of wheat.

(d) High-grade metamorphic crystallization

(i) At lower pressures

According to Walther and Wood (1984), the time necessary for a 0.1 cm-radius crystal to react or grow completely during moderate-grade thermal metamorphism is of the order of 10^2 yr. Overstepping of up to 50°C may be required, but because of the more rapid heat flow, metamorphic crystallization is potentially faster at lower pressures than at high pressures. Cashman and Ferry (1988) studied crystallization of magnetite, pyroxene and olivine in very high-temperature (1000°C) thermally metamorphosed basaltic rocks from Skye, and they found log-linear relationships in the crystal-size distribution curves, similar to those of volcanic rocks (section 4.1.2). Using reasonable growth rates they estimated growth times for magnetites at 0.03–32 yr., and they deduced nucleation rates of between $1\ \text{cm}^{-3}\ \text{hr}^{-1}$–$5\ \text{cm}^{-3}\ \text{min}^{-1}$. These growth and nucleation rates are similar to those of basalt, and it seems therefore that metamorphic and igneous rates of crystallization converge at high temperatures close to melting. As noted by Ridley (1985) and Cashman and Ferry (1988), the faster the heating rate the faster reactions are and the greater the number of nuclei formed. Such rocks are likely to be relatively fine grained.

(ii) At higher pressures

The ultimate result of prograde regional metamorphism is either anatexis (melting), discussed in section 5.3.6, or a dry refractory product such as pyroxene granulite. Formation of orthopyroxene from biotite involves a loss of volatiles which presumably assists the transport of reaction products during its formation. Once a totally water-free granulite is developed one might expect diffusion to be a rate-limiting factor in any further reaction; however, the lack of volatiles is to some extent counterbalanced by the increasing effectiveness of solid-state diffusion as well as the prolonged duration of metamorphism (very slow heating and cooling rates). Thus garnets are homogenized by volume diffusion at $T > 640°\text{C}$

(Yardley, 1977; Spear *et al.*, 1990), the smallest grains most easily, Ostwald ripening reduces the number of fine grains and coarsens the texture (discussed above), and annealing modifies grain boundary geometry. The end products tend to be even grained and coarse; most minerals are anhedral as part of the general mosaic, with dihedral angles reflecting relative surface energies (Hunter, 1987 – refer to section 4.2.2). The tendency to an even texture accounts for the coarse, sugary appearance of many granulitic rocks.

5.1.2 Retrograde metamorphic crystallization

Apart from the incipient effects of very low-grade metamorphism, apart from the metamorphic relics protected and preserved as inclusions in poikiloblasts, and apart from compositional zoning in minerals like garnet, the effects of **progressive** metamorphism are usually thorough and more-or-less complete. We have examined some of the reasons: devolatilization reactions provide a good means of diffusing reaction products; reaction rates increase progressively with temperature; devolatilization reactions are endothermic which means that the continued heat flow helps the overstepping of equilibrium conditions. The end product of progressive metamorphism represents the thermal climax of metamorphism, and the preceding *PT* history may not be easy to read.

In contrast, **retrogressive** metamorphism is seldom thorough and complete. Reactions involving the addition of volatiles are exothermic, so overstepping by decreasing *T* below the equilibrium conditions is necessary; since reaction rates also decrease with *T*, successful reaction is less likely than a prograde one. However, the principal obstruction to retrogressive reaction is the lack of fluids, already lost to the system during progressive metamorphism. Even if a nearby source of fluids exists, the completely crystalline nature of the rock may severely impede fluid ingress and hence fluid availability for reaction.

Paradoxically, the fact that retrogressive reactions are not thoroughgoing means that we often have a good record of them. Published *PTt* curves (section 5.5.2) usually lack detail of progressive metamorphism, and rely on regional information, whereas a complete retrogressive history may be recorded in one small area, even within the one thin section. Harley *et al.* (1990) describe Al-rich orthopyroxene–sillimanite–garnet–quartz–K-feldspar granulites which are progressively retrogressed from 1000°C and 11 kbar down to 750°C and 4.5 kbar. The first reactions involved sillimanite and garnet or orthopyroxene, and produced rims of cordierite surrounding and replacing the garnet or pyroxene, and symplectites of cordierite and sapphirine replacing sillimanite (Fig. 5.10A). Further reaction involved biotite formation at the expense of K-feldspar,

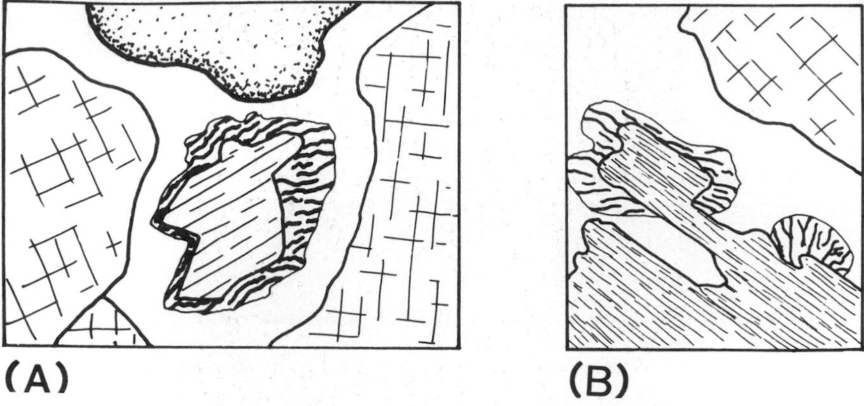

Fig. 5.10 Schematic representation of textures described by Harley *et al.* (1990). (A): Cordierite (clear) mantling orthopyroxene (crossed lines) and garnet (stippled), and cordierite + sapphirine symplectite replacing sillimanite (lined). Grains approximately 1 mm across. (B): Cordierite + orthopyroxene symplectite between biotite and plagioclase + quartz (clear). Older orthopyroxene also shown. Same approximate scale as (A).

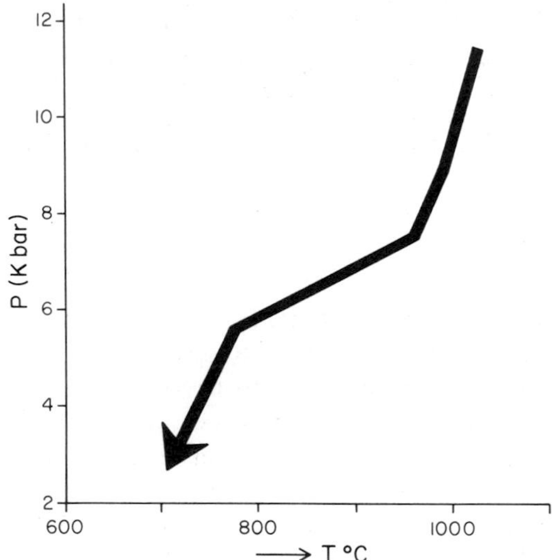

Fig. 5.11 *PTt* path of retrogressive metamorphism based on textural evidence and geothermobarometry, as described by Harley *et al.* (1990).

pyroxene and sapphirine, and later still, biotite was rimmed by symplectites of second-generation orthopyroxene and cordierite by reaction with plagioclase and quartz (Fig. 5.10B). The *PTt* evolution represented by the reactions (Fig. 5.11) can be reconstructed only because reactions were not complete because of limited diffusion, and because reaction products were often restricted to the grain boundaries.

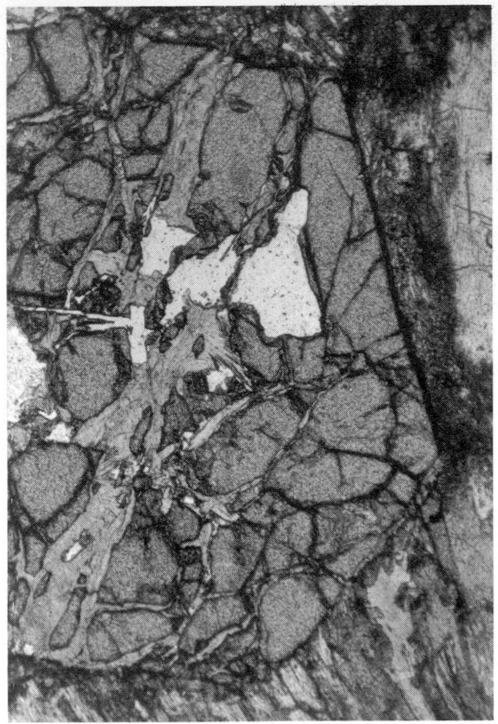

Fig. 5.12 Garnet partially replaced along a series of irregular cracks by chlorite. (Length of view measures 3.3 mm.)

At lower grades of metamorphism, the most familiar evidence for retrogression includes chloritization of biotite and garnet, uralitization of pyroxene, and sericitization and saussuritization of feldspar; these alterations have already been discussed in section 3.5. In addition, unmixing and symplectite formation such as myrmekite generally record a partial adjustment to falling temperatures (contrary to the general rule, some myrmekite represents a prograde reaction – section 3.5.2).

Retrogressive alteration products such as chlorite and tremolitic hornblende may form complete pseudomorphs of higher-temperature minerals, but partial alteration usually proceeds first along grain boundaries, along weaknesses such as cleavage planes or fractures, the latter being particularly common in hard minerals lacking cleavage such as garnet (Fig. 5.12). In the case of alteration of primary igneous minerals, it is not easy to distinguish incipient metamorphic effects (either retrogressive or progressive) from deuteric effects. Extensive or thorough-going retrogression is only likely in or adjacent to fault and shear zones where the fracture system acts as a pressure sink and zone of high fluid flow, and where shear-zone strain may induce a total recrystallization and modification of texture, eliminating evidence of the higher-temperature precursors.

5.1.3 Polymorphic transformations

Common examples of polymorphic transformations are found in the silica minerals, Al_2SiO_5, and calcium carbonate (the ideal stability fields for Al_2SiO_5 are shown in Fig. 5.13). It is well known that most polymorphic transformations, such as kyanite to sillimanite or quartz to tridymite, proceed very sluggishly, particularly during retrograde metamorphism. Diffusion cannot be the rate-limiting factor because no transport is necessarily required. The principal reasons for lack of reaction are a combination of very small free-energy differences in polymorphic systems, difficulties in nucleation, and in some cases, slow growth rates.

The small differences in free energy (related to small differences in entropy) mean that a large overstepping of equilibrium conditions is necessary to induce nucleation. If nucleation is also a problem, transformation may be prevented from taking place, even during progressive metamorphism. For example, seldom does one see kyanite or andalusite directly replaced by sillimanite. The structures of these polymorphs are utterly different so that when direct transformation does occur, reaction products form incoherent mosaics, as noted by Mezger *et al.* (1990) for sillimanite after andalusite.

The difficulty of nucleating sillimanite on either andalusite or kyanite may be resolved by its epitaxial nucleation on a sheet-silicate (Kerrick, 1987). This introduces the factor of diffusion, which is necessary to transport material between the two aluminium silicates, and which may therefore become a rate-limiting factor; it also points to the complexity of reaction in non-direct polymorphic transformations. Thus Carmichael (1969) shows that it is necessary to invoke ionic reactions involving K^+, H^+ and water, for the transformation to be achieved. Two of the reactions are:

$$\text{3 kyanite} + \text{3 quartz} + 2K^+ + 3H_2O = \text{2 muscovite} + 2H^+$$
$$\text{2 muscovite} + 2H^+ = \text{3 sillimanite} + \text{3 quartz} + 2K^+ + 3H_2O$$

and these reactions show why kyanite is replaced by muscovite as muscovite is replaced by sillimanite (Fig. 5.14A). The overall effect is kyanite = sillimanite, but the actual process of ionic transfer is more complex (Fig. 5.14B). Many metamorphic reactions are likely to proceed by similar complex ionic transfers; thus Bell and Rubenach (1983), for example, describe the production of andalusite from staurolite, which proceeds by muscovite replacing staurolite and andalusite replacing muscovite.

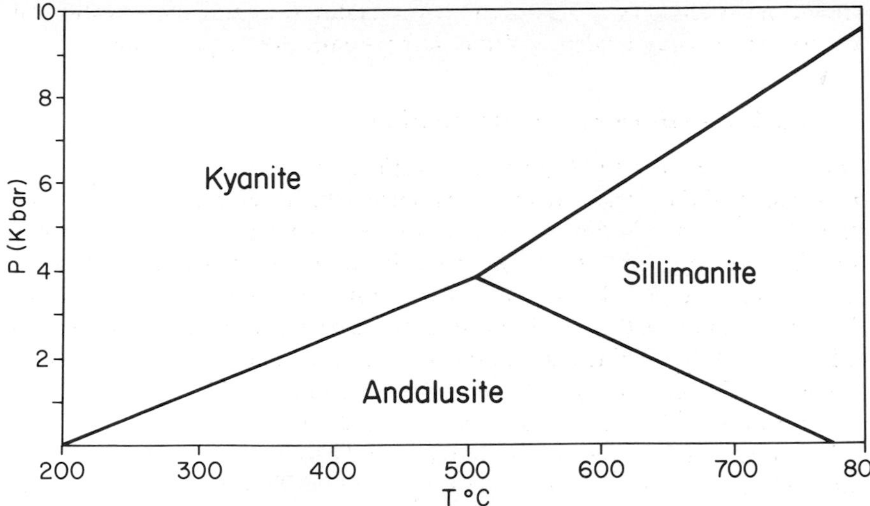

Fig. 5.13 Phase diagram for the aluminium-silicate polymorphs. Based on Holdaway (1971), with permission of the *American Journal of Science*.

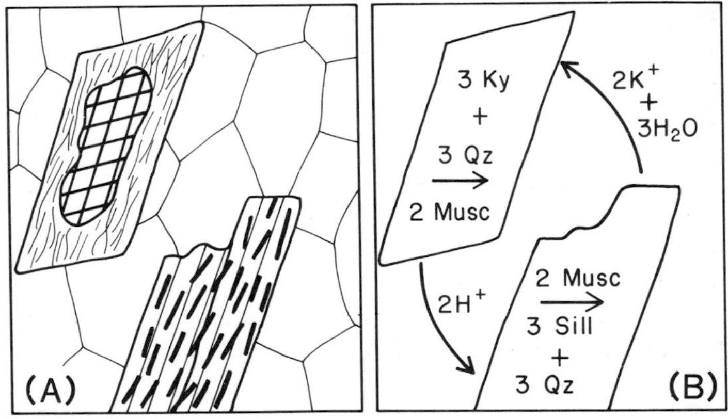

Fig. 5.14 (A): Kyanite to sillimanite reaction achieved indirectly by replacing kyanite by muscovite (top left) and muscovite by sillimanite (bottom centre). (B): The ionic transfer involved, as described by Carmichael (1969).

The fact that sheet-silicates seed sillimanite introduces a serious problem in assessing *PT* conditions. Unstable polymorphs that nucleate easily may grow outside their stability fields if nucleation of the stable polymorph is less easy. This happens with sillimanite (Kerrick, 1987), and it probably happens with the high-*T* polymorphs of silica in the sedimentary environment.

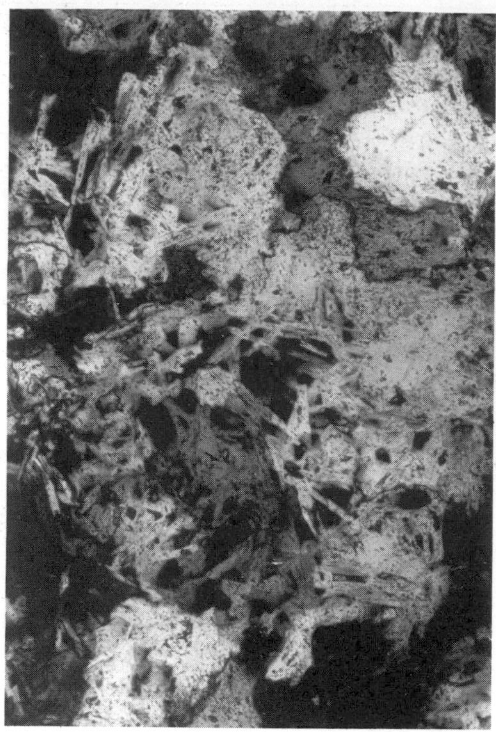

Fig. 5.15 Quartz sandstone in which quartz grains were replaced by tridymite plates during baking by basalt; the tridymite was then replaced by quartz during cooling. Oxford, NZ. (Length of view measures 0.85 mm.)

Transformation is even more difficult during a falling temperature regime because atomic movement rates are declining during overstepping; of course this explains why all of the silica and aluminium-silicate polymorphs (except high-T quartz) are found at the Earth's surface. Direct coherent transformation is possible in the special case of high to low quartz, and possibly aragonite to calcite, but in general it is not possible, and the general textural result is a mosaic of the new polymorph. Examples include euhedral tridymite plates (in high-T hornfels) replaced by anhedral quartz mosaics (Fig. 5.15), and the partial replacement of coesite by quartz (Fig. 5.16). The coesite is particularly intriguing, and according to Chopin (1984) represents subduction of sediments to extreme depths of >90 km. He believes that all quartz changed to coesite, but most converted back to quartz during uplift. Only parts of those grains included within garnet survived. He notes that quartz partially replacing coesite has a characteristic radiating pattern, and that the volume expansion

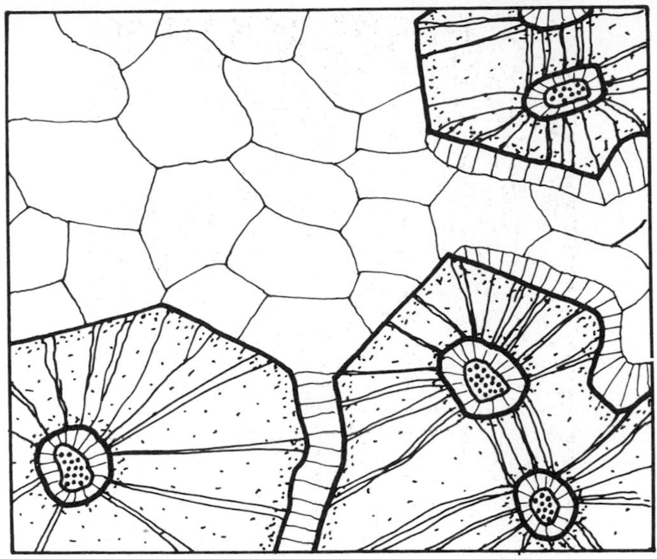

Fig. 5.16 Schematic representation of garnet–quartz rocks with coesite, described by Chopin (1984). Coesite inclusions (coarse stipple) in garnet are partly replaced by radiating quartz. Cracks in garnet radiate from the coesite inclusions, as explained in text. Radial quartz growths rimming garnet also suggest replacement of coesite. Coesite inclusions are of the order of 0.1 mm across.

accompanying the change of coesite to quartz caused cracking of the garnet (Fig. 5.16). A similar radiating pattern of quartz around the outer margins of garnet crystals is the main evidence that all the matrix quartz had also been transformed.

An important petrological indicator is transformation of aragonite to calcite. High-pressure aragonite is commonly found in blueschist-facies rocks, yet aragonite is metastable at the Earth's surface and can be easily changed to calcite, as for example in sea shells during diagenesis. However, this involves solution of aragonite and precipitation of calcite, and it is clear that aragonite in dry, completely crystalline rocks does not change so readily. Carlson and Rosenfeld (1981) show that growth is the rate-limiting factor controlling transformation. Nucleation is easy because calcite can replace aragonite in a coherent topotactic arrangement; this is indicated optically by the common orientation of discrete calcite grains replacing aragonite. In any case, the myriad grains of calcite found around the margins of some aragonite indicate that nucleation was not difficult. Experimentally determined growth rates near the stability field boundary at 250°C are about 10^3 mm Ma^{-1}, whereas at 100°C they reduce dramatically to 10^{-3} mm Ma^{-1} (Fig. 5.17A); calcite growth is then so slow that

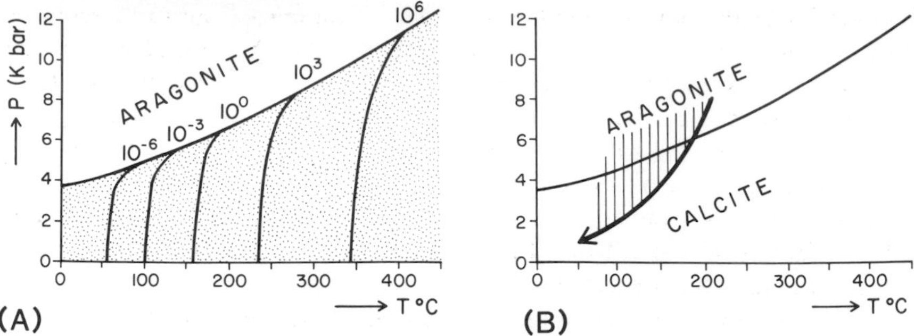

Fig. 5.17 (A): Aragonite is stable above stippled area, calcite within stippled area. Curved lines within stippled area give rates of growth (mm Ma^{-1}) for calcite crystallizing from aragonite. (B): Slow calcite growth rates mean that aragonite (following a *PTt* path within the shaded area) will often survive uplift. Based on Carlson and Rosenfeld (1981), © 1981 by The University of Chicago (*Journal of Geology*).

aragonite crystals survive uplift. Carlson and Rosenfeld (1981) conclude that aragonite indicates uplift into the calcite stability field at $T < 175°C$ (Fig. 5.17B).

5.2 RECRYSTALLIZATION (*SENSU STRICTO*)

Recrystallization (*sensu stricto*) describes the replacement of an existing mineral by other grains of the same mineral. It may refer to one grain replaced by a number of smaller grains or to a polycrystalline area where the number and/or shape and/or orientation of grains changes. Recrystallization (*sensu lato*) is also used to describe the total transformation of mudstone, for example, to garnet schist, a process that involves both crystallization and recrystallization (*s.s.*). The sense in which the term is used is usually self-evident from the context. This section is about recrystallization in the strict sense.

The most common geological examples of recrystallization involve quartz, olivine and calcite, and in these cases the principal driving forces are energy reduction by eliminating dislocations and defects in mineral grains, and reduction of surface energy by reducing grain-boundary area. If recrystallization involves a slight change of chemistry, as in plagioclase which may change composition by 1–10% An during recrystallization (Vernon, 1975; White, 1975), the driving forces may include differences in chemical free energies. In fact, there is no absolutely sharp dividing line between crystallization and recrystallization.

Temperature, strain and stress all have a vital influence on recrystallization. Recrystallization is essentially the movement of existing grain boundaries or the development of new grain boundaries, and this requires diffusion and adjustment of material either side of the boundary. In dry material this is highly temperature dependent, usually requiring $T >$ 0.3–0.5 melting T (Drury and Urai, 1990). Fluid films along grain boundaries may also promote recrystallization. Strain involving plastic deformation stores dislocation energy which drives recrystallization. In general, there is a negative correlation between recrystallized grain size and differential stress. The relationship is complex in detail, as discussed below under dynamic recrystallization.

The following terms are used in connection with recrystallization:

Subgrain: An area within a deformed mineral grain which has a slightly different orientation from other adjacent subgrains (Fig. 5.18A). The distinction between grain and subgrain is usually obvious in practice, but the degree of misorientation between subgrains defies definition (Urai *et al.*, 1986); some workers have found 15°, others <5° as a useful boundary.

Triple-point junction: A well-recrystallized mosaic has a similar geometry to a foam. The ideal polyhedral shape of one cell in a foam, or one grain in a mosaic, can be described as a truncated octahedron with slightly curved faces and edges, a 'minimum-area' form called a tetrakaidecahedron (Smith, 1964). The typical thin-section view has triple-point junctions where three slightly curved boundaries meet with dihedral angles of approximately 120° (Figs 5.18B, 4.39B and 4.40Af). If more than one mineral constitutes a mosaic, then the dihedral angles depend on the relative surface energies of adjacent minerals (Hunter, 1987, and refer to section 4.2.2)

Dislocations: Plastic deformation of minerals such as quartz, olivine and calcite can be analysed in terms of slip along certain lattice planes, bending of the lattice, and the production and movement of dislocations. This subject is discussed further in Part Three, but it is useful to introduce the concept of dislocations here. The main types are edge and screw dislocations (Nicolas and Poirier, 1976), and Fig. 5.19 shows how plastic deformation (slip) is achieved by their movement. Edge dislocations represent a small imperfection in the lattice structure which in three dimensions defines a line. One can envisage two kinds of edge dislocation of different sign, and they cancel each other out if they come together (shown in section view in Fig. 5.20). The strain of a deformed grain, represented by numerous dislocations, can be reduced, provided dislocations of opposite sign cancel each other out in this way. If grains were bent during deformation, dislocations of one sign will dominate; during recrystallization these tend to move into a common plane (Fig. 5.21) which then delineates the boundary between two subgrains.

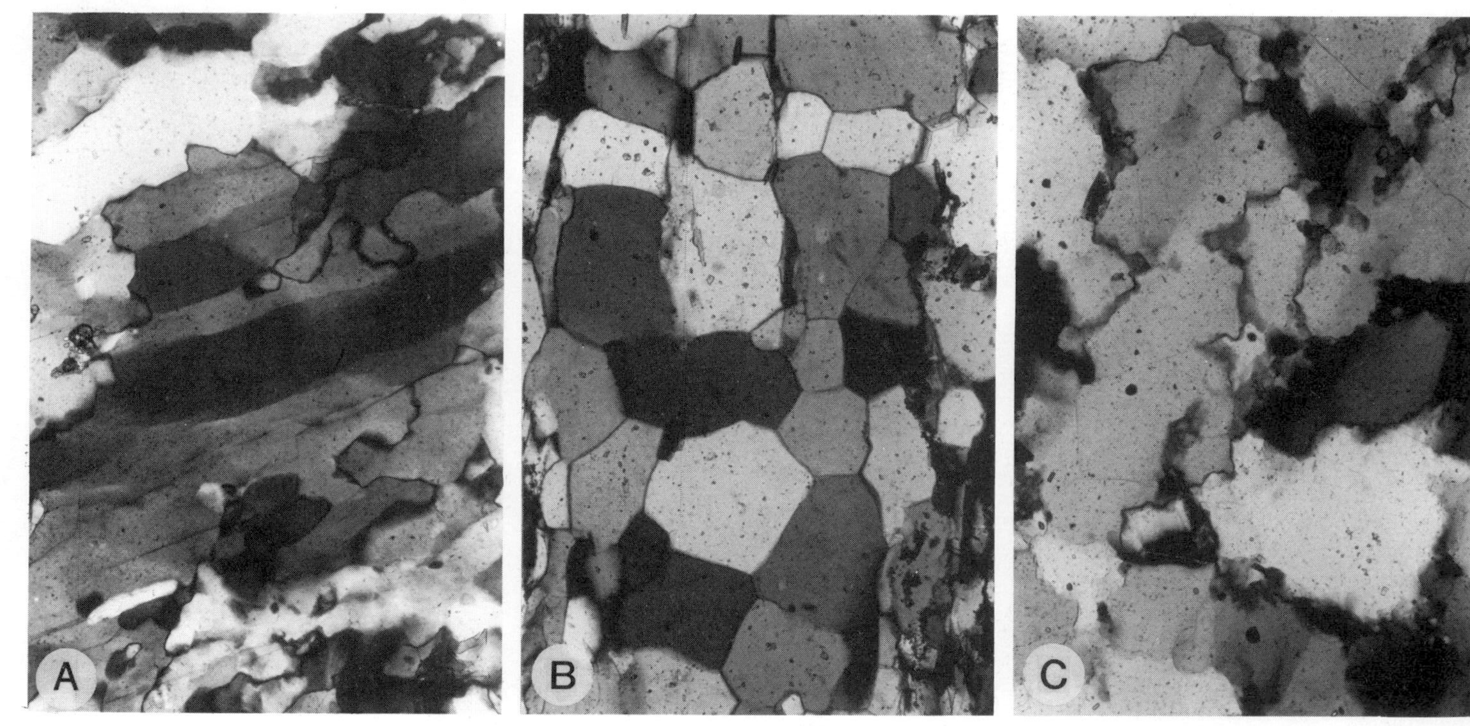

Fig. 5.18 (A): Subgrains in quartz (several sharply defined elongate subgrains are clearly visible in the large central grain). (B): Well-recrystallized quartz mosaic with dihedral angles close to 120°. (C): Serrated grain boundaries typical of dynamic recrystallization. View lengths measure 0.85 mm (A), (B) and (C).

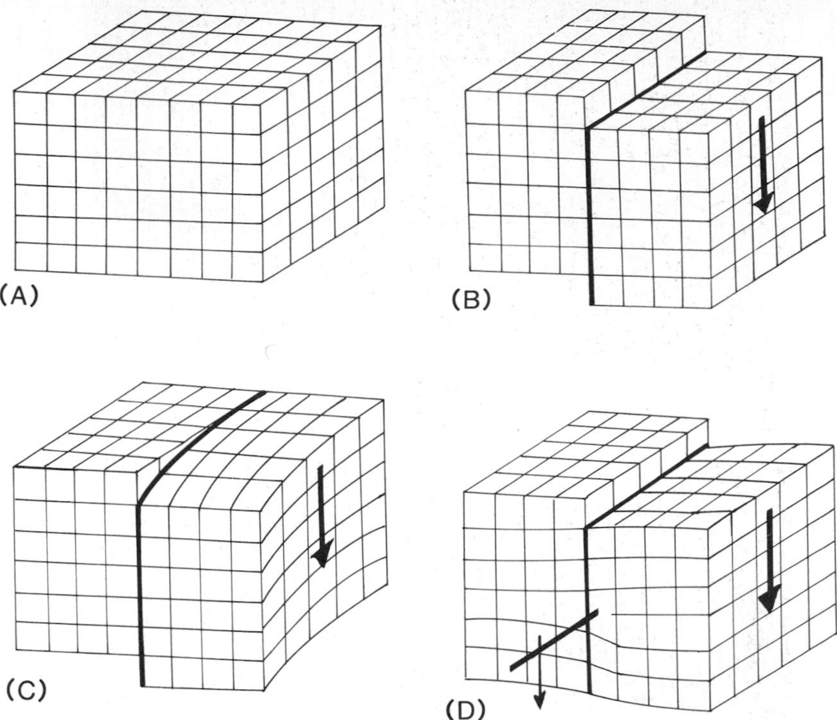

Fig. 5.19 Crystal transformed from (A) to (B) by slip involving either movement of screw (C) or edge (D) dislocations. In (C) and (D) the arrow indicates the slip direction; in (D) the line indicates the orientation of the edge dislocation which moves progressively from top to bottom in the transformation of (A) to (B).

Each subgrain will be left relatively strain free, unless deformation continues to propagate dislocations.

Static recrystallization: This is due either to an increased heat flow (promoting grain coarsening), a reduction in lattice defects after deformation, or some combination of the two. The reduction in lattice defects without grain coarsening is often termed **primary recrystallization**, and the grain-coarsening is termed **secondary recrystallization**. Ideally, static recrystallization produces a stable mosaic characterized by 120° dihedral angles. The terms annealing and post-tectonic recrystallization are sometimes used instead.

Dynamic recrystallization: This accompanies deformation and is often characterized by the perpetuation of strain features, elongate grains, subgrain structures, and serrated grain boundaries (Figs 5.18C and 5.39D), although at high temperatures, stable mosaics similar to those produced by static recrystallization may develop. Dynamic recrystalliza-

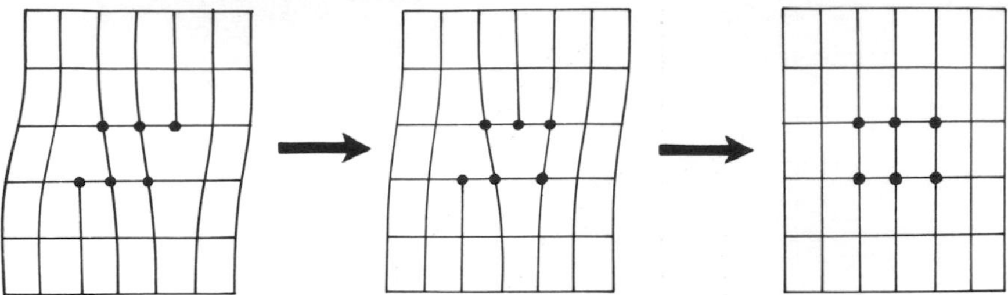

Fig. 5.20 Two edge dislocations of opposite sign move towards each other and cancel each other out. The progressive positions of six lattice points are given for reference.

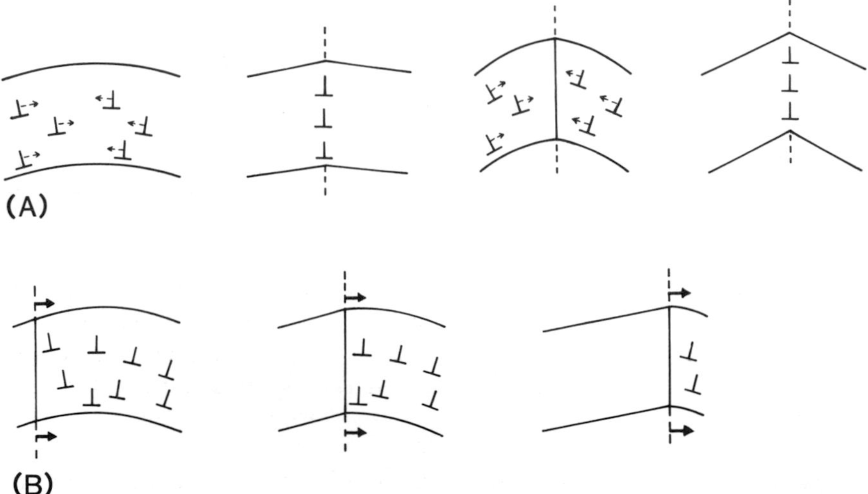

Fig. 5.21 (A): Edge dislocations migrate into the one common wall of dislocations which then separates two strain-free subgrains. Further bending generates more dislocations which migrate to the same wall; hence the subgrains progressively rotate with respect to each other. (B): A wall of dislocations (= subgrain boundary) sweeps through a bent grain, progressively eliminating dislocations. Based on Urai *et al.* (1986). The inverted T represents an edge dislocation similar to the right-hand one in Fig. 5.20.

tion is also known as syntectonic recrystallization or hot-working. Dynamic recrystallized grain and subgrain size decrease with increased differential stress, as shown by Chester (1989) for halite (Fig. 5.22). Note that grain size also varies according to the recrystallization mechanism, and according to Ross *et al.* (1980), olivine subgrain sizes increase with temperature in a wet environment as well as decreasing with increased differential stress. Modification of the stress-related grain size also occurs if temperature increases cause post-deformation static

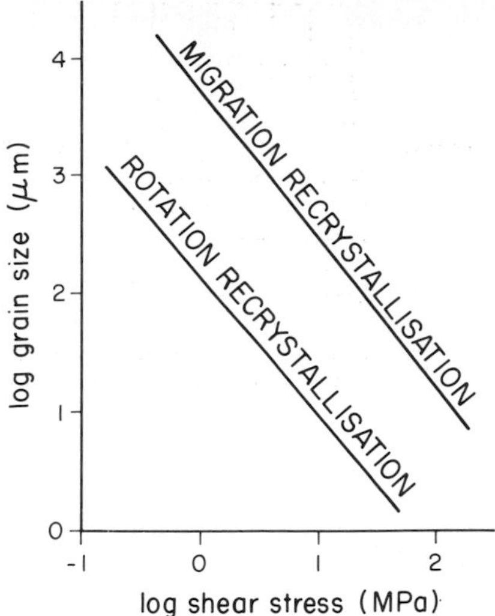

Fig. 5.22 Relationship between mean recrystallized grain size (for rotation and migration recrystallization mechanisms) and shear stress for halite. With permission from Chester, 1989. Copyright: Pergamon Press PLC.

recrystallization, and some grains or subgrains may adjust in size to progressive destressing; in this respect, Ord and Christie (1984) believed quartz grain (not subgrain) size to be relatively unaffected by destressing, whereas Ross *et al.* (1980) believed olivine subgrain size to be the best palaeostress indicator.

It is not always easy to distinguish the effects of static and dynamic recrystallization, especially if the former are modified by minor late deformation, or the latter by late annealing.

The two main mechanisms involved in recrystallization are discussed below.

5.2.1 Rotation recrystallization

Drury and Urai (1990) define this as the formation of new high-angle grain boundaries. Rotation recrystallization is therefore concerned only with the formation and evolution of subgrain mosaics, and because these develop diverse orientations from the one original grain, the impression is that subgrains have rotated relative to each other.

Subgrain formation follows heterogeneous mineral deformation (lattice bending) and involves the reorganization of dislocations. Urai *et al.* (1986) consider two cases. In the first (Fig. 5.21A), a subgrain boundary forms by movement of dislocations of like sign into one zone. Further bending may generate more dislocations, which again move towards the same assembly area, in this case a fixed subgrain boundary. Through this process, subgrains progressively rotate relative to each other. In the second case (Fig. 5.21B), instead of dislocations moving towards subgrain boundaries, the boundaries sweep through the grain, collecting dislocations *en route*.

Reasons for heterogeneous deformation include the lack of sufficient slip systems to satisfy von Mises' criterion (see Part Three), and the presence of more than one mineral of differing ductility. Thus quartz may be bent around rigid feldspars in low-grade metamorphic rocks (Fig. 2.7C). Homogeneous deformation is more likely at high temperature, as mineral plasticity and the number of operative slip systems increase with temperature. Therefore rotation recrystallization is most characteristic of low–medium-grade metamorphism.

5.2.2 Migration recrystallization

Drury and Urai (1990) define this as grain-boundary migration. If subgrains develop, the misorientation between them is slight, unlike the sharply defined subgrains of rotation recrystallization. Included here too is twin-boundary migration (Fig. 5.23). Material thoroughly recrystallized by this process (even in a dynamic environment) is characterized by equidimensional grains and dihedral angles of 120°. Imperfectly recrystallized material displays serrated and irregular grain-boundaries, and in the dynamic environment, inequidimensional grains.

The speed of grain-boundary migration depends on a number of factors, principally temperature, lattice orientation and the presence or absence of impurities. It is known from metals that boundaries are most mobile when lattices differ in orientation either side of a boundary by approximately 40° (Urai *et al.*, 1986). Grains with almost the same orientation have the least mobile boundaries. Mobility also increases with temperature, related to the ability of atoms to diffuse and/or readjust in position either side of a migrating boundary. Impurities slow grain-boundary migration, and this is seen in many rocks where quartz mosaic grain-size correlates strongly with mica grain size (Fig. 2.22). The micas effectively 'pin' the boundaries, preventing further migration and grain coarsening. Only if the driving forces (driven by deformation and aided by higher T) are sufficiently high will boundaries migrate around impurities incorporating them as inclusions (Fig. 2.9). At low T, there is a 'catastrophic jump' from slow to fast migration, whereas at high temperature there is no such jump. These factors are summarized in Fig. 5.24.

Fig. 5.23 Twin lamella within a plagioclase crystal in metatuff (hornblende–hornfels facies), illustrating the phenomenon of twin-boundary migration. Bluff, NZ. (Length of view measures 0.85 mm.)

The ideal effect of migration recrystallization is to create or maintain a stable equidimensional mosaic, even if accompanied by continuous deformation; grain size will be coarser with higher temperature, especially in a static environment, and in the dynamic environment, grain size is inversely related to differential stress (discussed above). Pure migration recrystallization in a dynamic regime depends on homogeneous deformation, and that is only likely at high temperature for most silicates.

5.2.3 General recrystallization mechanisms

Rotation and migration recrystallization are but two end members in a spectrum of recrystallization types. More generally subgrain evolution is accompanied by a degree of grain-boundary migration, and by new grain formation.

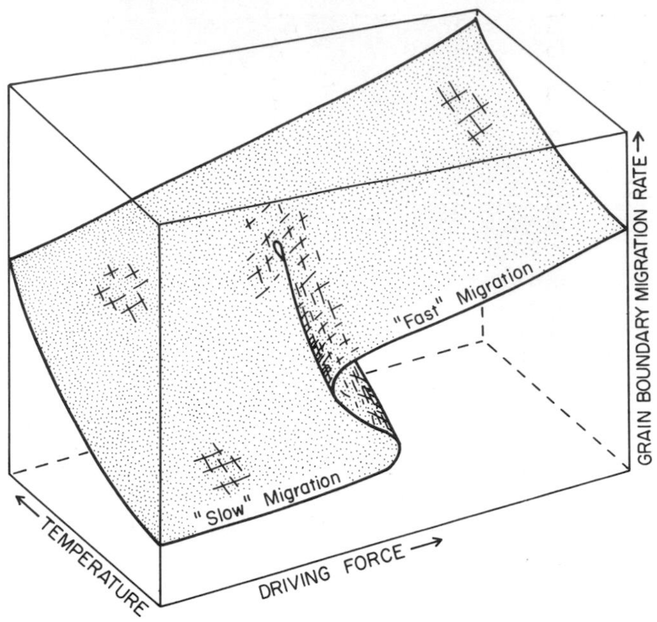

Fig. 5.24 Summary of relationships between temperature, driving force and grain-boundary migration rate. See text for discussion. With permission from Urai *et al.* (1986, Fig. 11), copyright by the American Geophysical Union.

Nucleation of new grains in the classical sense is deemed unlikely by Drury and Urai (1990) unless a chemical change is also involved, as in feldspar recrystallization. Instead, new grains develop either by grain-boundary bulging or from small rotated subgrains. New grains that develop in isolation within larger parent grains may be euhedral, as described for quartz (Hobbs, 1968) and olivine (Mercier, 1985). Grain-boundary bulging occurs where serrated or irregular (possibly irregularly pinned) boundaries develop (Fig. 5.25). If a bulge becomes sufficiently large, and especially if it develops a subgrain boundary at its neck and rotates due to deformation, it may quickly grow as a separate grain. Development will be assisted if the lattices of neighbouring grains are at the optimum angle for fast migration. Small, perhaps very small rotated subgrains may rapidly develop into fast-growing grains if lattice misorientation reaches the optimum angle, provided that such grains can break away from any impeding impurities.

The relationship between mineral preferred-orientations and recrystallization is explored in Part Three.

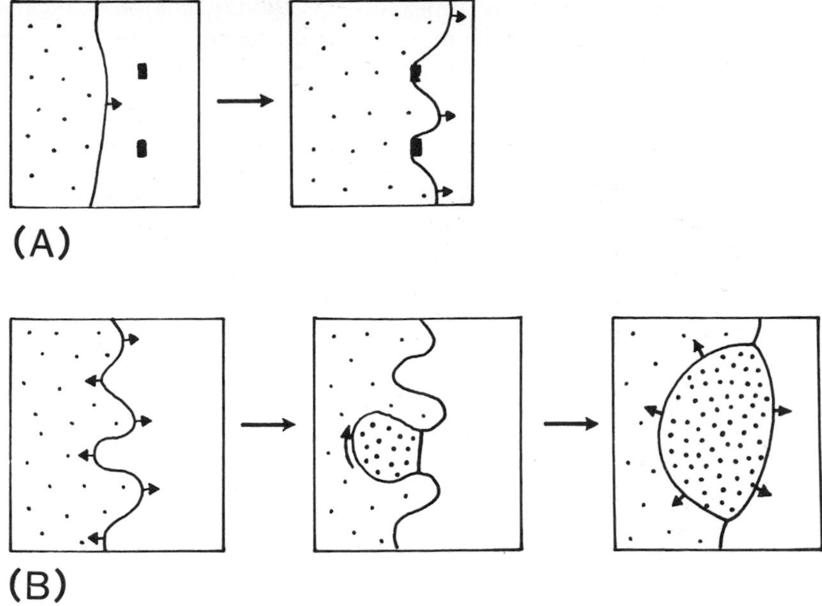

Fig. 5.25 (A): Bulges in grain boundary developed by pinning of migrating grain boundary with inclusions. (B): Serrated boundary developed by two grains bulging into each other. If a bulge area rotates during dynamic recrystallization into a subgrain with favourable orientation for fast migration, it forms the nucleus of a new rapidly growing grain. Unlike Ostwald ripening during static recrystallization (Fig. 5.7), grain boundaries migrate away from their centres of curvature during dynamic recrystallization.

5.3 METAMORPHIC LAYERING, CRENULATION CLEAVAGE, VEINS AND PRESSURE SHADOWS

Metamorphic petrology overlaps in scope with many other branches of geology, especially mineralogy, geochemistry and structural geology. The boundary with structural geology is particularly difficult to define. In this book I have opted to devote Part Three to mineral preferred-orientation studies, a branch clearly overlapping with structural geology as indicated by the oft-used title 'structural petrology'. In Part Two, I exclude discussion of large-scale structures, but the subjects of crenulation cleavage, metamorphic layering, veins and pressure shadows are very much an integral part of petrology, and subjects that can be illuminated by thin-section observations.

One of the biggest problems with metamorphic layering is recognizing it in the first place. It is commonplace for sedimentary and igneous layering to be inherited by a metamorphic rock, and even in migmatitic terranes, layering is commonly thought to be inherited (Johannes and Gupta, 1982). Primary layering may be significantly modified during metamorphism, and then it is a moot point whether or not it should be termed metamorphic layering. As is often the case in geology, sharp lines between categories seldom exist.

Recognition of sedimentary layering depends to some extent on common sense. A sequence of layers comprising diverse lithologies such as marble, quartzite and mica schist, can be identified clearly with an original sedimentary sequence. But in thick sequences of relatively monotonous lithology, bedding should only be labelled as such if unequivocal sedimentary structures are still recognizable.

5.3.1 Grain-size sorting and transposition

One of the great debates of the 1960s and 1970s concerned slaty cleavage: could it be produced by the movement of water expelled from unconsolidated sediment during deformation? This was the idea of Maxwell (1962) who pointed to clastic dykes and flame structures, seemingly parallel to slaty cleavage, suggesting water-induced grain flow during folding. Much of the original evidence has now been reinterpreted or shown to be ambiguous; clastic dykes that originate before folding will rotate during strain towards the same orientation as the cleavage, for example. It is a mistake, however, to throw Maxwell's idea out completely, as some have done, and a particularly clear example of cleavage-parallel layering produced by water-induced grain flow is described in Shelley (1975). Figure 5.26 illustrates the centimetre-scale layering, pervasively developed in Ordovician slates, and exposed on a shore platform in New Zealand. Clastic quartz grain shapes indicate an absence of pressure-solution effects. Folded bedding and sedimentary structures are well preserved in the area, and juxtaposition of secondary layering and bedding is sometimes seen, but where layers extend for metres parallel to cleavage, bedding is completely destroyed. The layering is similar to water-escape structures described by Laird (1970) but on a larger scale, and it does not precede folding. The only explanation seems to be grain-size sorting in channels where water escaped during folding.

More generally, primary sedimentary layering is transposed into new orientations during deformation. In slates, flame structures become more pronounced and rotate towards slaty cleavage during progressive strain (Fig. 5.27A); clastic dykes also rotate towards this orientation. In tightly folded and sheared sequences, primary bedding will be transposed into parallelism with cleavage (Bishop, 1972a), even though the larger-scale enveloping surfaces remain at an angle (Fig. 5.27B).

Fig. 5.26 Layering (parallel to slaty cleavage) which represents a grain-size sorting during water expulsion. Greenland Group, near Greymouth, NZ. (View length measures 35 mm.)

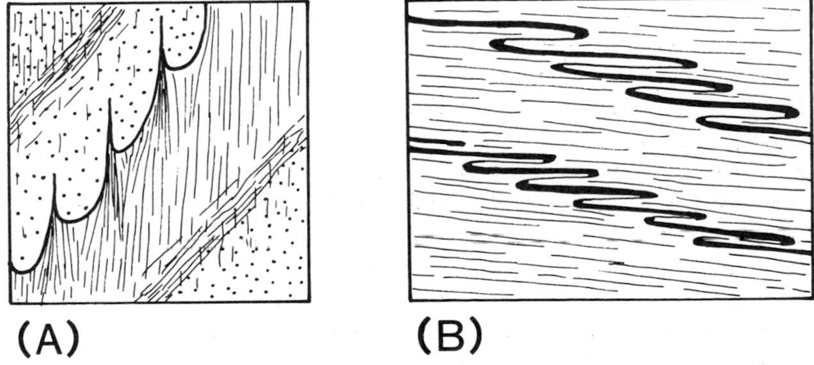

Fig. 5.27 Bedding subparallel to foliation. (A): Flame structures rotated due to strain during cleavage formation. (B): Transposition structures due to tight folding, shearing and formation of rootless folds. Note that enveloping surfaces of bedding are not parallel to schistosity.

5.3.2 Mechanical segregation

The idea that ductile minerals can be segregated from non-ductile ones by laminar shear has been a popular one to explain metamorphic layering (Spry, 1969). The idea goes back to Schmidt (1932), who based it on the supposed production of layering in metals by some such process. In fact the metallurgical analogy is false (Shelley, 1974). The idea that the Bagnold effect might produce mechanical segregation of porphyroblasts in schists and mylonites (just as phenocrysts are segregated in dykes – see section 4.1.7) was examined by Shelley (1974), and rejected for lack of evidence.

More recently, Toriumi (1986) has advocated mechanical segregation of garnet in schists from Japan. He notes that garnets in thermally metamorphosed rocks and low-grade schists have random distributions but are clustered in highly-sheared oligoclase-zone schists. He believes that non-ductile garnets collided during shearing, and that rapid overgrowth caused them to adhere to each other, preventing separation during further shearing. Toriumi (1986) does not consider other possible origins of segregation such as solution transfer (discussed below), and the strength of his case is therefore difficult to judge.

5.3.3 Layering produced by strain of existing features

We have already noted that planar features such as clastic dykes (and of course bedding) rotate progressively during strain towards parallelism with the cleavage or schistosity. The same phenomenon occurs with veins. Thus Norris and Bishop (1990) describe a deformed conglomerate where early quartz and quartz–albite veins cut through competent pebbles and the more ductile matrix; in the latter, veins rotated towards the schistosity and became boudinaged (Fig. 5.28A). Many quartz-rich layers in schists are slightly discordant to schistosity, and probably result from rotation of veins previously cross-cutting at a higher angle (Fig. 5.28B).

If more-or-less equidimensional objects such as pebbles in a conglomerate or mineral grains in an igneous rock become severely strained, layering develops. Thus Norris and Bishop (1990) measure a 70% shortening of pebbles in the Otago Schists, the result being a layered schist where distinction between flattened pebble and other forms of layering becomes difficult (Fig. 5.29). The layering in fine-grained mylonite commonly results from the strain of pre-existing minerals grains. Thus the extreme plastic deformation of quartz in massive granite results in extremely long, thin ribbons (Figs 2.6 and 2.7). At the same time, shredding of mica (by slip along cleavage planes) produces thin layers of mica debris. An example of such layering is found in the ultramylonites of the Moine

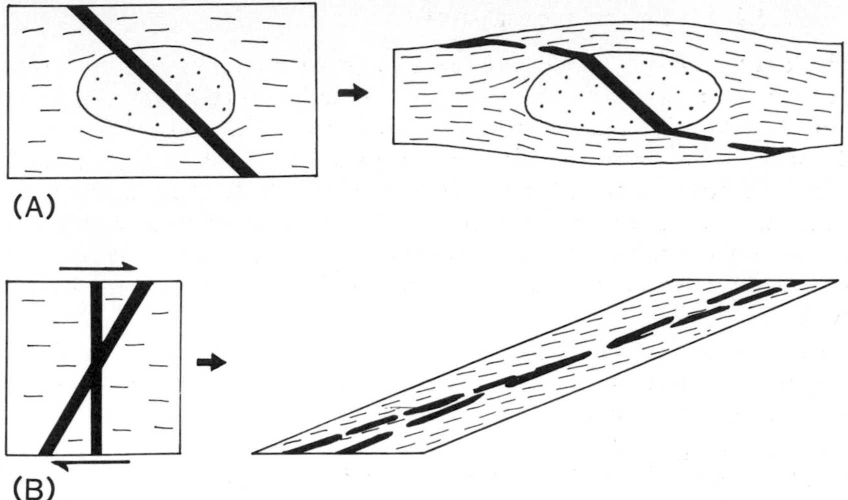

Fig. 5.28 Schematic representations of how veins may rotate towards parallelism with schistosity. (A): Vein cuts competent pebble and incompetent matrix; heterogeneous pure shear causes rotation and boudinage of veins in the matrix. (B): Two pre-existing veins are thinned, rotated and boudinaged during simple shear.

thrust system. Dixon and Williams (1983) describe how a coarse granitic gneiss was transformed: quartz was plastically deformed and dynamically recrystallized, the ultra-fine grain size being related to a high differential stress; muscovite was shredded and smeared out into fine-grained layers; metasomatism caused sericitization of all feldspar, this mica being added to the shredded mica layers (Fig. 2.7D). The result is a very fine-grained and finely layered quartz–muscovite rock.

The smearing out of mica is a kind of cataclastic flow, although the well-known flexibility of mica plates means that the deformation is not entirely one of fragmentation. Feldspar is a very brittle mineral at low to medium grades of metamorphism, and Tullis and Yund (1987) describe its cataclastic flow. Augen-shaped relics become elongated and develop undulatory extinction as a result of breakage along the cleavage planes and microfaulting on a submicroscopic scale. Macroscopically, deformation seems to be ductile, and the elongate grains resemble plastically deformed quartz ribbons. It is only on very close examination that angular, cleavage-bound fragments of diverse, but very fine size, prove cataclasis to be the causative mechanism. Augen are surrounded by a fine-grained veneer of dynamically recrystallized material. Cataclastic flow of brittle minerals like feldspar may take place over a wide range of temperature. Gapais and White (1982) observe that dynamically recrystallized quartz grains are

Fig. 5.29 Deformed conglomerate in which the pebbles are strained into thin plates which mimic other forms of secondary layering parallel to schistosity. Chlorite zone, Haast Schist, Haast Pass, NZ. Photograph kindly provided by R. J. Norris.

inequidimensional and bounded by prism and rhomb planes. They suggest that grain-boundary sliding of such an aggregate will align the grain long-axes so that *c*-axes end up close or parallel to the extension direction. Such grain-boundary sliding cannot occur without accompanying cataclasis, plastic deformation or diffusion. In the low-T example of Gapais and White (1982), it was not likely to be the latter.

Superplasticity also involves grain-boundary sliding, but in this case the grains are polygonal and even in grain size. First recognized in metals, it enables extremely fine-grained material (<0.01 mm) to be extended *ad infinitum* without necking (boudinage). It occurs at relatively high temperatures (>0.5 melting T). If one tries to imagine a mosaic of polygonal grains sliding past each other (without cataclasis), it is mechanically impossible unless accompanied by a process of intragranular diffusion allowing continuous grain shape changes to take place; such diffusion is possible at high temperatures, and is only effective because of the extremely fine grain size. The strain rate must not be too high otherwise normal plastic deformation becomes dominant. Characteristics of superplasticity are grains that remain equidimensional throughout deformation, and lack lattice preferred-orientation.

Boullier and Gueguen (1975) describe three natural examples of superplasticity. One is a mylonitized olivine–orthopyroxene peridotite: olivine porphyroclasts are set in a matrix of fine recrystallized olivine, and orthopyroxene porphyroclasts are surrounded by thin veneers of ultra-fine recrystallized orthopyroxene. The latter has been drawn out into very thin layers in either direction (Fig. 1.22). The temperature of deformation may have been as high as 1400°C, and the textures strongly support the interpretation of superplasticity. Another example contains thin layers of ultra-fine amphibole set in fine-grained recrystallized plagioclase; deformation took place at 600°C (amphibole would break down at 900°C).

Other workers have suggested that quartz and feldspar become superplastic when the dynamically recrystallized grain size is sufficiently fine (Allison *et al.*, 1979; Behrmann, 1985). Behrmann advocates quartz superplasticity because of an observed rapid loss in lattice preferred-orientation below a certain grain size. The major problem in advocating superplasticity in mylonites is that quartz usually has a very strong lattice preferred-orientation suggesting plastic rather than superplastic deformation.

5.3.4 Segregation due to stress- or strain-induced solution transfer

Solution seams and stylolite formation are well-known phenomena in sediments where the more soluble materials (e.g., carbonates) are removed from interfaces at a high angle to the maximum principal compressive stress, particularly in clay-rich layers, and precipitated in veins or as a cement in the clay-poor layers. Consequently, trivial differences in clay/carbonate content become exaggerated, and indeed some layered clay–limestone sequences are thought to be secondary rather than primary.

In crenulated slates and schists, preferential solution of quartz takes place within the steeper limbs (Fig. 5.30) where sheet-silicate cleavages are at a higher angle to the shortening direction; relic quartz grains become increasingly inequidimensional, and precipitation takes place within the less-steep limbs or in more distant fracture systems. Dark, almost opaque, solution seams mark where all the quartz has been removed and where the insoluble carbonaceous and sheet-silicate material is concentrated (Gray, 1979). Gray also describes how particular layers (sedimentary) or veins are truncated by such seams (Fig. 5.31), the apparent displacements being the result of solution not microfaulting.

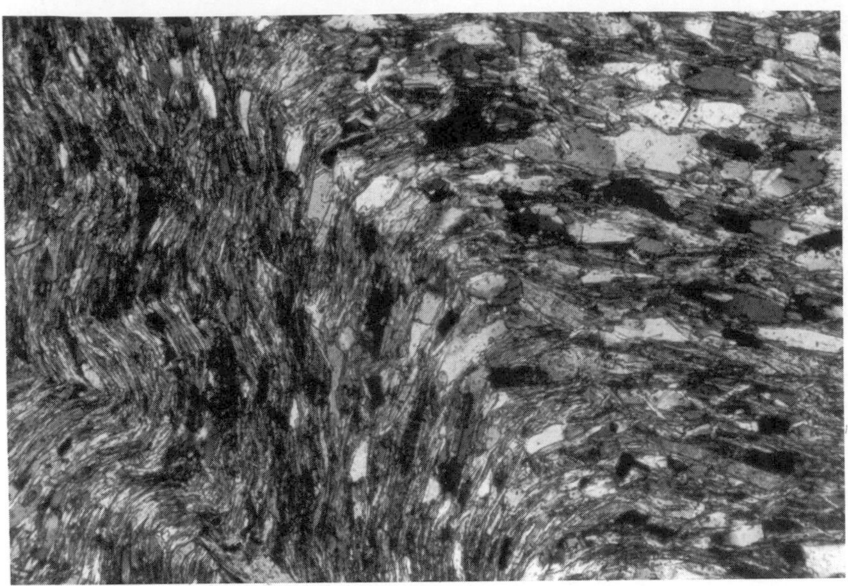

Fig. 5.30 Crenulated schist. Preferential solution of quartz and enrichment in mica has taken place along the steeper limbs (left of photo). The steeper limbs have themselves been crenulated, as illustrated in Fig. 5.55D. Palaeozoic schist, Springs Junction, NZ. Length of view measures 0.85 mm.

Textural zones in the Otago Schists of New Zealand (section 2.5.3), first advocated by Hutton and Turner (1936) and refined by Bishop (1972b), describe transitions from non-schistose material (zone I) through schists lacking segregation layering (zone II) to schists with conspicuous layering (zones III and IV). The changes in the Otago Schists usually take place within the lower grades of metamorphism, and the causative process is solution transfer (Norris and Bishop, 1990). Thus quartz is dissolved out of sheet-silicate-rich layers or out of layers where sheet-silicate cleavages are at a high angle to the shortening direction, and precipitated in low-pressure areas, either in fracture systems, or to the side of large, non-ductile, clastic grains forming strain shadows (section 5.3.5). Progressive strain during solution transfer only serves to exaggerate layering further, and to rotate initially cross-cutting veins into parallelism with schistosity (Fig. 5.28).

Stress- versus strain-induced solution

Stress is a force which may cause a strain which is the change in shape and/or size of the rock. Solution transfer has been explained alternatively by either stress or strain. The common observation that grain boundaries

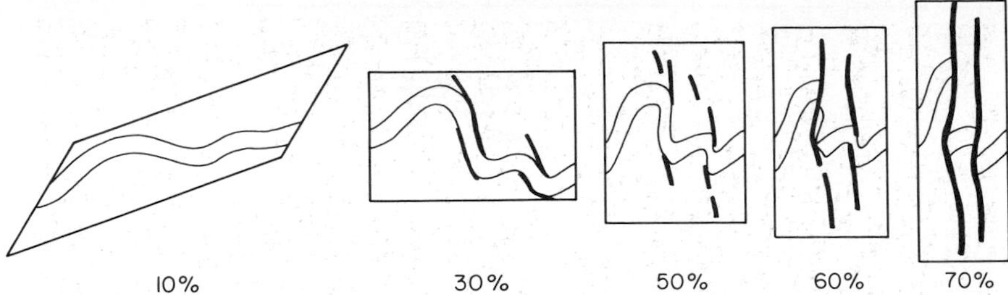

Fig. 5.31 Progressive development of solution seams and apparent offsets of buckled quartz veins due to solution. Reproduced from Gray (1979) by permission of the *American Journal of Science*.

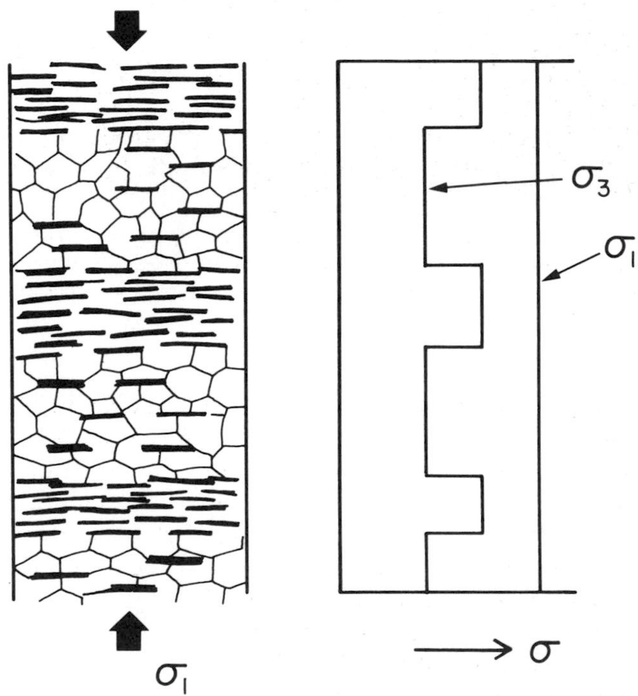

Fig. 5.32 The changing values of σ_3 relative to σ_1 in a column of rock with alternating layers of mica and quartz-rich rock may cause segregation of quartz from mica. Based on Robin (1979).

at a high angle to the maximum principal compressive stress are more susceptible to solution than other boundaries led to the term pressure solution. The idea that stress controls the degree of solution is a natural one, and it has been elaborated on most recently by Robin (1979), Sawyer and Robin (1986) and Gratier (1987). Robin's explanation involves the concept of stress heterogeneities in heterogeneous layered rocks. Imagine a

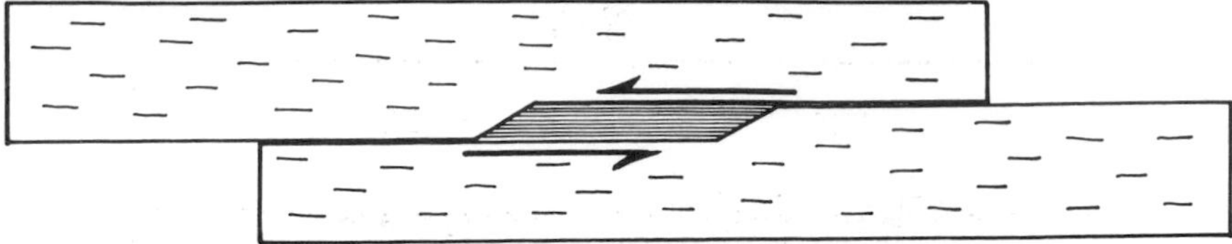

Fig. 5.33 A step on a shear plane creates a vein parallel to the shear plane, usually filled at first with fibrous crystals precipitated from solution.

layered quartz-rich/mica-rich rock subject to differential stress (Fig. 5.32). The value of σ_1 is constant, but the differential stress and value of σ_3 varies from layer to layer if strain is laterally constrained. In all layers, silica tends to dissolve along interfaces at a high angle to σ_1, but precipitation is dominantly in quartzose layers because of the higher differential stress there. Thus quartz-rich layers become even richer in quartz, and mica-rich layers richer in mica. Sawyer and Robin (1986) observe that quartz veins tend to develop at the interfaces between contrasting (sedimentary) layers, especially at low grades, and this is because on interface is the site of the sharpest gradient in σ_3, and also the site of differential movement and consequent dilation. Another mechanism advocated by Sawyer and Robin (1986) is precipitation in low-*P* pull-apart zones where shear movement jumps from one plane to another (Fig. 5.33); these sites are often characterized by fibrous growths in low-grade rocks. Several mechanisms are combined to account for the progressive thickening and extension of some veins: thus σ_1-parallel fracture filling is accompanied by accretion at layer interfaces (Fig. 5.34).

Gratier (1987) adds one further idea. Imagine a rock with its cleavage perpendicular to σ_1 (Fig. 5.35A), and imagine the tensile strengths perpendicular and parallel to cleavage to be 1 and 30 mPa respectively. If the rock is subjected to a differential stress less than 30 mPa, then the pressure needed to open a fracture parallel to cleavage ($P > \sigma_1 + 1$) is less than that required to open one perpendicular to cleavage ($P > \sigma_3 + 30$). Precipitation in the fracture produces a vein parallel to cleavage which may grow progressively by a combination of boudinage and more precipitation in the low-*P* zones between boudins (Fig. 5.35B).

The alternative idea of strain-induced solution transfer is advocated by Bell *et al.* (1986) and Bell and Cuff (1989). We have already seen that solution of quartz takes place mainly in the steeper limbs of crenulations (Figs 5.30 and 5.31); this process leads to the progressive development of crenulation cleavage, especially in pelitic sediments, as shown in Fig. 5.36. Strain is actually partitioned between M-domains (high non-coaxial shear strain and sheet-silicate rich – **M** for mica) and Q-domains (low strain and

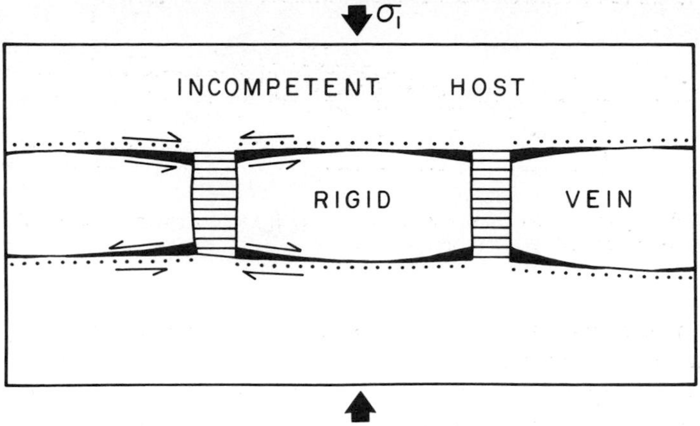

Fig. 5.34 Combination of mechanisms whereby rigid quartz–feldspar veins perpendicular to σ_1 may thicken and extend. Boudinage creates spaces that fill with quartz and feldspar precipitates. Sliding of host against the vein near fracture sites creates further dilation and precipitation, and if an excess of quartz and feldspar is available it will tend to precipitate at the vein host interface according to the mechanism described in Fig. 5.33. Based on Sawyer and Robin (1986).

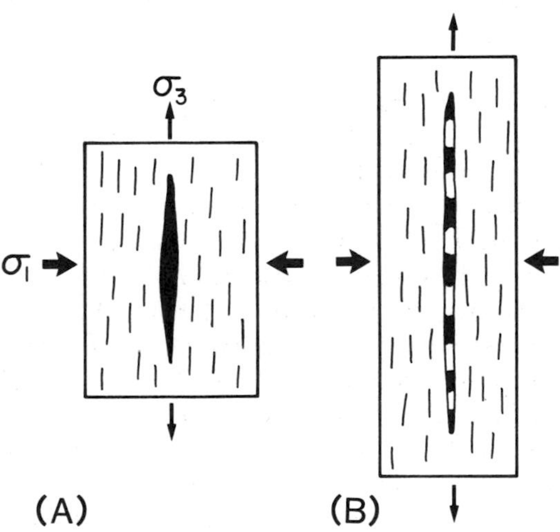

Fig. 5.35 Formation of veins perpendicular to σ_1, as described by Gratier (1987). Explanation in text.

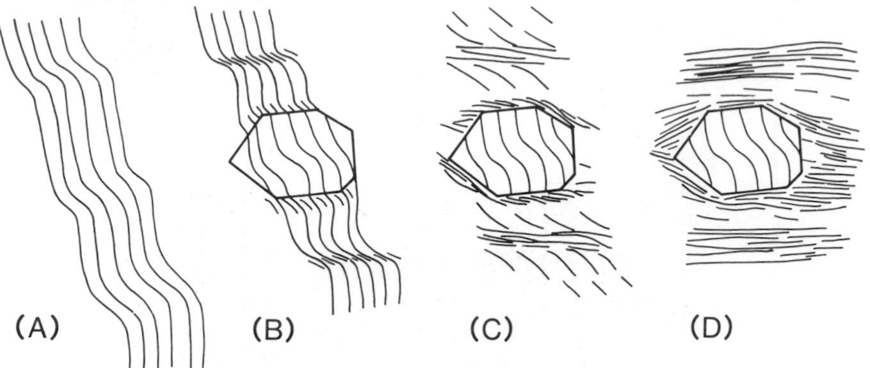

Fig. 5.36 Open crenulations (A) tighten in (B), and solution of quartz causes a concentration of sheet-silicates, and incipient crenulation cleavage to develop. A rigid porphyroblast overgrows and preserves the early stage of crenulation. Concentration of sheet-silicates due to solution increases (C), producing a new cleavage or schistosity. The M-domains increase in size, and evidence of early crenulations is found only in porphyroblasts (D).

Quartz-rich). Owing to progressive inhomogeneous shear, the M- and Q-domains are not parallel sided but anastomose (Fig. 5.37). The M-domains grow progressively at the expense of Q-domains until the new cleavage eventually dominates the structure (Fig. 5.36).

All minerals are more strained in M-domains than Q-domains, and the non-coaxial shear strain is accommodated most easily by sheet-silicates slipping on {001}, whereas quartz and feldspar become dislocated. Owing to this greater strain, Bell *et al.* (1986) suggest that quartz (and other non-sheet-silicate minerals) are most susceptible to solution in M-domains. The Q-domains are more competent than M-domains and therefore tend to deform by microfracturing; consequently, they are sites where fluids migrate, assisting porphyroblastic growth as well as segregation of quartz from sheet-silicates. The end product is metamorphic layering parallel to cleavage or schistosity; evidence of the original crenulation may be lost (Fig. 5.36) except where preserved in porphyroblasts (section 5.5.1).

Bell and Cuff (1989) suggest that most material dissolved is transported out of the rock rather than into Q-domains; if this is the case, pelitic rocks become even more Al-rich during metamorphism. Nevertheless, in the Otago Schists, for example, significant precipitation does occur in Q-domain strain shadows and in quartz-rich veins. The relative proportions of material precipitated in schistosity-parallel rather than cross-cutting veins, and the quantity of solute totally lost to the system are subjects requiring further investigation.

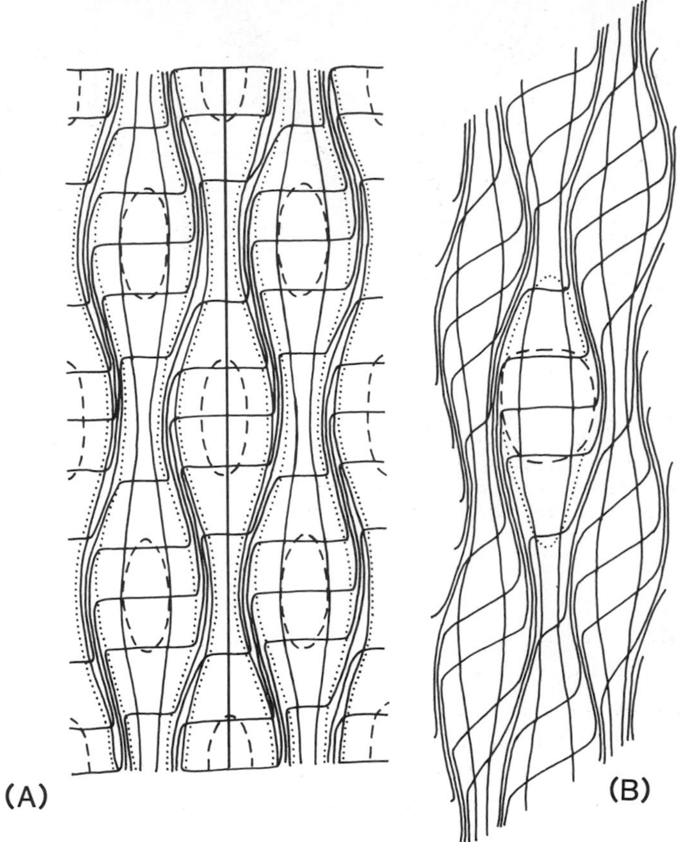

Fig. 5.37 Deformation partitioning and development of M- and Q- domains in strain-field diagrams for a non-coaxial progressive bulk inhomogeneous shortening (plane shown contains axes of shortening and extension). (A): Progressive shortening plus shearing occurred between the dotted lines (M-domains). Q-domains undergo no strain in the areas enclosed by dashed lines, and progressive shortening dominates between dashed and dotted lines. (B): A rigid porphyroblast has grown in one of the areas contained by dashed lines in (A), and further deformation is repartitioned about it. The porphyroblast itself does not rotate despite progressive shearing of the matrix. Reproduced with permission from Bell (1985).

5.3.5 Veins and pressure (strain) shadows

It has already been noted that material transported in solution may precipitate in the 'shadow' of porphyroblasts (the sides facing the direction of extension or the minimum principal compressive stress direction, σ_3), and in veins. Some veins represent an essentially non-dilational replacement (or metasomatism), as judged from the lack of displacement of marker horizons either side (Fig. 5.38), but the majority represent

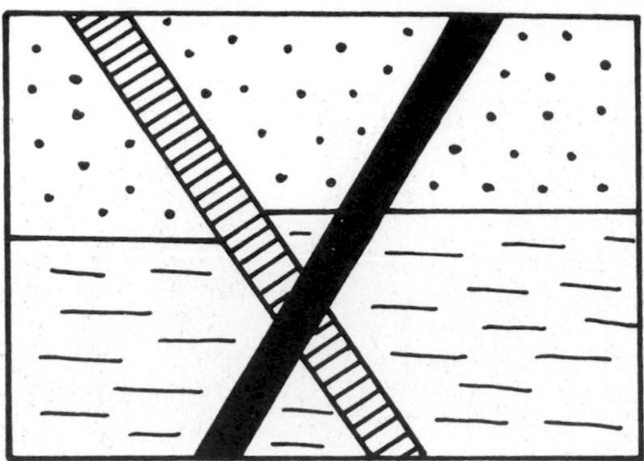

Fig. 5.38 An early vein (lined) displaces a marker horizon due to dilation. In contrast a later vein (black) of replacement origin does not displace either the marker horizon or the early vein. The lack of displacement can only be judged as such if at least two 'markers' of different attitude are intersected by the vein.

dilational fissure fillings. We have already seen that stepped shear-surfaces create open fissures, and these may parallel existing layering and schistosity, or cross-cut, possibly as a conjugate set. Other cross-cutting veins, perhaps the majority, represent tensional fissures, often created by the fluids themselves; this is because dehydration reactions can induce fluid pressures that are greater than lithostatic pressure, causing 'hydraulic fracturing' and dilation, even in compression and when the rock is otherwise ductile (e.g., Shelley, 1968). Gratier (1987) suggested that some tensional fissure fillings form perpendicular to σ_1 (section 5.3.4), but the most common attitude is likely to be perpendicular to the extension direction or σ_3.

The fillings of pressure shadows and veins, especially at low or very low grades of metamorphism, tend to be fibrous (Figs 5.39 and 5.40). The fibre lengths may or may not correspond with some rational crystallographic direction. In complexly deformed schists and higher-grade rocks, such fibres are unstable due to their high surface area/volume and irrational crystallography, and fillings are instead a mixture of granoblastic mosaic and euhedral crystals (of sheet-silicates, for example), thus mimicking the textures of the rock as a whole. Low-grade fibrous veins can be categorized as follows:

1. **Syntaxial veins**. The fibres nucleate on existing crystals either side of the fissure, and grow progressively inwards as the fissure opens (Fig. 5.40A). Fibres may thicken and thin depending on the supply of material and competitive growth among fibres. Syntaxial veins are therefore:

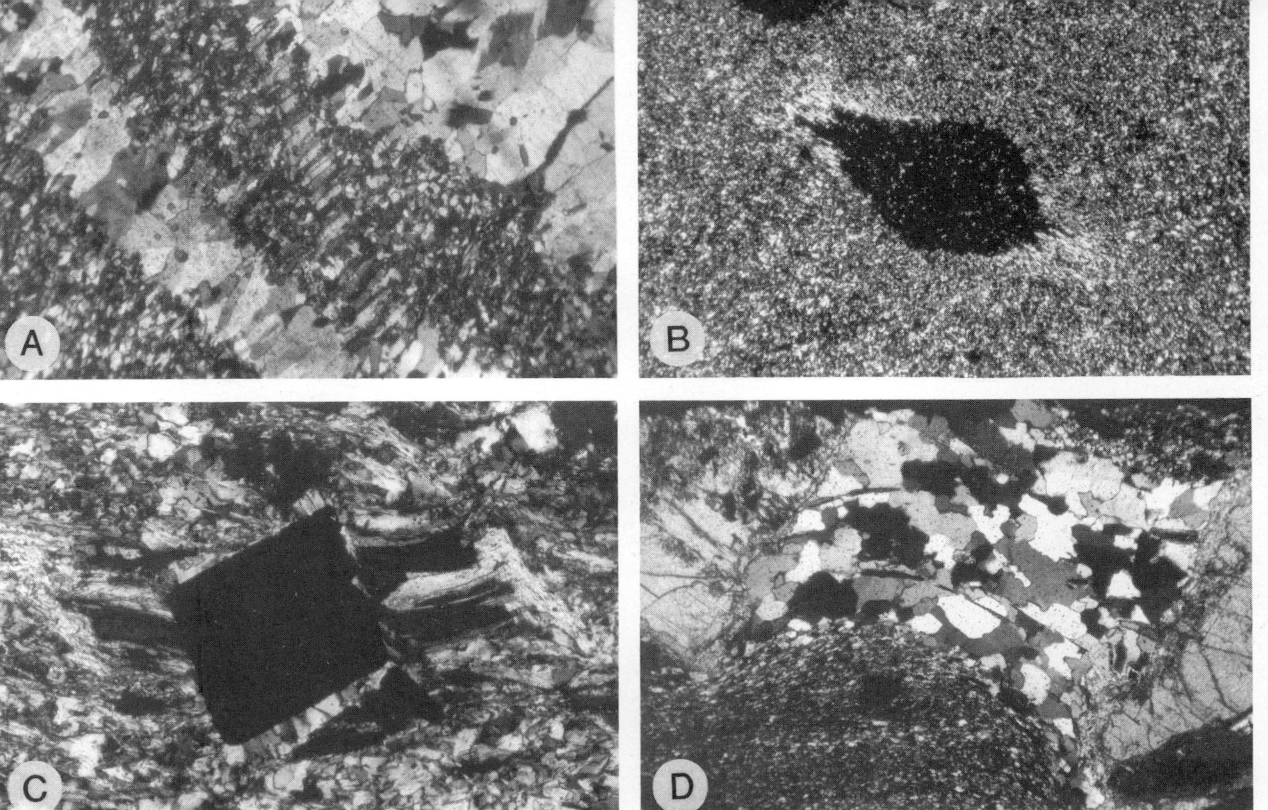

Fig. 5.39 (A): Fibrous vein fillings of quartz (especially well shown in the smaller central vein). Morgat, France. (B): Fibrous mica pressure-shadow around a siderite crystal. (C): Fibrous quartz, muscovite and chlorite pressure-shadow around a pyrite crystal showing a mixture of long displacement-controlled fibres and shorter face-controlled fibres. Waingaro Schist, NZ. (D): The space created when an andalusite crystal was broken in two was filled with a quartz pressure-shadow which underwent later dynamic recrystallization (note the serrated grain boundaries). Buller Gorge, NZ. (View lengths measure 3.3 mm (A) and (B), 0.85 mm (C), 3.3 mm (D).)

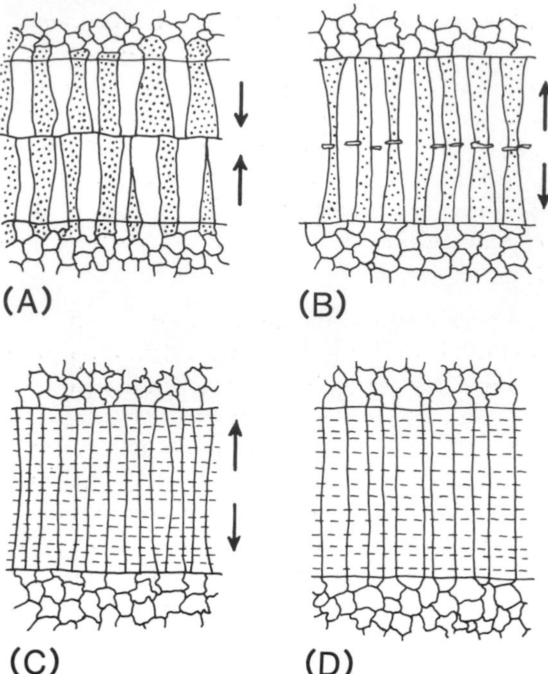

Fig. 5.40 (A): Syntaxial vein with central suture. (B): Antitaxial vein with debris from host rock at centre. (C): Antitaxial vein with numerous planes of wall-rock or fluid inclusions indicating a crack–seal mechanism of growth. (D): Stretched crystal fibre vein formed by nucleation on broken grains in the vein wall and propagation by a crack–seal mechanism. Arrows indicate the directions of crystal growth in (A) to (C); the growth direction in (D) may be variable. Discussion in text.

(a) of the same mineralogy as the host rock;
(b) made of fibres that generally do not match in size or species at the centre of the vein (the centre is always marked by a prominent suture).

2. **Antitaxial veins**. Fibres do not nucleate on the fissure wall, but instead grow continuously from the centre outwards (Figs 5.40B and C). In sharp contrast to syntaxial veins, the fibres in antitaxial veins run continuously across the centre in either direction, although some may thicken or thin due to competitive growth. Most antitaxial vein fillings have a different mineralogy from the host rock, which probably explains why they did not nucleate on existing crystals in the walls. The centre of antitaxial veins may be marked by debris from the host rock, trapped there as the fissure opened. A succession of opening events may be recorded by several planes of debris, sealed in by further growth: a process called crack–seal deformation by Ramsay (1980).

3. Various combinations of these two types of vein-filling process occur. Composite veins, for example, may be syntaxial at their margins (recording the first stages of opening), and antitaxial at the centre. In addition, 'stretched crystal fibre veins' (Ramsay and Huber, 1983) record the syntaxial-type nucleation of fibres on both sides of a fissure (on crystals broken in two), and the subsequent growth of the vein-filling by a crack–seal mechanism. The impression created is that individual crystals of the host rock have been stretched across the vein (Fig. 5.40D).

Fibrous pressure shadows show similar features to vein-fillings. The opening is caused by deformation of matrix around a rigid body such as a pyrite crystal (Figs 5.39B and C). In some cases, the fibres nucleate on crystals in the host rock and grow towards the rigid crystal. This is typically the case with quartz fibres around pyrite crystals where fibres may be either parallel to the displacement direction (called **displacement-controlled fibres** – Fig. 5.39C), or perpendicular to the rigid crystal faces (**face-controlled fibres** – Fig. 5.39C). In other cases, fibres nucleate on the rigid crystal and grow outwards in the displacement direction (another form of displacement-controlled fibre), as found with calcite pressure-shadows around fossil fragments.

Curved fibre growths in pressure shadows and veins hold valuable information on strain history, a subject dealt with thoroughly by Ramsay and Huber (1983).

5.3.6 Migmatite layering

Migmatites are rocks that superficially at least give the impression of a mixed igneous and metamorphic parentage. The terminology is provided in section 2.5.3. The 'metamorphic' part is usually a high-grade schist or gneiss. The nature of the 'igneous' part is more contentious: does it represent an external supply of magma injected as veins and layers, or is it a partial melt of the metamorphic rock, variously segregated in layers and/or injected as veins or is it simply metamorphic segregation layering of high grade, similar to that discussed in the previous section?

Debate on migmatite was an integral part of the 'granite controversy', and since the 'granitizers' lost that battle there has been a deeply held suspicion of invoking widespread metasomatism during migmatization. In general we lack hard data on this issue, mainly because of the difficulty of sampling migmatites for geochemical study (Olsen, 1985).

The notion that supposedly viscous granitic magma can be injected to form the leucosome layers in areally extensive migmatites is now rejected by most workers. The only exceptions may be immediately adjacent to plutons.

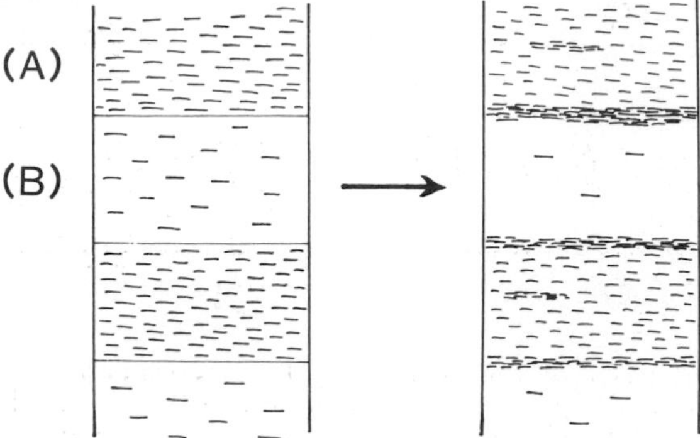

Fig. 5.41 Migmatization as described by Johannes and Gupta (1982). Relatively felsic layers in the original paragneiss (B) undergo partial melting and form leucosomes as mafic restite material is segregated and concentrated at the boundary of leucosome and mesosome. The more refractory layers (A) undergo less change and form mesosomes. However, with progressive increase in temperature, mesosomes may also undergo progressive partial melting with further segregation of mafic material.

The notion of partial melting is widely accepted, and Johannes and Gupta (1982) and Johannes (1988) believe that it proceeds almost isochemically, layer by layer (Fig. 5.41). In this model, neosomes represent layers that melted. Segregation of mafic restites towards neosome margins produces the striking biotite-rich melanosomes that border leucosomes. The mesosomes represent more refractory layers. Leucosomes do not necessarily form layers. For example, Pattison and Harte (1988) describe refractory hornfels interlayered with semi-pelitic material that partially melted. Deformation accompanying melting caused cracking of the hornfels, and the melt migrated into the opening fissures (Fig. 5.42A). McLellan (1988) also describes leucosome fissure-filling, but in this case it was accompanied by shearing (Fig. 5.42B).

The popularity of partial melting belies a number of serious problems. One involves the composition of leucosomes. Following the experimental work of Tuttle and Bowen (1958), it was widely expected that leucosomes would have compositions corresponding with the temperature minimum in the system Qz–Ab–Or–H_2O (Figs 3.18 and 4.19). The reality is often far from this, and trondhjemitic (plagioclase + quartz) compositions are very common. Sometimes these have cotectic compositions (Ashworth, 1976), but Yardley (1978) argued that the lack of K-feldspar indicates a subsolidus metamorphic origin. Ashworth (1985) shows that it is possible to have low-temperature trondhjemitic melts; in other cases K-feldspar has been replaced during retrogression by myrmekite and muscovite–quartz symplectites (discussed below). However, the serious problem remains

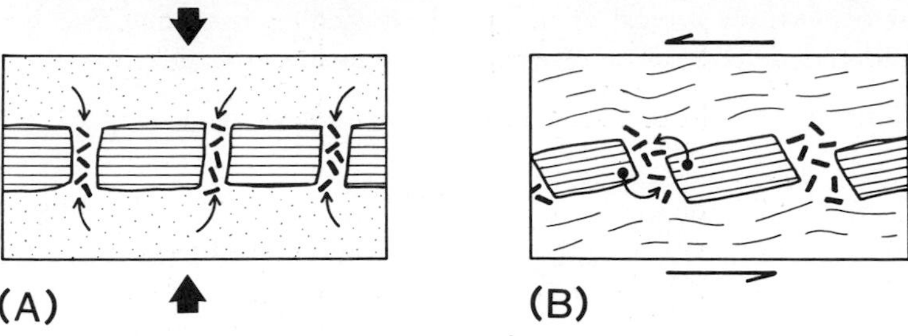

Fig. 5.42 (A): Migration of melt from melting layers (stippled) into the spaces between boudins of more refractory layers (lined), as described by Pattison and Harte (1988). (B): Collection of melt between boudins rotated during shear, as described by McLellan (1988). McLellan suggested that much of the melt is derived from melting the corners of boudins.

that bulk leucosome compositions cannot be easily used as an unequivocal argument for or against melting.

Another problem concerns plagioclase composition, which one might expect to be fractionated during melting; leucosomes should have the more Na-rich plagioclase (Fig. 3.10). This is by no means always the case, and again Ashworth (1985) provides a string of possible reasons including post-migmatization re-equilibration, unstable melting (where An_{15}, for example, simply melts to An_{15} rather than differentiating as expected from Fig. 3.10), and homogenization due to reactions with the melt on cooling (favoured by Johannes, 1985).

Owing to difficulties in applying the melting hypothesis, other workers such as Yardley (1978) have favoured subsolidus metamorphic segregation. Certainly the abundant segregation layers and veins found in many schist terranes are similar in many respects to those of supposed 'magmatic' origin. Sawyer and Barnes (1988) take a geochemical approach to solving this problem for some Canadian migmatites; they show that schistosity-parallel layers are metamorphic but cross-cutting layers were partial melts, and they can discriminate between them on the basis of Th/Y and Hf/Yb ratios.

More relevant to this book are textural criteria which may discriminate magmatic and subsolidus origins. To this end, Vernon and Collins (1988) contrast crystal shapes in leucosomes and mesosomes. In the latter, grains form polygonal mosaics, and except for {001} of biotite, crystal faces are rare. In the former, K-feldspar, plagioclase, cordierite and andalusite all show euhedral forms against quartz, and they view this as a typical igneous feature. They identify inclusion-rich cores to crystals in leucosomes as restite material identical to material in mesosomes; in the leucosomes the restite cores are overgrown by inclusion-free euhedral material. Ashworth and McLellan (1985) add euhedral zoning, especially oscillatory zoning, to the list of textures typical of magmatic crystallization; they also point to the following retrogressive reaction as diagnostic of a melt:

sillimanite + K-feld + plag + hydrous melt = musc + qtz + plag

The reaction products typically form myrmekite and muscovite–quartz symplectite replacement of K-feldspar. Carlsbad twinning should be added to this list (Gorai, 1951), and I would concur that all the above features strongly suggest an igneous environment. However, here are some words of caution: euhedral feldspar is found in greenschist-facies segregation layers (Shelley, 1989a); euhedral zoning is common in metamorphic garnet and oscillatory zoning does occur in metamorphic rocks (Philippot and Kienast, 1989). These features must be regarded as suggestive but not diagnostic indicators of magmatic crystallization.

McLellan (1983) tackles the textural question in a different way. She proposes that the distribution of grain contacts between the minerals present can be used to categorize various origins.

1. An 'aggregate relationship' (i.e., grains of one mineral tend to cluster together) typifies metamorphic segregation, because in a solid state, nucleation is strongly heterogeneous.
2. A truly 'random relationship' typifies growth from a melt or hydrothermal fluid (but not a grain-boundary fluid).
3. A 'dispersed distribution' has more like/like contacts than random, less than aggregate distributions, and is thought to typify high-grade rocks affected by annealing where the initial aggregate distribution becomes dispersed during textural equilibration.

The practical matter of how to measure such distributions is dealt with by McLellan (1983) and Ashworth and McLellan (1985). In any such analysis, one must bear in mind that migmatite textures may represent subsequent metamorphic events, not only the process of migmatization.

Textural observations can provide direct clues to the reactions involved during migmatization. We already know from section 4.2.3 that water has an important effect in reducing the melting temperature in granite systems. Johannes (1985) notes that mineral assemblages in many migmatites indicate temperatures between 630°C and 730°C, and at 730°C and 5 kbar, for example, at least 1.4 wt. % water is required to produce 24% melting in suitable composition rocks. In view of the contention of Johannes that only neosome layers melt significantly, one can see that melting 40% of 20% of the total rock requires only 0.48 wt. % water, provided that water generated by reactions in refractory layers migrates to melting layers.

Water is released from mica during production of aluminium silicate (usually sillimanite) and K-feldspar. In most migmatites, swarms of tiny sillimanite prisms are found embedded in mica (Fig. 5.43), and the usual

Fig. 5.43 Swarms of sillimanite (fibrolite) embedded in mica, and with incoherent relationship. West Coast, South Island, NZ. Length of view measures 0.85 mm.

interpretation is that this represents prograde water-generating reactions such as:

$$\text{musc} + \text{qtz} = \text{K-feld} + \text{sillimanite} + H_2O$$

However, in many cases the sillimanite forms 'folded' clusters, incoherent with respect to mica, and the texture may represent the reverse retrograde reaction with mica overgrowing and replacing sillimanite.

Dougan (1983) showed how grain-contact distribution relationships can establish the precise reactions involved. A general examination of some migmatites suggested that leucosomes were generated from melanosomes by the reaction:

$$\text{biot} + \text{musc} + 3\ \text{qtz} = \text{garnet} + 2\ \text{K-feld} + 2H_2O$$

However, in the melanosomes he found that biotite–muscovite contacts increase in frequency and quartz–plagioclase, quartz–muscovite and biotite–quartz contacts decrease in frequency as the degree of leucosome removal increases. This indicates instead the following alternative water-generating reactions:

$$\text{musc} + \text{qtz} = \text{K-feld} + Al_2SiO_5 + H_2O$$

$$\text{biot} + 2\ \text{qtz} + Al_2SiO_5 = \text{K-feld} + \text{garnet} + H_2O$$

Some migmatites represent high-temperature dry melting; others again depend on an external supply of water. Thus, Pattison and Harte (1988) work out for the Ballachulish migmatites that, even with total biotite dehydration, insufficient water to explain the melting observed would be produced; they propose an external source from adjacent cooling quartz diorite which devolatilized at 700°C. Finally on the subject of water, Yardley (1978) points out that water-rich melts will freeze rapidly if the migmatite rises in the crust; leucosomes in such a case should be fine grained.

5.4 SHEAR-SENSE INDICATORS

Plate-tectonic theory has heightened geologists' interest in determining movement directions in tectonized rocks. The aim is to measure the rotational component of finite strain, and to determine the sense of rotation and the orientation of rotation axes. Studies of mineral preferred-orientations are important in this regard, as discussed in Part Three. Other methods involve the following microstructures, all of which are visible in the field or thin section, and which are described in turn below:

1. *S–C* planes (especially in mylonites)
2. Porphyroclast(blast) fracture patterns
3. Rolling structures, δ-type porphyroclast(blast) systems and asymmetrical microfolds
4. σ-type porphyroclast(blast) systems
5. Asymmetrical pressure shadows
6. Deformed vein systems

Much of the recent research has focused on narrow shear-zones and mylonitic rocks, but the criteria may be applied to a wide range of metamorphic rocks as well as flow and shear structures in igneous rocks.

The best way to view the structures is in a section perpendicular to the shear zone (or, in practice, perpendicular to a prominent foliation) and parallel to the movement direction (the direction of maximum elongation usually represented by a strong mineral lineation). Thin sections should be cut with this orientation.

It is generally assumed that the rotational component of finite strain is a non-coaxial plane strain, geometrically a simple shear (like a pack of cards deformation), with no shortening or extension along the intermediate strain axis (Fig. 7.15). Intuitively, this is how one would expect natural rocks to deform in a shear zone, but deviations undoubtedly occur, and some of the consequences of non-ideal simple shear are noted below.

5.4.1 *S–C* mylonites

Shear-zones in granitoids are often characterized by the development of two anastomosing foliations called *C* and *S* planes (Berthé *et al.*, 1979; Lister and Snoke, 1984). The *S* planes are at a high angle to the maximum principal compressive stress direction, and represent mainly a non-rotational shortening; the *C* planes develop parallel to or within a small angle of the shear-zone wall, more or less at 45° to the directions of shortening and extension, and represent a severe degree of strain including a significant simple-shear component. Both planes anastomose, and a strong stretching lineation is usually developed perpendicular to the line of intersection of the *C* and *S* planes (Fig. 5.44). The fact that *C* and *S* planes are both present in thin shear zones cutting homogeneous granite proves that they can develop more-or-less simultaneously. However, *C* and *S* planes resemble intersecting foliations of differing age (the *C* planes look as if they are later than *S*). Such a sequence of events is commonly the case in other rocks, and for mylonites Shimamoto (1989) suggests that *C* planes do develop late during any one shearing event.

There are in fact *C* planes and *C* planes, depending on their exact

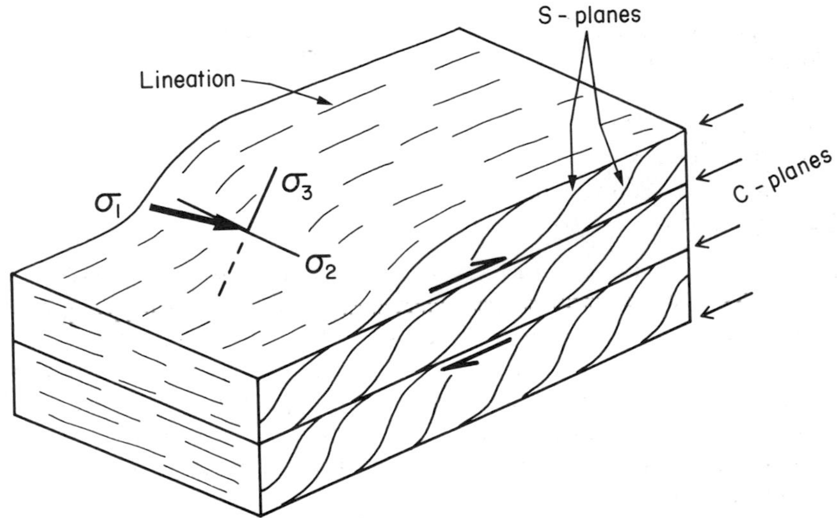

Fig. 5.44 Block diagram to illustrate *C* and *S* planes.

attitude in relation to the shear-zone wall. Strictly speaking, *C* planes are parallel to the shear-zone wall, but other shear surfaces (often called *C′* planes), which are synthetic Riedel shears at a small angle to the shear-zone boundary (Fig. 5.45), may develop instead (the antithetic Riedel shears are much less common). In experimental deformation of salt, Shimamoto (1989, fig. 7) presents histograms for the orientations of *C* and *C′* planes that show the two essentially merging one into the other; clearly, any distinction between types of *C* planes is difficult, all the more so if the shear-zone boundary wall itself is not exposed.

In thin section, *S–C* granite mylonites are characterized by ribbons of quartz, porphyroclasts of feldspar and mica (which occupy the spaces delineated by pairs of *C* and *S* planes), and tails of cataclased or recrystallized feldspar and mica that extend from the porphyroclasts into the *C* planes. In some deformed quartz–mica schists (called type-II *S–C* mylonites by Lister and Snoke, 1984), the *S* planes are not prominently developed, but the *C* planes are by virtue of the tails extending along them from mica porphyroclasts called mica fish (Fig. 5.46). The mica fish are conspicuous in the field by virtue of the strong reflection from the mica cleavages, all tilted in one sense relative to the shear plane and movement direction.

5.4.2 Porphyroclast(blast) fracture patterns

Crystals such as feldspar, unable to deform plastically, fracture along cleavage planes in order to accommodate some of the strain. In some cases, fracturing is so pervasive that the crystals appear to have been plastically deformed (Tullis and Yund, 1987). In other cases, a clear fracture pattern is produced, and the direction of movement on any one fracture system can be understood by reference to Fig. 5.45. Thus a fracture within 20° or so of the antithetic and synthetic shears will become normal faults, whereas those in the acute angle between *C* and *S*, and at a high angle to *S* will become reverse faults. Bodily rotation of the porphyroclast may cause the fracture planes to rotate into new orientations so that a normal fault becomes reverse, and vice versa.

5.4.3 Rolling structures, δ-type porphyroclast(blast) systems and asymmetrical microfolds

A rigid crystal will rotate as the matrix is sheared, provided that there is little or no decoupling of matrix and porphyroclast(blast). If the matrix is layered or if tails or pressure shadows extend either side of the porphyroclast, then the rotation will be marked by the layers or tails being rolled around the rigid crystal to form what is called a δ-type por-

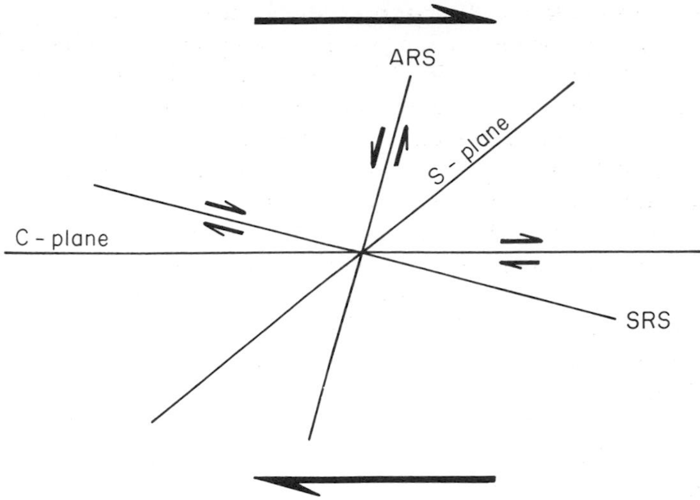

Fig. 5.45 Relationship of *C* and *S* planes to synthetic Riedel shear planes (SRS) and antithetic Riedel shear planes (ARS).

Fig. 5.46 'Mica fish' with tails of cataclased and/or recrystallized mica extending along *C* planes in mylonite from the Constant Gneiss, NZ. Length of view measures 0.85 mm.

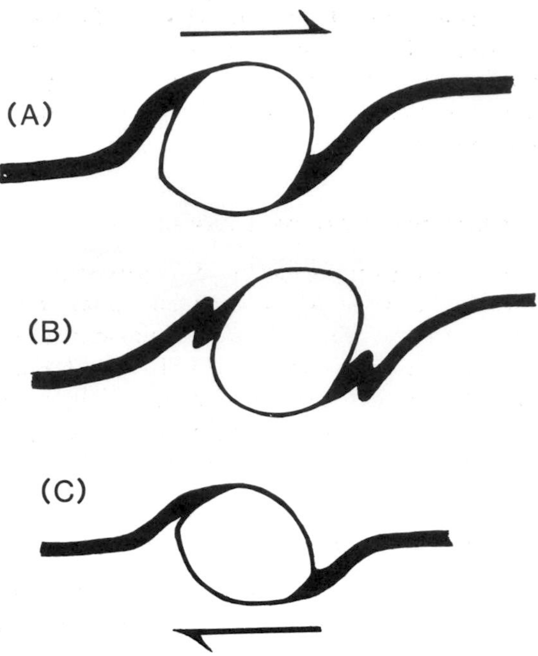

Fig. 5.47 Some δ-type porphyroclast systems in dextral shear. (A): Stair-stepped (not in the same plane) tails typical of ideal simple shear. (B): Asymmetrical microfolds developed in tails. (C): Non-stair-stepped (in the same plane) tails typical of non-ideal simple shear. Note also the typical orientation of the long axis of the clast for non-ideal simple shear.

phyroclast structure (Passchier and Simpson, 1986); note the double curvature that characterizes the deformed layers or tails (Fig. 5.47). In some cases, the matrix develops asymmetrical microfolds either side of the crystal (Fig. 5.47B), and similar structures about phenocrysts in flow-layered volcanic rocks provide information on flow directions (Vernon, 1987).

Hanmer (1990) has investigated how the structures differ in ideal and non-ideal simple shear. In the latter, the layer either side of the δ-structure is in the same plane, and inequidimensional porphyroclasts rotate until they reach a stable position with long-axes parallel to the shortening direction (Fig. 5.47C). In ideal simple shear, the layer on either side is stair stepped (Fig. 5.47A).

5.4.4 σ-type porphyroclast(blast) systems

If the rigid object behaves passively during shearing of the matrix, the pressure shadow or tail is not in any way rolled around the crystal, but simply develops a stair-stepped structure (Fig. 5.46) without the double curvature that characterizes δ-type systems. The σ-type structures are most commonly developed in association with *S–C* mylonites, and mica fish are just one example. It is possible for σ-type structures to evolve progressively into δ-types.

5.4.5 Asymmetrical pressure shadows

Fibrous pressure-shadow growths that are displacement controlled (section 5.3.5) may curve in response to non-coaxial strain (Fig. 5.39C), and record the position of the direction of extension through time. This method of strain analysis is treated thoroughly by Ramsay and Huber (1983). The major factor that must be determined is whether the fibres grew from the matrix towards the rigid crystal as space opened up between them, or the other way round (section 5.3.5).

5.4.6 Deformed vein systems

Brittle–ductile shear zones are characterized by tensional gashes approximately 45° to the shear-zone wall and parallel to the direction of shortening. The early-formed cracks rotate during progressive shearing (Fig. 5.48A), but the tips of the cracks continue to propagate parallel to the direction of shortening. The end result is an array of sigmoidal veins, whose geometry indicates the movement direction. Sigmoidal veins may also originate by mechanical interaction of fractures formed during brittle–elastic cracking (Olson and Pollard, 1991). The geometry of the two types differs (Figs 5.48A and B).

Pre-existing veins of whatever origin buckle, often with ptygmatic form, if within approximately 40° of the shortening direction. If within 40° of the extension direction they will be thinned or boudinaged instead. Such behaviour is not confined to shear zones, but the distinctive feature of a shear zone is that some veins may rotate from the shortening quadrant into the extensional zone; in other words, already-folded veins that rotate past the perpendicular to the shear-zone wall are subsequently boudinaged (Fig. 5.48C).

5.5 METAMORPHIC HISTORIES

5.5.1 Inclusion textures in porphyroblasts

Porphyroblast inclusion textures do not represent the most stable configuration because surface area increases with the number of inclusions.

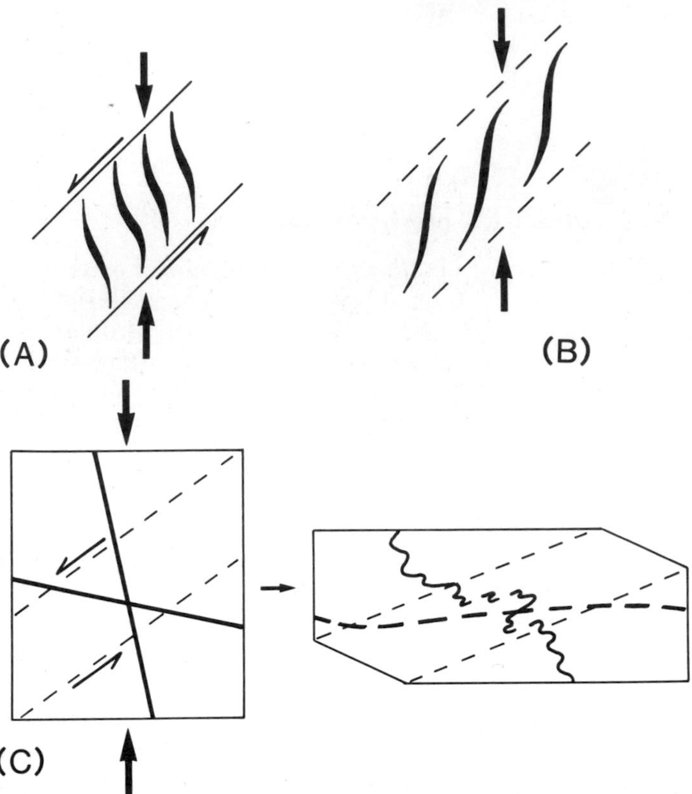

Fig. 5.48 (A): Sigmoidal veins in a brittle–ductile shear zone. (B): Sigmoidal veins due to brittle–elastic deformation and mechanical interaction of neighbouring fractures. The dashed lines delimit the deformation zone, and may develop into a shear zone. (C): Two veins in a block of rock undergoing shortening and shear-zone development. One vein simply boudinages; the other buckles at first, but may subsequently undergo boudinage, especially in the shear zone.

Similar textures in other environments result from rapid growth and/or a high growth/nucleation rate ratio. In metamorphic rocks this is not necessarily the case. Consider garnet growing according to the reaction:

$$\text{musc} + \text{chlorite} = \text{biot} + \text{garnet} + \text{qtz} + H_2O$$

Various stages of garnet growth are shown in Fig. 5.49. The rock is at all times a completely solid mesh of crystals, and therefore the quartz grains not used in the garnet-forming reaction are not likely to be 'swept aside' as if in a liquid, but will instead be trapped as inclusions. This will generally be the case if the rate-limiting factor is the actual rate of dissolution or the diffusion rate, or when the included grains form a rigid three-dimensional

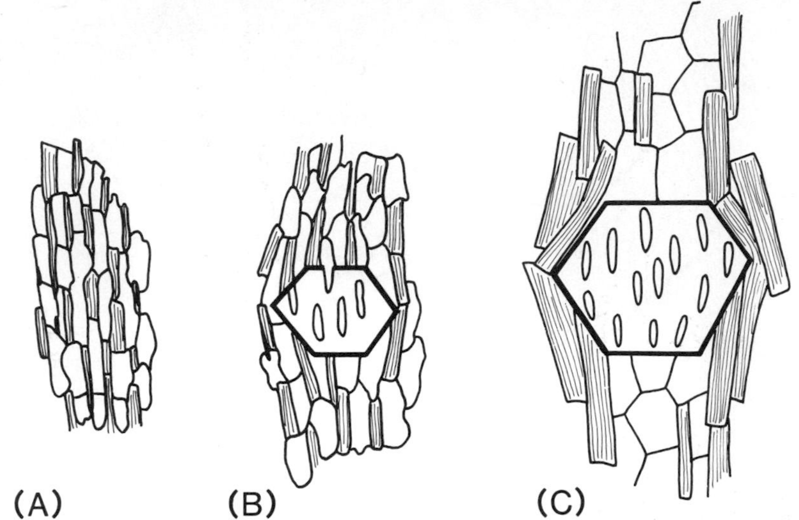

Fig. 5.49 Schematic development of a garnet porphyroblast with quartz inclusions in a chlorite–muscovite–quartz schist, as discussed in text.

framework. Only if growth rate is the rate-limiting factor will poikiloblastic development be inhibited.

Subsequent to porphyroblastic growth, the matrix around the garnet may be deformed and/or recrystallized and/or coarsened (e.g., Fig. 5.49C), and herein lies the value of poikiloblastic textures: they preserve a record of grain shape and orientation, and an indication of grain size at the time of porphyroblastic growth.

Some words of caution are needed, however. Grain size may change during the process of inclusion. For example, quartz is a product of the garnet-forming reaction given above, so that one could imagine the included quartz grains growing larger than in the original matrix. In other situations, the included phase may be a reactant but one present in excess of that required; included grains then become smaller than the original. In other cases, grain size results from complex interrelated reactions proceeding simultaneously. A few moments thought must also be given to distinguishing poikiloblastic textures from those formed by unmixing, replacement or coupled growth (Table 3.1). Fortunately, distinction is usually easy.

Progressive regional metamorphism typically involves a complex of structural and mineralogical change. In pelitic rocks, crenulation of early planar features evolves into a new cleavage or schistosity, which in turn may be crenulated, and then develop into a new schistosity, and so on (Fig. 5.50). The schistosity sequence may be labelled S_1, S_2, S_3, S_4, etc.

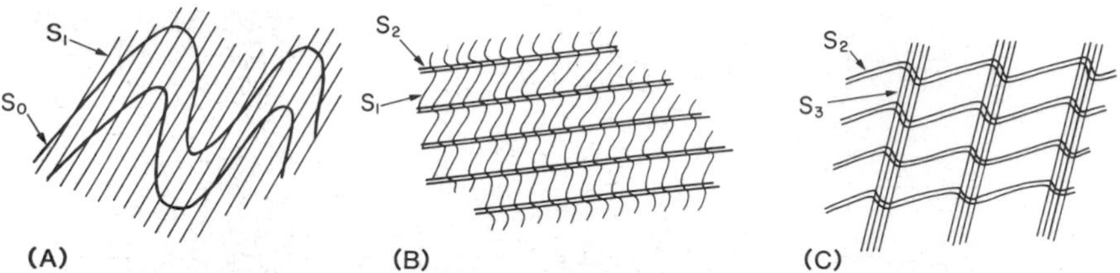

Fig. 5.50 (A) to (C): Sequential folding and cleavage or schistosity development starting with folding of bedding S_0 and developing in turn S_1, S_2 and S_3.

Bedding is S_0. Porphyroblasts that grow at any one stage of the structural development may preserve a record of it. The included schistosity is called *Si*, and the evolved schistosity of the matrix *Se*. The object of a historical analysis is to equate *Si* with, for example, S_2, and therefore perhaps *Se* with S_3. To be done properly, this requires thorough integration with structural field-work. On a practical note, oriented samples should be collected in the field (see Part Three). The principal thin section should be cut perpendicular to fold structures (and parallel, or sometimes perpendicular, to linear structures); if not done, the geometry of inclusion textures will be difficult to interpret.

Emphasis in the past has been placed on determining whether a porphyroblast is pre-tectonic, syn-tectonic, or post-tectonic, and some of the principal criteria used are shown in Fig. 5.51. Such emphasis, however, is misplaced. First of all an understanding of M- and Q-domain evolution during crenulation shows many criteria to be ambiguous at best. Secondly, the idea of pre-, syn- and post- is often artificial in terms of a structural development involving several phases of schistosity formation; for example, a crystal may be pre-S_3 but syn-S_2 and post-S_1. Thirdly, some criteria depend on the idea that porphyroblasts rotate during deformation, and there is increasing evidence that they do this very rarely.

The idea that hard, rigid porphyroblasts remain more-or-less fixed in spatial orientation during deformation, and that rotational forces are taken up by the matrix, was first mooted seriously by Ramsay (1962). Similar-type folds are produced by penetrative deformation of the matrix, and Ramsay's idea means that porphyroblasts with inclusion trails pre-dating the folding should remain fixed in orientation (Fig. 5.52) – (obviously this does not refer to concentric-type folds produced by bulk rotation of layers). Many workers now find this to be the case (Fyson, 1975, 1980; Bell and Johnson, 1989; Johnson, 1990); indeed inclusion trails do not change orientation over hundreds or thousands of square kilometres, regardless of the complexity of deformational history.

Even within a single similar fold, S- and Z-shaped inclusion trails

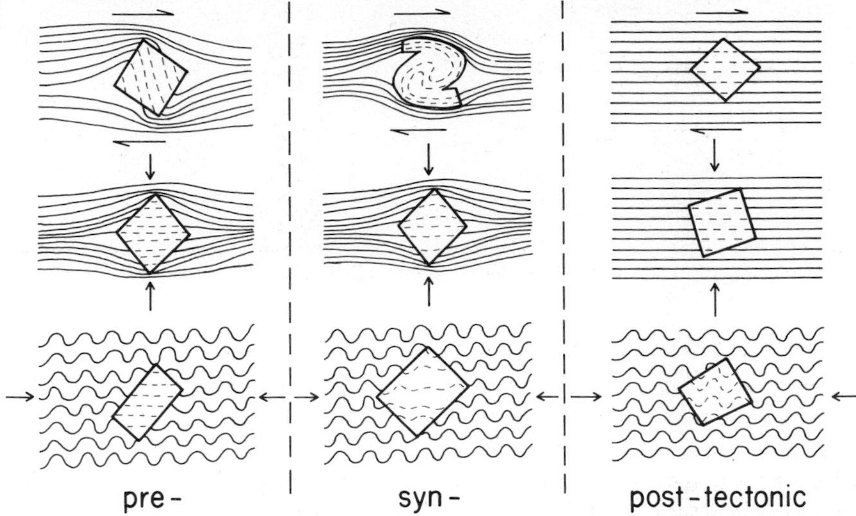

Fig. 5.51 Relationships between porphyroblasts, external foliation and inclusion patterns for pre-, syn- and post-tectonic porphyroblast growth, as proposed by Zwart (1962).

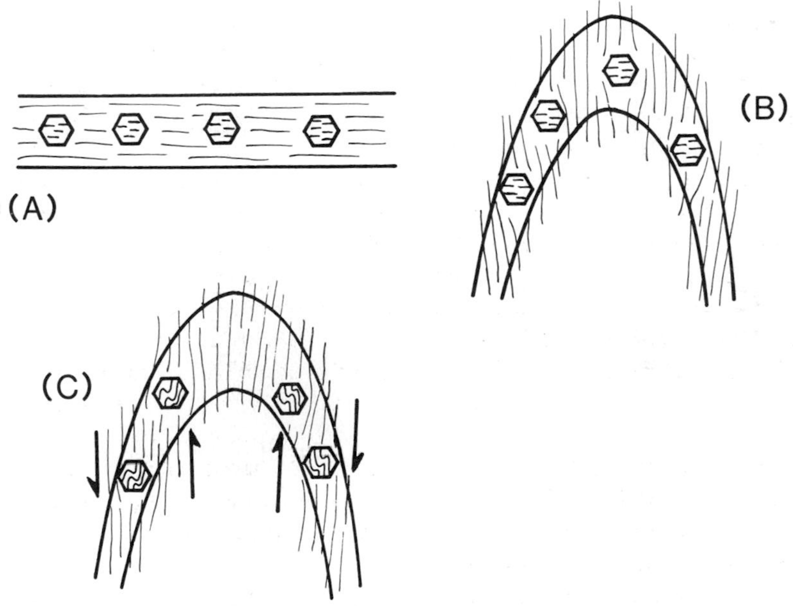

Fig. 5.52 (A) to (B): Idea of Ramsay (1962) that porphyroblasts do not rotate during similar-type fold formation. (C): Inconsistency between observed S- and Z-shaped inclusion patterns in similar-type folds, possible rotation senses (arrows), and predicted patterns of Fig. 5.51.

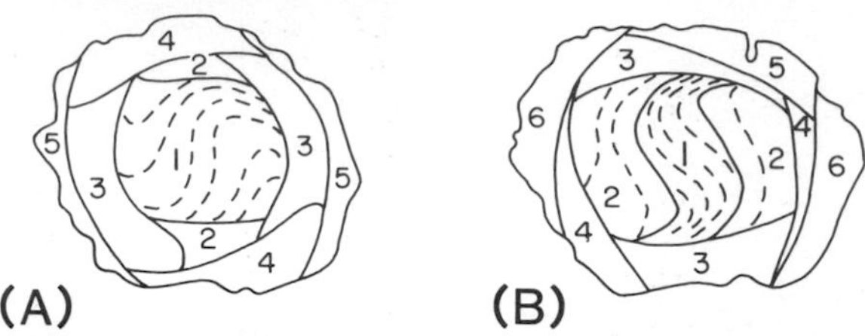

Fig. 5.53 Two examples of the interpretation given by Bell and Johnson (1989) of garnet inclusion patterns that superficially appear to be spiral shaped. Numbers 1 to 6 refer to successive growths that truncate earlier inclusion patterns.

cannot be easily interpreted in terms of rotation. Thus in Fig. 5.52C, the Z- and S-patterns on the left and right limbs are the reverse of that expected during syn-tectonic rotation (Fig. 5.51); the simpler explanation is that they represent porphyroblastic overgrowth of crenulations (which perhaps no longer survive in the matrix). If the trails 'spiral' through an angle of >180°, rotation seems likely, but according to Bell and Johnson (1989), very few of the 'spiral' inclusion patterns documented (e.g., Rosenfeld, 1970) really exist. Most seem to represent successive phases of inclusion which only give a false impression of spiralling (Fig. 5.53). Indeed, Bell and co-workers argue that spiralling is only feasible in an ideal (and rare) simple-shear deformation.

In terms of the typical anastomosing M- and Q-domain pattern of crenulated schist, porphyroblasts are most likely to grow in Q-domains. Consider the following scenarios (Fig. 5.54):

1. Garnet grows in a Q-domain and includes a straight trail. M-domains continue to encroach on Q-domains so that the final result is a truncation of *Si* by *Se*. The impression is of a pre-tectonic (*Se*) garnet, yet *Si* and *Se* developed together, and the garnet is syn-tectonic.
2. Garnet overgrows trails that curve from Q- to M-domains. Subsequently M-domains tighten around garnet, and if garnet is dissolved in the M-domain, *Si* is truncated.
3. If Q-domains are large relative to garnet crystals, one may mistakenly take the crystal to be post-tectonic, and this example shows how important is the scale of observation.
4. Shortening of pre-existing schistosity adjacent to a growing porphyroblast may produce a 'millipede' texture (Bell and Rubenach, 1980).

These figures are from sections perpendicular to fold axes. Sections parallel to a fold axis will often display relatively straight trails with *Si* more-or-less parallel to *Se*. Such garnets are easily misinterpreted as post-

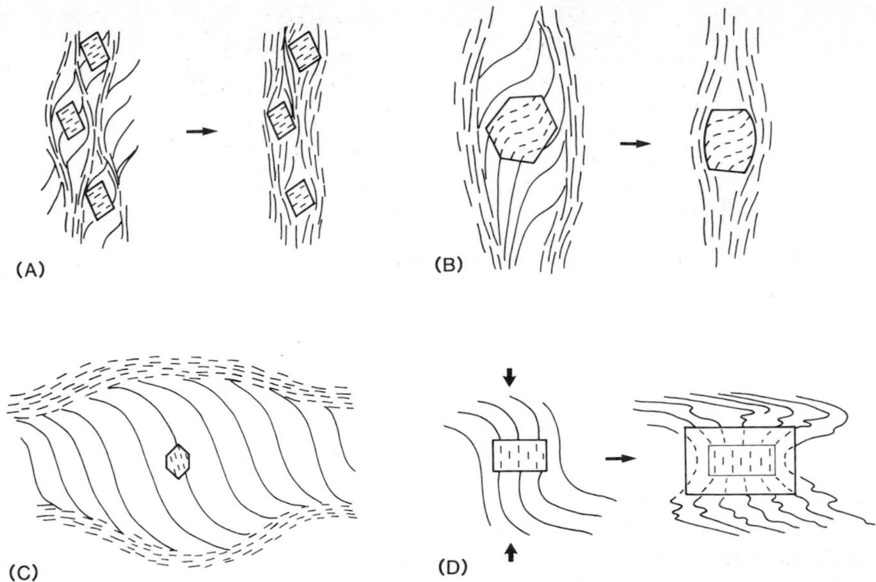

Fig. 5.54 Syn-tectonic porphyroblasts with various relationships between inclusion patterns and external foliation developed during the evolution of M- and Q-domains. The (A), (B), (C) and (D) are explained in the text by discussion of scenarios (1), (2), (3) and (4), respectively.

tectonic, and it illustrates the importance of knowing the orientation of a thin section relative to a structure.

Bell *et al.* (1986) contend that all poikiloblastic growth is accompanied by deformation. During thermal metamorphism it is true that a large pluton can cause significant deformation during emplacement (spotted slates and schists in thermal aureoles are commonly crenulated), and porphyroblast growth implies fluid movement and some degree of microfracturing. Nevertheless, this contention seems to conflict with common observations of post-tectonic porphyroblasts. Vernon (1989) notes that judgement on this requires a complete appreciation of the scale of observation. 'Post-tectonic' may be valid on the thin-section or outcrop scale, for example, but not in the wider geological perspective.

5.5.2 Determining metamorphic histories

(a) Sequences of mineral growth and deformation

Inclusions in porphyroblasts record the general characteristics of the metamorphic fabric of the time (including spatial orientation), and the fact that matrix continues to evolve, perhaps accompanied by further

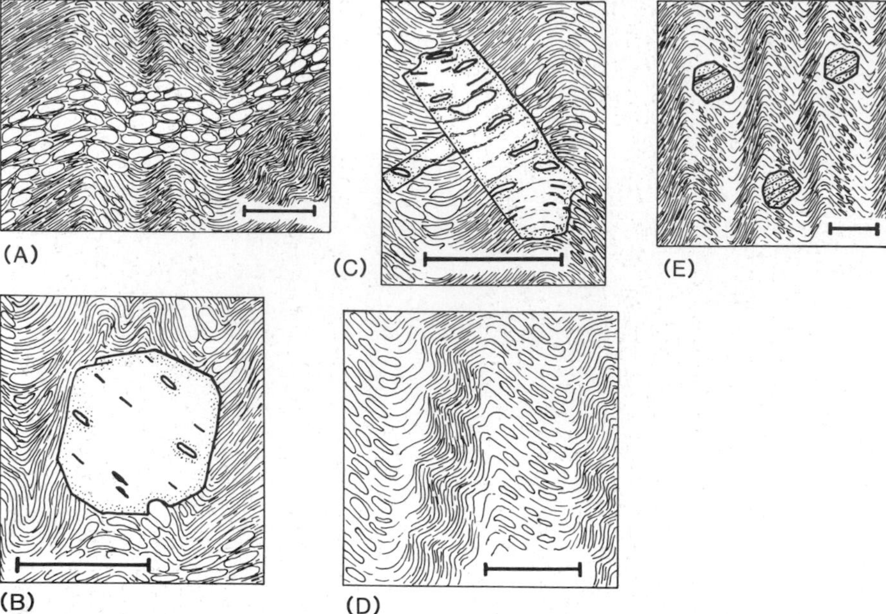

Fig. 5.55 Sketches to illustrate textural features that elucidate the historical development of a pelitic garnet schist from the Palaeozoic near Springs Junction, NZ. Discussion in text. Scale bars represent 0.5 mm.

porphyroblastic growth, means that a sequential record of metamorphic events may be preserved. Following the pioneer work of Zwart (1960) and Johnson (1961), the 1960s were a time when every metamorphic rock was analysed in terms of $D_1, D_2, \ldots, F_1, F_2, \ldots, S_1, S_2, \ldots, M_1, M_2, \ldots$ etc. (D = deformation, F = folding, S = schistosity, M = mineral growth). The following example (Fig. 5.55) is such an analysis of a garnet schist from the Palaeozoic of New Zealand:

1. Original sedimentary features include clastic quartz and compositionally distinct sedimentary layering (*So*). Much quartz remains as discrete unrecrystallized grains encased in newly crystallized mica. Quartz grain size is coarser in the quartz-rich layers (Fig. 5.55A).
2. Biotite and muscovite define a schistosity (S_1) at a small angle to *So* (the wedge-shaped fine-sand layer in Fig. 5.55A, for example, has a consistent obliquity to schistosity). The S_1 is probably parallel to the axial-plane surface of folds (F_1) not visible in thin section. Quartz grains are elongate parallel to S_1 due to grain rotation and/or plastic deformation and/or pressure solution.

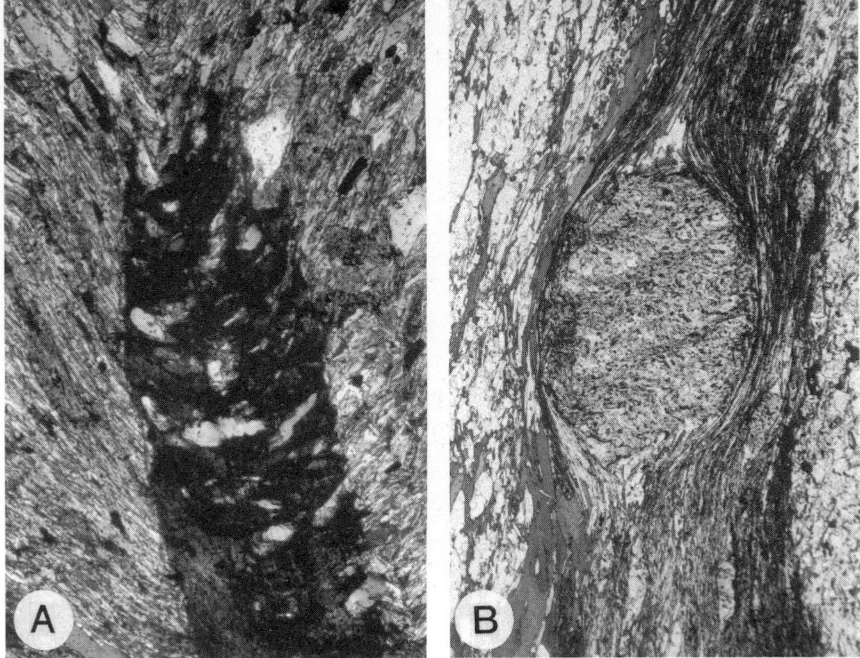

Fig. 5.56 (A): Chlorite porphyroblast that overgrew F_2 crenulation in Palaeozoic schist, Springs Junction, NZ. Note curved inclusion trail concordant with external fold, and compare with Fig. 5.55C. (B): K-feldspar porphyroblast in Haast Schist, NZ, that includes trails of fine-grained crenulated carbonaceous slate, discordant with the external coarser schist fabric. (Lengths of view measure 0.85 mm (A) and 3.3 mm (B).)

3. Garnet porphyroblasts represent the thermal climax of metamorphism and enclose S_1 inclusion trails (Fig. 5.55B).
4. The S_1 was crenulated, and biotite, muscovite and quartz buckled, during F_2. Micas did not recrystallize during F_2, and crenulations clearly truncate S_1 trails in garnet (Fig. 5.55B).
5. Chlorite porphyroblasts (more-or-less random in orientation) overgrew F_2 crenulations (Figs 5.55C and 5.56A).
6. One limb-set of the crenulations was further crenulated (F_3) (Figs 5.55D and 5.30), and chlorites were slightly buckled at the same time.
7. Cracks of constant orientation affect all garnets (Fig. 5.55E). The relationship of this microfracturing to other deformation phases is unknown.
8. Neat as this pigeon-holing may seem, the reality is more likely an overlapping of events. Thus, quartz was preferentially dissolved in the steeper limb-set of F_2 crenulations establishing Q- and M-domains (Figs 5.55A, C, D and E), and the fact that garnets are restricted to Q-

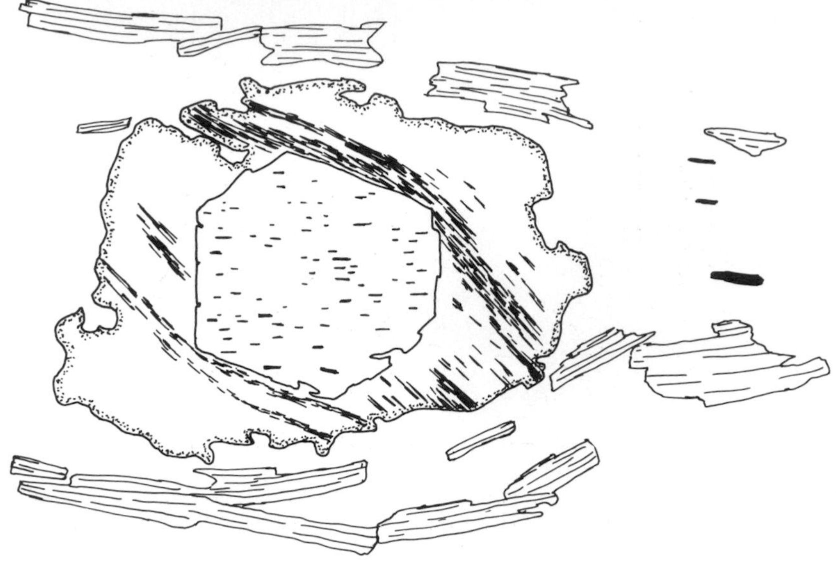

Fig. 5.57 Sketch from photograph showing two stages of garnet growth and three stages of foliation development. Foliation existed prior to the first euhedral growth stage and is preserved as inclusions. Foliation was reformed (or garnet rotated) and wrapped around the euhedral garnet before the second stage of garnet growth. The external foliation is discordant to the second foliation, and has been wrapped around the garnet. Source of photograph unknown.

domains (Fig. 5.55E) suggests that they developed at a very early F_2 stage, not simply pre-F_2. Chlorite porphyroblasts grow in both Q- and M-domains, but the fact that they tend to be richer in quartz inclusions than associated M-domains (Fig. 5.55C) suggests development at a late F_2 stage, not simply post-F_2. These observations also lend support to the proposition of Bell *et al.* (1986) that all porphyroblasts are syntectonic, even though the fact of it may not be immediately apparent on the scale of observation.

Perusal of recent issues of the *Journal of Metamorphic Geology*, among others, will provide numerous examples of such analysis. In essence, they rely mainly on the common sense developed in all geologists trained to think in terms of continuous or discontinuous successions of events. The relationship between porphyroblast inclusion trails and matrix fabric is often the major clue to sequential development (Fig. 5.56), but clues may also be found in reaction rims or replacement textures (Fig. 5.10); sometimes a single porphyroblast gives evidence of multiple structural episodes (Fig. 5.57). A fascinating and related subject is strain analysis

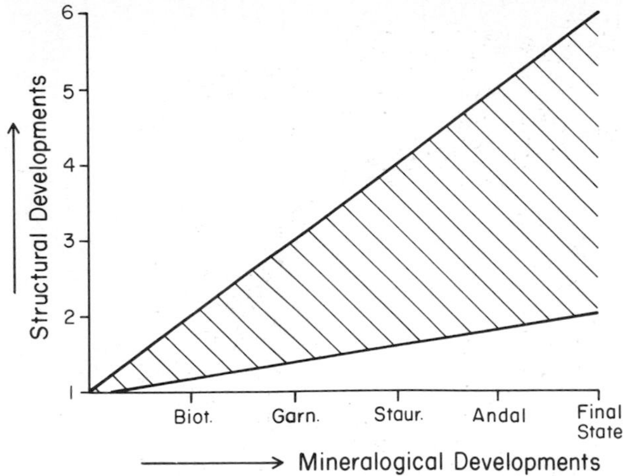

Fig. 5.58 The non-synchronous nature of mineralogical and structural developments is illustrated here, as described by Bell and Rubenach (1983). Rocks in the area that they describe may fall anywhere within the shaded area.

using complex strain-shadow patterns, discussed by Ramsay and Huber (1983, session 14).

The significance of sequential analysis within the larger geological perspective is often difficult to assess. Compare, for example, a thermal aureole that records several pulses of deformation and mineral growth, all within a geologically short period of time, with a gneiss that records a protracted series of, say, Precambrian events overprinted by Mesozoic events. Another point for consideration was raised by Bell and Rubenach (1983) who established that the sequence of porphyroblastic growth was always the same in the Robertson River Formation, Queensland, but that the stage of structural development varied in space and time (Fig. 5.58). What this means in general is that S_3 in one area, for example, might be synchronous with S_4 in another.

(b) *PTt* curves

In the 1980s, the main development in metamorphic petrology was analysis of the variation of pressure and temperature with time. Earlier, Miyashiro (1961) had been instrumental in getting geologists to think in terms of metamorphic facies series and geothermal gradients. Subsequently, it was realized that metamorphism of any one rock is a response to a complex series of changes in *PT* conditions, depending on tectonic circumstance. A common sequence of events is:

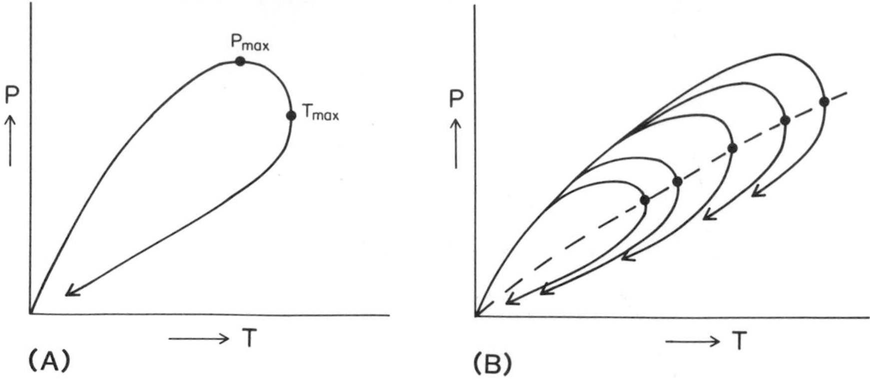

Fig. 5.59 (A): Typical *PTt* curve for rocks subducted, then subjected to uplift. Discussion in text. (B): Nested *PTt* curves for rocks subducted to different depths then uplifted. The dashed line joining points of maximum *T* gives a 'false' geothermal gradient.

1. Sediments are progressively buried and tectonized in an accretionary prism.
2. Some sediment is transported to great depth along the subduction zone where the geothermal gradient is exceedingly low.
3. The subduction system eventually ceases, usually stepping oceanwards. Consequently the older accretionary prism and subducted sediments are uplifted and eroded, and normal geothermal gradients gradually restored.

According to England and Richardson (1977), this sequence of events results in a *PTt* curve such as Fig. 5.59A. Rocks subducted to different levels within a regional metamorphic belt follow a series of nested curves (Fig. 5.59B). At low–moderate grades of metamorphism, mineral assemblages normally reflect thermal maxima (Fig. 5.59B), and one can see how a totally false geothermal gradient could be deduced from a simple *PT* analysis of a succession of schists. Thompson and Ridley (1987) and Spear *et al.* (1984) pursue further interrelationships between *PTt* curves and tectonic histories. Spear *et al.* (1984), for example, discuss what happens when a hot nappe is emplaced on top of cold rocks, or when a cold nappe moves over hot rocks. Anticlockwise *PTt* curves exist too, as recently described for granulites in Australia by Clarke *et al.* (1990).

Determination of a *PTt* curve depends on a combination of textural analysis, as discussed in this book, and electron microprobe analysis which allows an accurate determination of composition of coexisting minerals and zoning profiles (essential for geothermobarometry). The latter subject is beyond the scope of this book, and the reader is referred to key papers

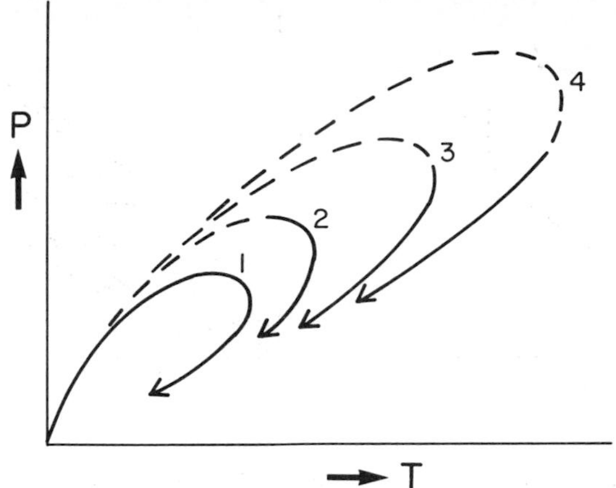

Fig. 5.60 Solid lines represent *PT* histories most likely to be preserved in the mineralogical/textural record. Curves 1, 2, 3 and 4 represent *PTt* histories for very low-grade, low-grade, moderate-to-high-grade, and very high-grade metamorphisms.

by Spear and Selverstone (1983), Essene (1989), Spear (1989) and Spear *et al.* (1990). Two particular problems need to be guarded against:

1. Homogenization of minerals like garnet in the upper amphibolite or granulite facies involves an 'internal metasomatism' which seriously affects the usefulness of methods proposed by Spear and Selverstone (1983) which presume crystal fractionation. Crystals <1 mm in size may even be homogenized in the upper greenschist facies (Jiang and Lasaga, 1990).
2. Retrogressive exchange reactions within the higher grades of metamorphism that affect rim compositions do not give an accurate indication of *PT* changes (Selverstone and Chamberlain, 1990).

Students wishing to study examples of *PTt* curves should peruse Daly *et al.* (1989), recent issues of the *Journal of Metamorphic Geology*, the *Journal of Petrology* and other petrological journals.

Reactions during progressive metamorphism tend to go to completion, and therefore it may be difficult to find evidence for the progressive stages of a *PTt* curve. Most evidence comes from zoned crystals such as garnet (Spear and Selverstone, 1983) and amphibole (Yardley, 1982), where early-formed mineral cores are protected from later reaction. Retrogressive reactions are not likely to go to completion at low–moderate grades, but at high grades, the enhanced rates of solid-state diffusion, combined with the very long times involved, may successfully eliminate the record of the highest temperatures achieved. Fig. 5.60 illustrates those parts of *PTt* curves most likely to be preserved or eliminated.

Part Three

Mineral Preferred-Orientations

6

Introduction

6.1 SHAPE PREFERRED-ORIENTATIONS (SPOs) AND LATTICE PREFERRED-ORIENTATIONS (LPOs)

Two main categories of mineral preferred-orientation are recognized:

SPO: Shape preferred-orientation describes an anisotropic spatial arrangement of the long- and/or short-axes of inequidimensional mineral grains.

LPO: Lattice preferred-orientation describes an anisotropic spatial arrangement of the crystal lattices in a population of mineral grains.

Some mineral preferred-orientations consist only of an LPO or SPO; others contain elements of both. Familiar examples include:

1. Sheet-silicates in schist exhibit an interrelated SPO and LPO: long-axes of platy grains are preferentially aligned parallel to schistosity (they usually define it), and {001} planes (which contain the long-axes) are aligned in the same direction.
2. Elongate grains of quartz in slate (formed by pressure solution – section 5.3.4) exhibit an SPO only; the lattice orientations represent the more-or-less isotropic arrangement of the original sedimentary grains.
3. The regular mosaics of high-grade quartzite often exhibit an LPO only, usually due to plastic deformation, but with grain shapes thoroughly modified by recrystallization.

6.2 ORIENTED SPECIMENS

A prerequisite for preferred orientation studies is an oriented specimen. In the field, strike and dip directions should be marked on a flat surface (Fig. 6.1B), preferably the foliation, and the readings recorded, together with other relevant information such as the orientation of associated folds, lineations, bedding, layering, etc. Specimens should be oriented *in situ*. Nevertheless, a lot of time can be wasted if one finds that a marked

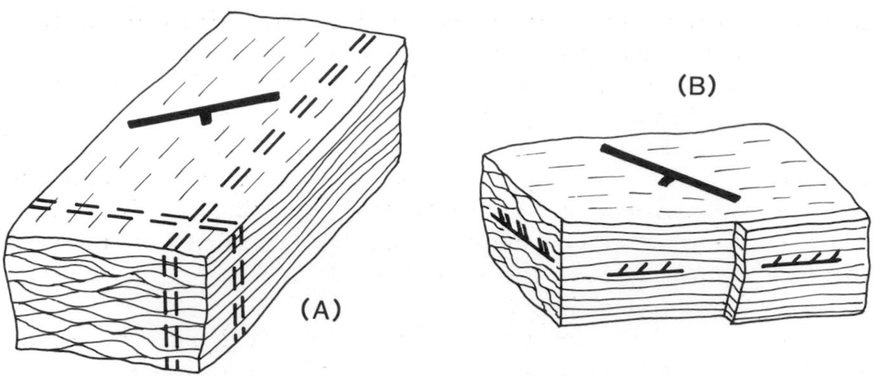

Fig. 6.1 How to orient specimens. In (A), the bold mark on the foliation surface represents the strike direction, the small tick, the direction of dip. Two positions for cutting slices perpendicular to foliation and perpendicular and parallel to lineation are shown. After cutting, (B), the cut surfaces are marked as shown (discussed further in text).

specimen cannot be extracted without severe damage. It is sensible, therefore, to loosen potential specimens, and to reposition them carefully before the marking is done.

Thin sections must be marked so that they can be related in orientation to the hand-specimen. This is what I do:

1. Immediately after a slice has been cut from the rock, both hand specimen and slice are dried, and then fitted back together. A line (possibly parallel to the trace of a foliation), with some oblique ticks, is marked on the cut surfaces of slice and hand specimen, so that they are exactly parallel on both pieces. It is easier to do this if the slice is broken or cut into two smaller pieces so that a ruler can cover both slice and hand specimen (Fig. 6.1B). It is not always easy to mark surfaces accurately, and it may be necessary to use clamps, or to put guide-marks at some point along the join of the two surfaces. If the inner surface of the slice is marked, the oblique ticks must be a mirror image of those on the hand specimen.
2. The unmarked surface of the slice is polished in the usual way before mounting on glass. If Canada balsam is used, then it is best to draw (on the polished surface) a mirror image of the existing mark in indian ink. Subsequent preparation of the section is not affected, and the mark stays in a permanent position relative to the section when it slides around during cover-slip mounting, etc. If an epoxy-resin is used, and especially if automated thin-section machinery is used, it is best to make a parallel mark at the end of the glass slide (using a diamond inscriber) after the slice has been permanently fixed in position.

It is normal practice to cut at least two sections perpendicular to foliation: one parallel, the other perpendicular to any lineation (Fig. 6.1). Even if one is not going to study mineral preferred-orientations in detail, this is good practice because the general appearance of textures and structural features can differ radically in these sections. For example, the texture of many dyke rocks of the Lyttelton volcano (Shelley, 1985a) would be described as trachytic in sections parallel to the flow direction, but lacking trachytic texture in sections normal to the flow direction. The reality is that feldspars have an LPO and SPO with poles to (010) forming a girdle about the flow direction (Fig. 6.2).

A section parallel to foliation is often useful, providing another perspective on the general texture. For a foliated rock that lacks obvious lineation and contains platy subhedral minerals such as feldspar or a sheet-silicate, this section is the one to make first. It often provides a means of locating a lineation, and once found, the other two sections perpendicular to foliation can be cut in appropriate orientations. This is discussed further under U-stage techniques (section 6.3.2). There is no easy way to locate a weak grain-shape lineation, and one should note that a section slightly oblique to foliation can exhibit a lineation along the intersection of foliation and section; this is not the penetrative mineral lineation being sought.

6.3 MEASURING PREFERRED ORIENTATIONS

A preliminary assessment of some preferred orientations can be made by eye. It is, after all, difficult to avoid seeing the sheet-silicate SPO and LPO in typical schists, but it is always a rewarding exercise to go a little further, and to examine the fabrics in properly oriented samples. As indicated above (Fig. 6.2), this can radically change one's idea about the spatial arrangements of minerals. Fabrics and textures are indeed three-dimensional, and should be described as such.

Apart from the fairly obvious SPO of inequidimensional minerals such as mica and feldspar, the less-obvious LPO of quartz is also readily detected by using a sensitive-tint plate. The trace of the quartz *c*-axis is the slow direction, and the line where the plane of *a*-axes intersects the thin section is the fast direction; therefore, if the majority of grains go blue, the *c*-axes are aligned NE–SW (in thin-section view). If the majority of grains display a high retardation, then the *c*-axes lie preferentially at a small angle to the section plane; if a large number display a low retardation, the *c*-axes lie at a high angle. Using such information from two or three sections, it is possible to build up a very good picture of the three-dimensional LPO. This simple approach may not work when mosaics are a mixture of quartz and feldspar: to separate the effects of the two minerals becomes extremely difficult.

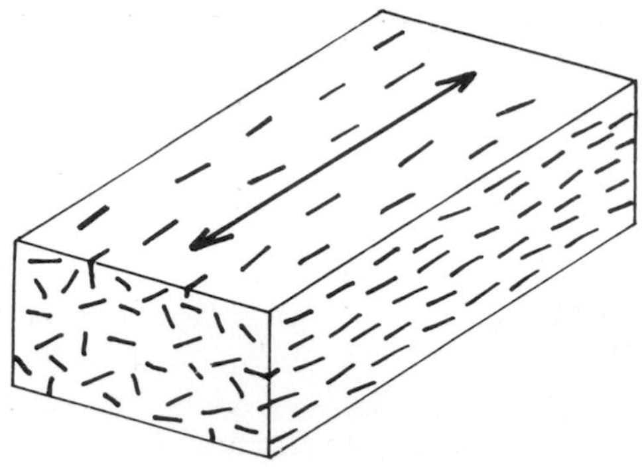

Fig. 6.2 Diagram to illustrate a common arrangement of feldspar laths in dykes. The arrow represents the trace of the flow line (the sense of direction of flow is given by the obliquity of the feldspar laths to the dyke wall – section 8.1.3). A section parallel to flow line and perpendicular to foliation shows a trachytic texture; some other sections do not.

Such simple techniques do not work for all minerals, and in any case, it is preferable to employ the U-stage or other techniques to quantify a preferred orientation's characteristics.

6.3.1 SPO measurement and aspect ratios

Measurement of an SPO can be usefully combined with measurement of grain shape using a method based on the *Rf*/ϕ technique (Dunnet, 1969). Grain lengths and widths are measured together with the angle (ϕ) between grain length and a fixed reference line (often chosen as the visible foliation). The aspect ratio (length/width) is then plotted against ϕ, and the preferred orientation of long-axes is marked by a concentration of points combined usually with the maximum aspect ratio (Fig. 6.3). The *Rf*/ϕ technique was originally devised for measuring finite strains from deformed grains; because grain shapes in metamorphic rocks do not usually represent finite strains, those details of the technique need not concern us here.

Measurements are best accomplished with a micrometer set in the eyepiece, using a mechanical stage, and by reading degrees off the graduated rotating stage. Measurements should be made in at least two perpendicular sections to assess the strength of the SPO in three dimensions and to appreciate the spatial variation of aspect ratios.

One problem is that length and width measurements in a section are not the same as actual lengths and widths of the three-dimensional grain (we are dealing with two-dimensional, possibly oblique views of grains). Cashman and Marsh (1988) conclude that there is no satisfactory way of

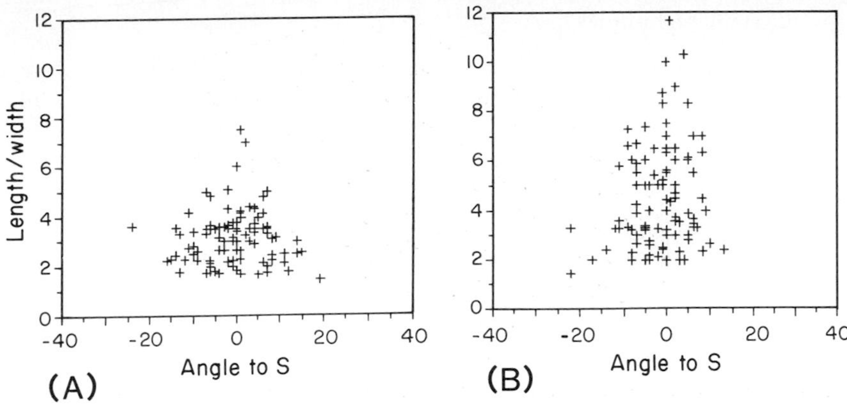

Fig. 6.3 SPOs represented by plotting grain aspect ratios against the angle between grain length and a reference plane (usually foliation). The example shows albite SPO in sections parallel to lineation and perpendicular to foliation for (A) quartz-rich layers, and (B) sheet-silicate-rich layers. Reproduced with permission from Shelley (1989a), copyright Pergamon Press PLC.

correcting two-dimensional grain-size data, but by examining at least two thin sections of different orientations we do obtain some insight into the three-dimensional variation. There is also the practical problem of deciding what is the long-axis of a not very inequidimensional grain. For example, the longest dimension of a rectangle is along its diagonal, yet one's instinct, and I believe the more correct approach would be to measure the length parallel to the long side, thereby dividing the grain in the most symmetrical way.

6.3.2 LPO measurement

There are two established techniques for measuring LPOs: optical methods using the U-stage, and X-ray diffraction using the texture goniometer. The X-ray technique (Wenk, 1985a) allows a complete determination of crystallographic data for some minerals, and for quartz, this compares with the fact that only the *c*-axis and the plane of the *a*-axes can be determined optically. The method has provided invaluably complete data for quartz and calcite, but does not cope well with some common minerals (e.g., feldspar), or with complex multimineralic assemblages typical of many metamorphic rocks. It works best with quartzite, marble and slate, and is especially useful for very fine-grained specimens, such as chert.

X-ray and optical techniques both have their advantages and disadvantages, and they complement rather than supplant each other. Further discussion of X-ray methods is beyond the scope of this book.

U-stage techniques

Optical methods rely on the use of the universal stage, which can rotate a thin section up to 50° or so from its normal horizontal position. The following terminology for rotation axes is a common standard:

Axes initially vertical	A1	inner stage
	A3	outer stage
	A5	microscope stage
initially horizontal	A2	N–S inner ring axis
	A4	E–W main control axis

The actual number of rotation axes varies with the model of U-stage. Only three are essential for the work described here, which means that A3 is seldom used: it should be kept clamped in the zero position.

Setting up a U-stage is dealt with in some mineralogy texts (e.g., Shelley, 1985b); workers using it for the first time should in any case refer to the appropriate equipment manual. The glass hemispheres set above and below the thin section are manufactured with a variety of refractive indices, and they should be chosen to match as closely as possible the refractive index of the mineral to be studied. Glycerine is used to seal the contacts between hemispheres, U-stage and thin section, and should be removed after use with a dry tissue and/or water.

Few mineralogy or petrology texts outline standard procedures with the U-stage, and the following notes may therefore prove useful.

(i) Recording measurements

U-stage readings are normally recorded on an equal-area lower-hemisphere projection, and it will be assumed that the reader is familiar with how to use such a projection from structural geology. Software is available to plot data through the keyboard of a computer, but until one is completely familiar with the procedures of structural petrology, it is preferable to plot data by hand. This is done on tracing paper placed over a fixed net marked with degrees mimicking those on the inner stage (for a Leitz U-stage, zero is at the south pole of the net with degrees up to 360 in the clockwise direction). The tracing paper has a reference mark placed initially over the zero point of the fixed net.

(ii) Measuring the orientation of planar features

Determining the orientation of cleavage cracks and twin-lamellae requires the use of only two axes. Clamp A3, A4 and A5 in their 'zero' positions, then proceed as follows:

1. Rotate the feature to be measured into the N–S position using A1.
2. Rotate about A2 until the feature is seen at its sharpest (so that it is vertical). If you use a microscope with the non-standard N–S polarizer orientation, it is advisable to rotate it into the E–W position when dealing with biotite; this will lighten its colour when the cleavage is N–S.
3. Record the readings of A1 (0–360°) and A2 (0–50° L or R). 'L' and 'R' records whether A2 was rotated up to the left or right.
4. To plot A1 (54°) and A2 (30° L) rotate the tracing paper until the reference mark lies over 54° on the underlying fixed net, then plot the point 30° along the E–W diameter from the left. This represents the pole to the cleavage or twin-lamella. If A2 had been (30° R), the plot would have been 30° in from the right side of the net.

A2 cannot be rotated more than 50°, and therefore there is a 'blind spot' in the centre of the net. For example, the cleavage orientation of sheet-silicates with {001} <40° from the plane of the thin section cannot be measured. The simplest way to deal with blind spots is to gather data from at least two sections perpendicular to each other.

Very inequidimensional platy grains can generate spurious pole patterns. This is because a thin section is more likely to cut grains that lie at a high angle to the section. In other words, a random orientation of micas would actually generate a spurious girdle of poles to {001} around the perimeter of the net. Such concentrations may be of the order of 3–4% per 1% area. There are two ways to deal with the problem. In foliated rocks, the preferred orientation will usually consist of some combination of pole maximum and girdle. Follow the procedure described below to find the girdle axis; the preferred orientation measured in a section cut perpendicular to the girdle axis will not contain spurious concentrations. In rocks lacking a strong preferred orientation, data should be corrected as described by Billings and Sharpe (1937).

Poles to {001} of sheet-silicates in metamorphic rocks and {010} of feldspars in igneous rocks often form a girdle distribution (Figs 7.11B, 7.23B and 8.6A), and the axis of the girdle is a lineation though not always an obvious one. It is desirable to examine such rocks in sections parallel and perpendicular to the girdle axis, but if the lineation is not obvious it may first be necessary to examine a section parallel to foliation. Most cleavages or twin lamellae make a small angle with the plane of this section, and hence occupy the 'blind spot'. The U-stage can be used to determine the orientation of the few planar features that do not fill the blind spot, and if these have a preferred orientation they define the position of the girdle. The grains sought are commonly obscured by plates parallel to the plane of the thin section, and the following procedure may be required:

1. Position A1 at zero, then rotate A2 slowly to the left and right through the full 100°. Grains with sharp planar features close to N–S, previously invisible, often appear during this procedure. Record their orientations (adjust A1 as appropriate for an accurate reading).
2. Reposition A1 at 20° intervals up to and including 160°, and repeat the above procedure for each position.
3. Repeat steps 1 and 2 for a succession of areas within the section until enough data are recorded.
4. Plot the readings as before. The girdle axis is on or close to the perimeter of the net at right angles to the maximum concentration of poles plotted.

(iii) Measuring the orientation of c*-axes of a uniaxial mineral*

We aim to set the *c*-axis parallel to A4, or failing that, to set it vertical. Either should be possible, so no 'blind spot' exists except where twin lamellae physically obscure the optics, sometimes the case with calcite (section 7.4). Proceed as follows:

1. With all other axes at zero, rotate about A1 to an extinction position. (In the special case of a grain with *c*-axis close to vertical, which stays in extinction all the time, find a position for A1 when a tilt about A2 maintains extinction, then proceed to step 3).
2. Tilt about A2, and if extinction is not maintained, go to the other extinction position about A1. The grain should now remain in extinction when rotated about A2.
3. Rotate A4 by some moderate amount.
   (1) If the grain illuminates: (a) Maintain the A4 rotation and restore extinction by rotating A2; check that extinction is now maintained when A4 is rotated. (b) Return A4 to zero; the *c*-axis is now either horizontal E–W or vertical.
   (2) If the grain does not illuminate, maintain the A4 rotation and rotate A2. (a) If the grain still does not illuminate, the *c*-axis has been set N–S and is close to horizontal in the thin section. Go to the other A1 extinction position and start again. (b) If the grain illuminates, then put the grain back into extinction, then return A4 to zero. The *c*-axis is now either horizontal E–W or vertical.
4. To check whether the *c*-axis has been set horizontal or vertical, rotate A5. If the grain remains in extinction, the *c*-axis is vertical. Return A5 to its zero position, and clamp.
5. Record the A1 and A2 readings, noting whether the *c*-axis is horizontal (H) or vertical (V). To plot the *c*-axis at A1 (120°) A2 (35° L and H), rotate the tracing paper so that the reference mark lies over 120° of the net, and plot the pole of the *c*-axis 35° in from the left side along

the E–W diameter. A2 (35° L and V) would plot 35° from the centre of the net towards the right side, (35° R and H) would plot 35° in from the right, and (35° R and V) 35° from the centre towards the left.

The following may allow one to anticipate some of the steps above:

1. A grain with an initially low relative retardation is likely to have its *c*-axis set vertical by the above procedure.
2. When a *c*-axis is close to vertical, the retardation reduces progressively as extinction is approached; for a *c*-axis that is close to horizontal, the retardation actually increases as the final extinction position is approached.
3. The common undulatory extinction pattern in quartz is made up of elongate subgrains subparallel to the *c*-axis. It saves time to set them E–W when embarking on step 1 above.

The following alternative procedure for finding the *c*-axis is sometimes useful when extinction positions are not sharp, often the case with calcite:

1. Tilt A2 by some set amount (say 20° L).
2. Rotate A1 until the grain extinguishes with the ordinary ray E–W (this is the fast direction for quartz, the slow direction for calcite; for the latter it is not possible to use the sensitive-tint plate, so set the highest RI, judged from relief, E–W, or the opposite if you use a non-standard N–S polarizer orientation).
3. Use A1 and A2 readings to plot the position of this ordinary ray.
4. Repeat the above procedure several times, setting A2 in a different position each time.
5. The several orientations of the ordinary ray should plot on a great circle, and the *c*-axis is the pole to that great circle.

(iv) Measuring the orientation of a biaxial indicatrix

We aim to set the optical directions X, Y and Z parallel to A4. Proceed as follows:

1. Rotate about A1 to an extinction position.
2. Rotate A4 by a moderate amount. The grain normally illuminates. If so, restore extinction by tilting A2.
3. Retain the tilt of A2, and restore A4 to zero.
4. If the grain illuminates further, repeat steps 2 and 3, and go on doing so until extinction is maintained throughout the entire rotation of A4.
5. Either X, Y or Z is now set E–W and horizontal. To determine which, rotate A5 45° anticlockwise. If it is Y, A4 is perpendicular to the optic axial plane, and rotation about it will usually bring at least one optic axis into a vertical or near-vertical position. This can be recognized

by extinction or a sharp drop in retardation. If it is not *Y*, use the sensitive-tint plate to determine whether it is fast (*X*) or slow (*Z*).
6. Record the A1 and A2 readings.
7. Find at least one other optical direction by repeating the above procedure, starting with an alternative A1 extinction position in step 1, or tilting A2 in the opposite direction in step 2.
8. *X*, *Y* and *Z* are always exactly 90° from each other, and only two need to be found by direct observation. The position of the other can always be determined using the equal-area net (it is the pole to the great circle that contains the other two).
9. Plot the data for *X*, *Y* and *Z* separately. To plot the pole A1 (67°) A2 (27° L), rotate the tracing paper so that the reference mark lies over 67° of the net, and plot the pole 27° in from the left side along the E–W diameter. The A2 (27° R) would plot 27° in from the right.

The above procedure does not always work easily. If an optic axis is close to vertical, an exact extinction position is difficult to achieve, and sometimes a trial-and-error approach seems necessary. Another common problem is that repeating steps 2 and 3 can lead the A2 axis into impossibly high tilts. In this case the solution is to start again and tilt A2 in the opposite direction. A good check on results is that optical directions must plot exactly 90° from each other. In fact, if at least one optical direction has been found with confidence, but the others are difficult to find, it often pays to construct the great circle to the determined pole, and by trial and error, experiment with possible positions for the other two directions: they must lie on that great circle.

(v) How many measurements, and where?

Before interpreting a mineral preferred-orientation, one should be sure that it really exists. Numerous researchers have commented on the statistics of the matter, but in practice the following are most important:

1. First of all decide what is to be measured. For example, if there is a fold, perhaps one should measure the preferred orientation for each limb separately. If a particular mineral occurs in a number of obviously different ways (for example, as veins **and** in the body of the rock, or as very fine-grained areas **and** coarse-grained areas), measurements for each type of occurrence should be made independently.
2. A systematic approach to covering a representative area must be taken. This means not missing out certain types of grains (it is easy to overlook very-low retardation grains, for example), and covering a large enough area. The exact approach will vary from section to section. For example, in a coarse-grained rock it may be necessary to

measure every single grain, perhaps photograph the section first, and tick off each grain as it is measured. In finer-grained rocks, it may be better to make systematic traverses over the section, measuring every single grain encountered along each traverse, but spacing out the traverses to cover a wide area. Always be on guard for a personal bias against measuring particular types of grain.

3. The best test of significance is reproducibility. When starting work on a particular suite of rocks, one should measure preferred orientations at least twice, preferably in different thin sections, and preferably in sections of different orientation. One soon learns which are the significant elements of the pattern and which are simply ephemeral.
4. One learns from experience how many data should be recorded. At first it does no harm to err on the high side, but it pays not to measure unnecessary numbers. A strong LPO of sheet-silicates or calcite may require no more than 100 or 150 readings whereas a weak quartz or feldspar LPO may require 300 or 400. A broad girdle pattern requires more data for definition than a single maximum.

It may be possible to construct several diagrams of sheet-silicate data in one day. Quartz and calcite (especially if twin data are to be processed) diagrams typically require a day's work, a feldspar diagram requires several days. Some of the procedures can be quickened significantly by computer processing of data, and there is a commercially available modification of the Leitz U-stage that enables a computer to sense the A1 and A2 stage readings – an added expense, but a great time saver.

Further details on procedures for particular minerals occur throughout Chapters 7 and 8.

6.3.3 Integrating SPO, LPO and CSD data

A disappointing aspect of much recent work on LPOs is that one is given little or no information on the relationships between LPO, SPO and grain-size data. Collecting data relevant to the SPO, LPO and CSD (see section 5.1.1) all at once is a lengthy procedure, but to do so in the first place saves a lot of time in the end. It is often a rewarding experience.

Two questions to ask are:

1. Do the most elongate grains have their lattices oriented in a particular way?
2. Do grains of a particular size have their lattices oriented in a particular way?

To answer these questions, the following can be done:

1. Plot grain lengths or grain length/width ratios against the angle

between a particular optical direction and the foliation or lineation (Shelley, 1989a).

2. Determine CSDs separately for grains that have particular lattice orientations, perhaps: (a) those contributing to maxima in the LPO pattern, (b), those which lie close to the maxima and (c), those that lie at a high angle to the preferred orientation (Shelley, 1977, 1980).

Such plots have an important bearing on interpretation of the LPO.

7

Metamorphic rocks

The terminology of metamorphic rocks hinges to a very large extent on whether or not a mineral preferred-orientation exists, and if so, what sort of mineral preferred-orientation. Slate, schist and gneiss are named directly after the preferred orientations that characterize them. It is therefore a matter of some importance that we understand the characteristics of those preferred orientations, and understand some of the mechanisms that produce them.

7.1 MECHANISMS

Three main categories of mechanism for producing LPOs and SPOs exist:

1. Plastic deformation.
2. Mechanical alignment of inequidimensional grains.
3. Grain formation and growth, including diffusion-aided anisotropic crystallization in a stress field, and recrystallization.

7.1.1 Plastic deformation

We are all familiar with the plastic behaviour of metals: copper and gold can be beaten into any shape, and nails all too often bend. We also know that most rocks are brittle under ordinary conditions. What many do not realize is that metals, like rocks, are made up of an aggregate of crystal grains.

Metal crystals deform plastically by a process known as translation gliding or slip. In essence, what happens is the crystal slips along certain planes (and along certain directions within those planes). One way of visualizing the difference between a plastic and non-plastic crystal is to compare a centimetres thick slab of wood with its processed version, a telephone directory. Both are made of the same material, but the latter will deform, and bend, simply because it has the potential to slip along the plane of the sheets of paper. If you draw cross-sectional shapes of crystals on the side of the directory, then deform it, you will get some idea of what plastic deformation is like.

The reasons that metals deform plastically include their simple structure, weak bonding and high symmetry. The most plastic rocks at low temperatures are also composed of relatively simple substances such as halite and calcite. Silicates are much more difficult to deform, and only become plastic at elevated temperatures, sometimes requiring water to hydrolyse and weaken the lattice bonding.

A slip system is specified using Miller indices for the slip plane and a letter or zone symbol for the slip direction: hence (001)[100] or (001)*a*. Many metals are cubic, in which case (001) is part of the cube form {100} which consists of three planes (100), (010) and (001); each plane contains two *a*-axes. If a cubic crystal slips on (001)[100], symmetry demands that it must also slip on all cube planes and along all *a*-axes. This would normally be summarized as $\{100\}\langle a\rangle$. In monoclinic and triclinic minerals, such as feldspar, the situation is very different. Symmetry is so low that slip on {010}[001], for example, means on just one plane and along one direction.

A single crystal, by itself, will deform on a slip system if the slip plane is close to the plane of maximum shearing stress (a moderate angle to σ_1), and if the trace of the slip direction is towards the extension direction. Of the three grains of Fig. 7.1, only *B* will deform simply by slip. Grain *C* will not deform at all, but *A* may be subject to kinking which could bring it into a grain *B*-type orientation. Note that slip rotates the direction perpendicular to the slip plane towards σ_1 (Fig. 7.1B): thus basal slip in quartz rotates the *c*-axis towards σ_1. Imagine now an aggregate of grains (Fig. 7.2). If the mineral is triclinic, say plagioclase, and slip is only possible on the one system (say {010}[001]), then the aggregate as a whole cannot deform; grains *A* and *C* prevent grain *B* from deforming. In contrast, a cubic metal with numerous possible planes and directions of slip (even with just the one slip system $\{100\}\langle a\rangle$) will deform relatively easily: most grains are likely to have one plane or another in a suitable orientation for slip. More precisely, homogeneous plastic deformation is possible only when slip can occur on at least five independent systems (von Mises, 1928), a requirement known as **von Mises' criterion**. 'Independent' means a slip plane and direction that can achieve a strain not possible with any combination of the other slip planes and directions.

von Mises' criterion has an important bearing on how rocks may or may not deform. For example, at low–moderate grades of metamorphism, quartz deforms by slip dominantly on $\{0001\}\langle a\rangle$. In a simple-minded way, one can visualize a mosaic of quartz deforming on this slip system to produce ribbons in the plane of extension (i.e., an SPO), and with *c*-axes rotated towards σ_1 (hence an LPO of *c*-axes). However, the reality is that

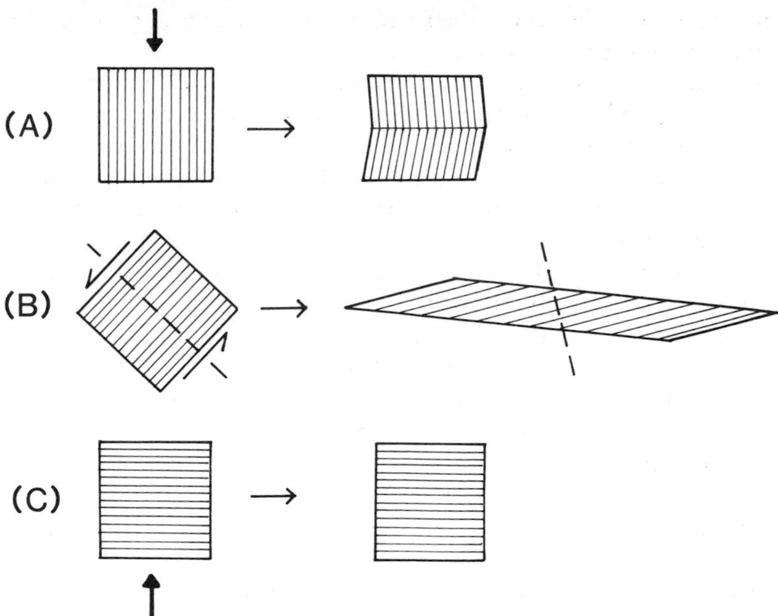

Fig. 7.1 Three grains, A, B and C, with slip planes (fine lines) in different orientations relative to the maximum principal compressive stress direction (bold arrows). Explanation in text.

von Mises' criterion is not complied with, and the deformed mosaic of quartz grains would develop impossible geometries with overlapping grains or gaps between grains (Fig. 7.3). Nevertheless, some SPOs and LPOs of quartz (section 7.3) strongly suggest basal slip to be the dominant mechanism of deformation, and this may be possible if the deformation is heterogeneous: for example, slip itself could be heterogeneous (bending of plastic grains about non-plastic ones), and it could be combined with pressure solution and grain boundary sliding.

It is convenient to visualize slip in terms of planes of movement, as discussed above. In terms of overall geometry the model is close to the truth, but in reality, crystals, even plastic ones, are far too strong to allow all the bonds along one plane to be broken simultaneously. It is actually the movement of dislocations that enables slip to take place (Fig. 5.19), and by way of analogy, imagine shifting the position of a large square of carpet in an even larger room. Simply pulling it across the floor is a difficult task, due to friction, but if one edge is folded up, the fold (equivalent to an edge dislocation) can be easily transferred from one side of the carpet to the other, shifting the carpet without having to overcome the friction. The planar shift we call slip is really a movement of dis-

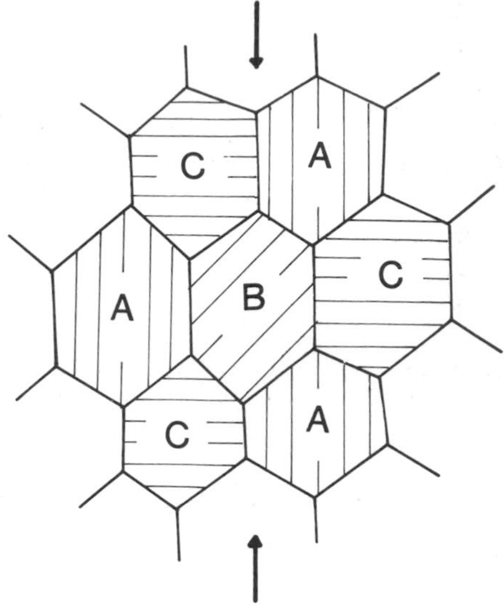

Fig. 7.2 Aggregate of grains with slip plane orientations shown as fine lines. Grain B is in a suitable orientation to deform plastically, but is prevented from doing so by A and C grains. Bold arrows represent the maximum principal compressive stress direction.

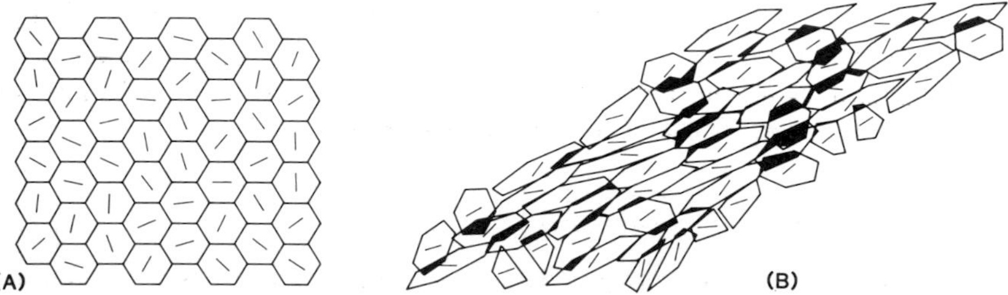

Fig. 7.3 (A): Aggregate of quartz grains with orientations of the (0001) planes given by lines. (B): Same aggregate after simple shear during which individual grains deformed by basal slip. Homogeneous deformation is not possible: hence, 'overlapping' grains, shown in black, and gaps between grains, left clear. Reproduced with permission from Etchecopar (1977).

locations from one side of the crystal to the other, and so the process is sometimes called dislocation slip.

At higher temperatures, dislocation slip is accompanied by a more active movement of dislocations known as climb, the same process that occurs during recrystallization (section 5.2), and the combination is called dislocation creep. More information on these important processes can be found in Nicolas and Poirier (1976).

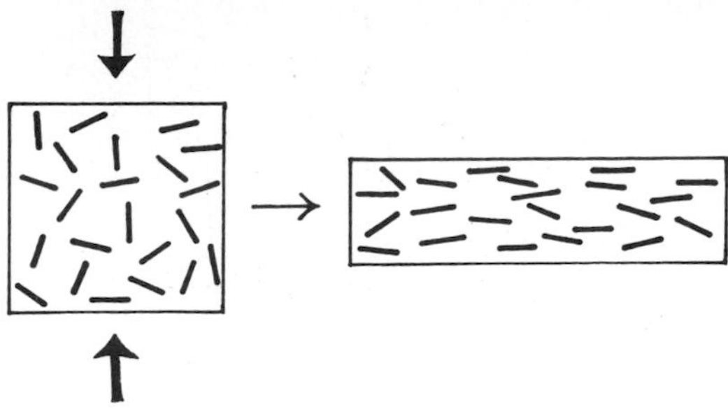

Fig. 7.4 The mechanical alignment of rigid inequidimensional plates or rods during strain of a matrix. Bold arrows represent the maximum principal compressive stress direction.

7.1.2 Mechanical alignment of inequidimensional grains

The idea that rotation of inequidimensional grains during strain can produce an SPO, and usually an LPO, was advocated by Sorby (1853). It is not difficult to visualize the process (Fig. 7.4): plates rotate towards being perpendicular to the shortening direction, and the long-axes of elongate crystals rotate towards the elongation direction. The ways in which the overall strain is achieved include some combination of grain-boundary sliding, plastic strain of the matrix, and a decrease in volume due to dewatering and/or pressure solution. In addition, plates and rods can be aligned in a flow of water through unlithified muddy sediments, or by cataclastic flow or the smearing out of inequidimensional fragments along shear planes. These ideas are pursued in section 7.2.

7.1.3 Grain formation and growth

The possibility that mineral preferred-orientations result from anisotropic grain growth has been neglected in recent fabric studies. Yet the potential is clear, as was clearly outlined by Cox and Etheridge (1983). One familiar result of competitive growth is vein quartz with its *c*-axes at a high angle to the vein wall. The reason is the propensity of quartz to grow fastest parallel to the *c*-axis. Thus, if quartz nucleates with random orientation on the vein wall, grains that can grow outwards most quickly outgrow the others (Figs 7.5 and 7.6). Such effects are likely to be even more marked with minerals like the sheet-silicates and amphibole which are much more strongly anisotropic in their growth habits than quartz.

Crack–seal deformation may also lead to a preferred orientation due to the anisotropic growth habits of inequidimensional minerals such as mica and hornblende. For example, crystals with long-axes parallel to the extension direction are most likely to be cut by fractures, and these are the

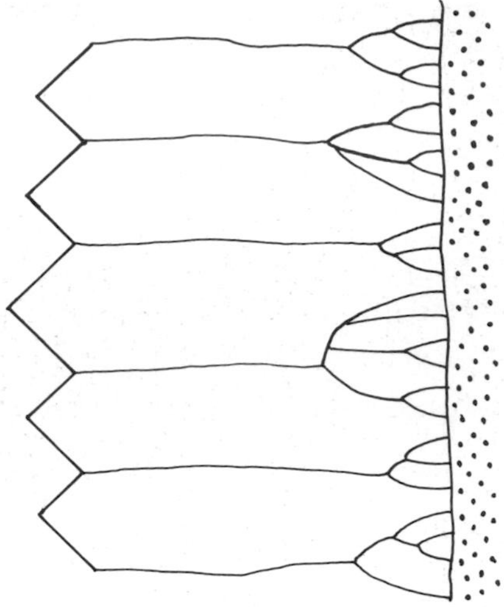

Fig. 7.5 Development of quartz *c*-axes preferred-orientation in a vein. Host rock is stippled. The large crystals with *c*-axes perpendicular to the vein wall outgrow early-formed grains of other orientations.

crystals most able to grow rapidly towards the extension direction (Fig. 7.7A). Crystals with long-axes parallel to the shortening direction are least likely to be cut by fractures (Fig. 7.7B); even if they are (Fig. 7.7C), they are not able to grow rapidly towards the extension direction. In addition, grains that are elongate parallel to the shortening direction may be reduced in size by pressure solution (Fig. 7.7B).

Growth processes may also enhance an incipient preferred orientation (Fig. 7.8). Plates or rods with the longest possible growth paths soon dominate. Grains with orientations of lesser abundance are naturally truncated by grains with the dominant orientation; in this way the preferred orientation is progressively enhanced.

At moderate-to-high grades of metamorphism, textural maturation and Ostwald ripening (section 4.2.2 and section 5.1.1) may further enhance a preferred orientation by eliminating altogether smaller grains that constitute the less-abundant orientations in Figs 7.7 and 7.8, for example.

Given that many metamorphic rocks are dominated by structures formed by solution transfer and precipitation, including segregation layers, pressure shadows and various crack–seal phenomena, it would be surprising if fabrics of competitive growth origin did not result. The matter will be considered further throughout sections 7.2–7.7.

Recrystallization normally reduces the strength of an existing SPO or

Fig. 7.6 Thin-section view of quartz vein with numerous small grains of variable orientation at wall, but a dominance of one orientation (*c*-axes perpendicular to wall) towards the vein centre. Quartz vein from Morgat, Brittany. (View length measures 3.3 mm.)

LPO; in some circumstances it leads directly to a significant new SPO or LPO, as discussed in sections 7.3 and 7.6.

7.1.4 P-, M- and G-type fabrics, and P-, M- and G-tectonites

Fabrics generated by the three distinctive categories of mechanism, described above, can be labelled **P-type**, **M-type** and **G-type** after **P**lastic deformation, **M**echanical rotation and **Gr**owth, respectively. The thrust of recent research has been the study of P-type fabrics in quartzite, marble and dunite, all essentially one-mineral rocks; consequently we have a good understanding of plastic deformation as it affects those three rocks. Not so clear is what happens in the more common metamorphic rocks made up of a complex assemblage of minerals. Are the deformation and orienting mechanisms for one mineral modified when that mineral is intimately

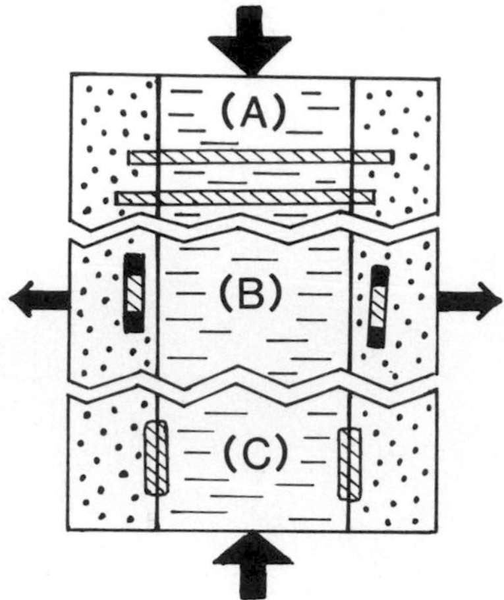

Fig. 7.7 Development of mineral preferred-orientation during crack–seal deformation. Host rock is stippled, crack–seal vein is lined, and shortening and extension directions are indicated by arrows. Explanation in text. In (B), the filled-in ends of crystals indicate parts susceptible to removal by pressure solution.

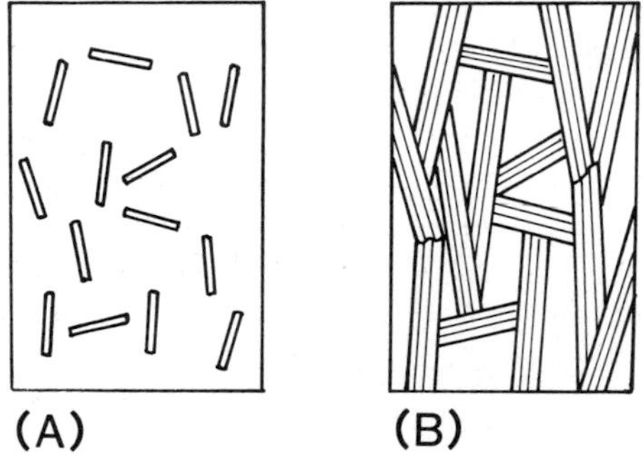

Fig. 7.8 A preferred orientation of inequidimensional grains (A) is enhanced by growth because grains of less-abundant orientations have shorter possible growth paths and become truncated (B).

associated with another? This question has seldom been investigated, but the answer is 'yes' according to Starkey and Cutforth (1978) and Walniuk and Morris (1985) who show that a positive correlation exists between quartz LPO strength and modal quartz content. As the quartz content approaches zero, so the *c*-axes pattern becomes random.

Another question is whether the interpretation of one mineral fabric sheds new light on the associated mineral fabrics. Shelley (1989a), for example, describes sheet-silicate, quartz and feldspar fabrics from a

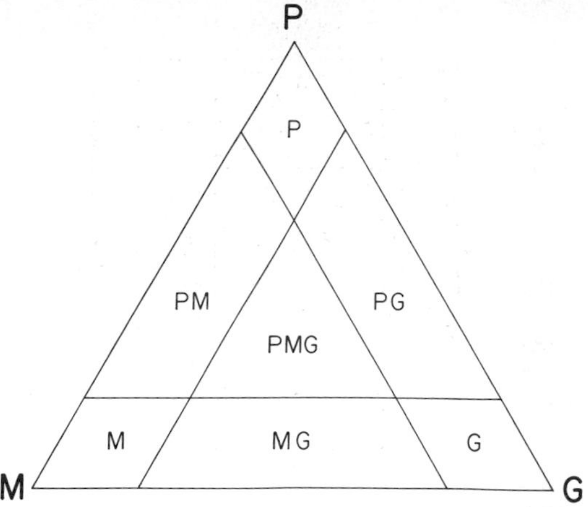

Fig. 7.9 PMG triangle which delineates fields of seven tectonite types defined in terms of fabric genesis. Explanation in text. Reproduced with permission from Shelley (1989b), copyright Pergamon Press PLC.

greenschist-facies schist. The quartz and feldspar are grown together in mosaics: the quartz fabrics appear to be P-type, the feldspar G-type. Many would label the sheet-silicate fabric as M-type. Are these three types compatible in intimate association?

To encourage such questions and a total, holistic examination of rock fabrics, Shelley (1989b) put forward a classification for tectonites according to their fabric types. The P-tectonites, M-tectonites and G-tectonites, are dominated by P-, M- and G-type fabrics, respectively. Tectonites of mixed fabric type can be plotted on a triangular diagram (Fig. 7.9) and classified as PM, PG, MG and PMG-tectonites. Fig. 7.10 illustrates where some common metamorphic rocks might plot on such a diagram.

7.2 SHEET-SILICATES

A sheet-silicate preferred-orientation is the most conspicuous element that defines foliation in the most common metamorphic rocks: slate, schist and gneiss. Accompanying minerals may also display preferred orientations, but it is the very strong anisotropy of the sheet-silicates that controls the fissility of slates and most schists.

Measuring sheet-silicate preferred-orientations

Sheet-silicate preferred-orientations are usually reported simply in terms of poles to the cleavage {001}. If the sheet-silicate is pseudo-uniaxial or has a

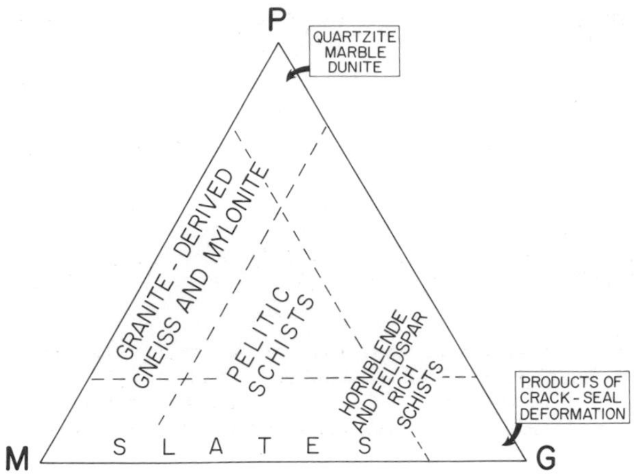

Fig. 7.10 Some common metamorphic rocks plotted on a PMG triangle. Reproduced with permission from Shelley (1989b), copyright Pergamon Press PLC.

very low birefringence (e.g., some biotite and chlorite), it is difficult to measure anything more than that. For those with a discernible 2V and low–high birefringence it is possible to measure the orientations of optical directions *Y* and *Z*. This can be done accurately with the U-stage in sections parallel to foliation, or in a semi-quantitative fashion on an ordinary flat stage using interference figures (isogyre positions give the approximate orientation of the optic axial plane which contains *Z*, and the perpendicular to the optic axial plane, *Y*). Muscovite has *Z* parallel to the *b* crystal axis; most biotite has *Y* parallel to *b*. According to Oertel (1985b), *a* and *b* crystal axes are not known to have a preferred orientation within the foliation of slates; I found no preferred orientation in a garnet schist (Shelley, 1977). A lack of preferred orientation is not unexpected because the {001} section of sheet-silicates is essentially pseudo-isotropic, but more observations are needed to confirm it as a general fact.

Measuring poles to {001} is a quick and easy procedure, as outlined in section 6.3.2, but pay some attention to the following:

1. It is easy to miss oblique grains that do not immediately display the cleavage with the U-stage axes at zero. To ensure that you do not do this, follow the second procedure given in section 6.3.2 for planar features, i.e., perform trial tilts of A2 for a systematic sequence of A1 positions.
2. It may be difficult to decide where one grain starts and another ends in a dense cluster of nearly parallel sheet-silicates. The best solution is to

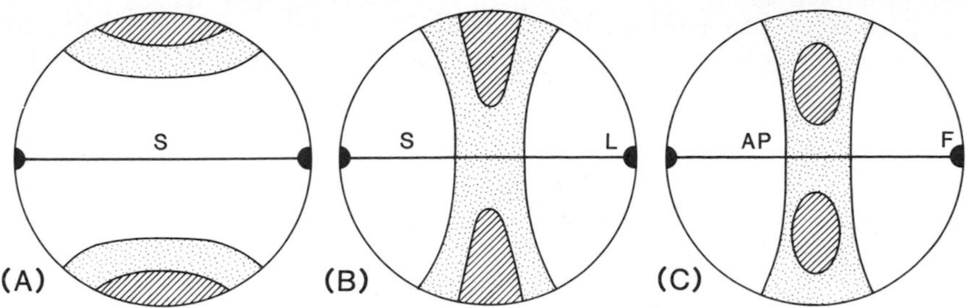

Fig. 7.11 Idealized contoured equal-area projection patterns for sheet-silicate (001) poles, S = foliation, L = lineation, F = fold axis, and AP = fold axial plane. (A): Maximum that often defines the position of foliation. (B): Girdle combined with maximum that together define foliation and lineation. (C): Double maximum and girdle that relate to a fold axis and two limbs.

measure grains at a systematic spacing interval (instead of measuring every grain along a traverse or within a specified area).

3. 'Blind spots' and possible spurious patterns generated by very in-equidimensional grains must be considered systematically. Methods are outlined in section 6.3.2: for foliated rocks, cut a section parallel to foliation, follow the second procedure for planar features, identify the position of any girdle of poles to {001}, then measure the girdle distribution in a section parallel to the girdle (usually perpendicular to a mineral lineation).

Interpretation of sheet-silicate preferred-orientations

Most sheet-silicate preferred-orientations consist of two main elements (Fig. 7.11): 1. a strong maximum of {001} poles, which defines the foliation; 2. a girdle or partial girdle which spreads out from that maximum. These two elements may be inextricably related, but in the first instance it is worth considering them separately.

Firstly, let us consider what might result from the deformation of a rock during a simple uniaxial shortening. Most rocks that contain sheet-silicates always contained sheet-silicates, so if the rock has undergone any strain at all, the possibility of M-type processes must be considered seriously. However, the perpetual crystallization and recrystallization of sheet-silicates throughout diagenesis and metamorphism mean that G-type processes need similar serious consideration: sheet-silicates grow most quickly parallel to their cleavage. Slip on {001} does occur, but there seems little evidence that P-type processes contribute to the preferred orientation.

The M-type processes depend on plates being able to rotate relative to one another during strain. Four M-type processes need to be considered:

1. Rotations in the transformation of muddy wet sediments to slate (or shale) are possible because up to 80% of the volume is water, and clay particles are not firmly attached to each other. In two reviews, Oertel (1985a and b) notes: (a) clay particles <1 mm size initially have a random orientation in muds; (b) the strengths of natural sheet-silicate preferred-orientations in slate can be directly correlated with measured strains; (c) the correlation of LPO and strain conforms with the expectations of March (1932) strain theory that assumes platy grains deform homogeneously with the matrix. The fact that a correlation exists does not necessarily confirm the proposed cause and effect, and Oertel is careful to caution readers on this point. It does, however, make M-type processes a viable possibility for shales and simple slate structures, and whatever the cause ultimately proves to be, LPOs in slates are a useful means of measuring finite strain.
2. During the escape of the large volume of water from mud, discrete water-escape structures may develop, and within those channels, sheet-silicates can be aligned by flow of water. Layering in some slates suggests a linkage between cleavage formation and water escape (section 5.3.1).
3. The removal of quartz and other minerals by solution transfer and the consequential concentration of sheet-silicates in seams (section 5.3.4) is a powerful means of developing a sheet-silicate preferred-orientation. The loss of volume within the seam causes an inevitable packing and alignment of the residual sheet-silicates; this is probably the principal process of LPO formation in slates with prominent solution seaming.
4. Mica or chlorite crystals are fragmented and smeared out along the *C*-planes of mylonitic rocks (sections 2.4 and 5.4) during cataclastic flow. Fragments are bounded primarily by {001}, tend to lie with {001} parallel to the *C*-planes, and impart the main foliation to such rocks.

The general observations in slates and schists that oblique sheet-silicate crystals are truncated sharply by {001} of adjacent subhedral grains (Figs 7.8B and 7.13B), and that sheet-silicates coarsen and change in composition with grade of metamorphism, indicate the importance of grain growth and recrystallization. The question is how much do growth processes contribute to the preferred orientation? If a rock is undergoing strain, crystals will naturally grow preferentially in directions perpendicular to the shortening direction and parallel to the extension direction. This is because spaces are opening up in the extension direction, perhaps at the margins of rigid grains, as a result of crack–seal deformation,

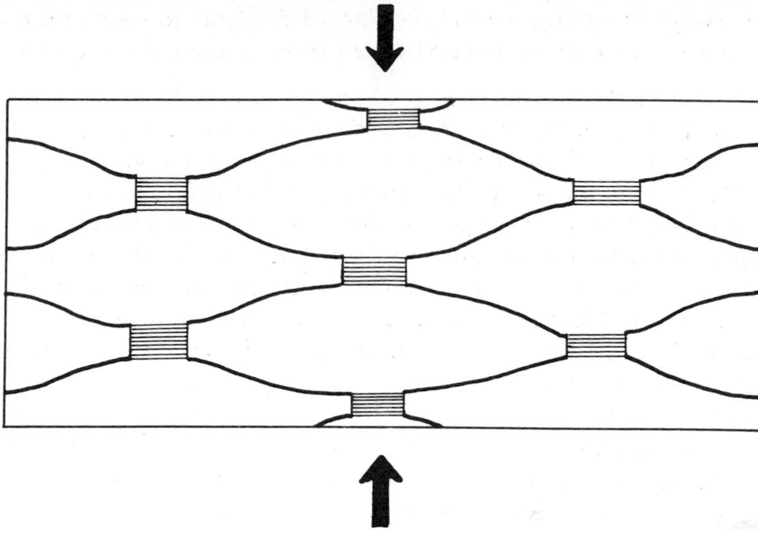

Fig. 7.12 Brick-shaped areas of fibrous or platy mineral growth that develop where pressure-solution seams converge and where extensional microfracturing (boudinage) is concentrated. Arrows indicate the shortening direction. Based on De Roo (1989).

or because of dilation in fractures and shear zones. In the direction of shortening, remaining spaces are closed, and pressure-solution (complementary to crystallization) takes place. Sheet-silicates grow fastest parallel to {001}, so those grains with {001} perpendicular to the shortening direction have a competitive advantage, thus producing a cleavage or schistosity. Grains with {001} at a low angle to the shortening direction tend to buckle, and this enhances their solubility and removal.

Evidence for G-type processes is found both naturally and experimentally. Etheridge and Lee (1975) describe a bimodal orientation of sheet-silicates in slate, some parallel to bedding, some to cleavage: this strongly contradicts the idea of rotation into the cleavage orientation. In slates and low-grade schists, sheet-silicates grow in pressure shadows and veins with a preferred orientation parallel to slaty cleavage or schistosity. Some grains may achieve their orientation by competitive growth, others probably nucleate on existing oriented grains of the host rock (Cox and Etheridge, 1983). Slates in New South Wales (De Roo, 1989) have micas that grew with a very strong preferred orientation in extensional sites where solution seams converge (Fig. 7.12). In high-temperature experiments, and in natural rocks, Etheridge and Hobbs (1974) and Williams *et al.* (1977) show that new crystals of mica replace existing deformed grains: 1. coherently, and 2. incoherently with {001} at a high angle to the shortening direction (Fig. 7.13). The incoherent replacement usually takes

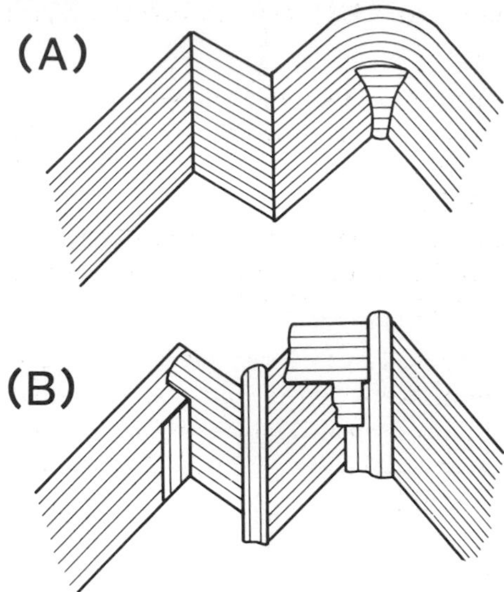

Fig. 7.13 (A): Mica crystal, or bundle of crystals, kinked and bent. (B): Mosaic of recrystallized micas that might replace (A). Replacement is partly coherent, outlining the fold, partly incoherent with (001) following kink-band boundaries. Note that grain boundaries in the sheet-silicate mosaic (B) are dominated by (001).

place along kink band boundaries (i.e., along the hinges of microfolds). In both cases the driving forces are a reduction in strain and chemical free energies. The incoherent grains form in an orientation similar to schistosities associated with natural folds; the grains develop and survive because they are not prone to further deformation, and because of their ability to grow quickly with that orientation.

The M- and G-type processes both tend to produce the same general preferred orientation of {001} at a high angle to the shortening direction, and the chief problem is deciding their relative importance. The M-type fabrics of the March kind seem most likely at very low grades of metamorphism, but even then grains are growing, and M and G-type processes probably act together. For example, a moderately oriented set of plates become better oriented by growth simply because those in the preferred orientation have longer possible growth paths (Fig. 7.8); plates parallel to the shortening direction are truncated. In mylonitic rocks, and rocks affected by solution transfer, M-type fabrics, but not necessarily of the March kind, are an important but not exclusive factor. In many cases the question to ask may be how much of an earlier M-type fabric has been inherited, how much of the fabric is due to neocrystallization or recrystallization?

The spread of poles to {001} as a complete or partial girdle can be explained in two ways:

1. If the strain departs from a simple uniaxial shortening, poles to {001} spread towards the intermediate strain axis. An L-tectonite will therefore have a prominent girdle with the girdle axis parallel to the stretching direction (Fig. 7.11B).
2. Folding of existing foliation surfaces spreads poles to {001} in a girdle about the fold axis. The girdle is likely to contain two maxima, one for each limb (Fig. 7.11C).

There is a surprising lack of investigation of sheet-silicate preferred orientations in natural rocks, despite its obvious importance. Common cleavage and schistosity are such familiar phenomena, even to the layman, that we tend to forget our inability to specify mechanisms of formation with any degree of certainty.

7.3 QUARTZ

Quartz preferred-orientations are common, easily recognized, and diverse in character. They range from the SPO (no LPO) of slates, the extreme SPO (with LPO) of mylonites, and the weak or non-existent SPO but moderately strong LPO of common schists and quartzites. The last fifteen years have seen a concentrated research effort devoted to understanding the ways in which quartz deforms plastically. Two different but complementary approaches have emerged. Etchecopar (1977) and Etchecopar and Vasseur (1987) propose heterogeneous deformation, getting around von Mises' criterion by involving other strain mechanisms such as pressure solution and grain-boundary sliding; recrystallization also plays an essential role in the fabric simulations of Etchecopar and Vasseur (1987). On the other hand, Lister *et al.* (1978), Lister and Williams (1979) and Lister and Hobbs (1980) propose homogeneous deformation and compliance with von Mises' criterion so that a complex of slip systems is entirely responsible for strain and preferred orientation development. Whichever approach is closer to reality, there is no doubt that many SPOs and LPOs of quartz are readily explained by plastic deformation, and the great volume of work on the subject has provided a great impetus to fabric work in general. Often lost sight of is the fact that some quartz LPOs and SPOs develop entirely by G-type processes, and that many fabric patterns develop only through the joint action of G- and P-type processes.

Measuring quartz *c*-axes preferred-orientations

Owing to the lack of cleavage and visible twinning, and because of its uniaxial nature, the only direction that can be identified optically is the

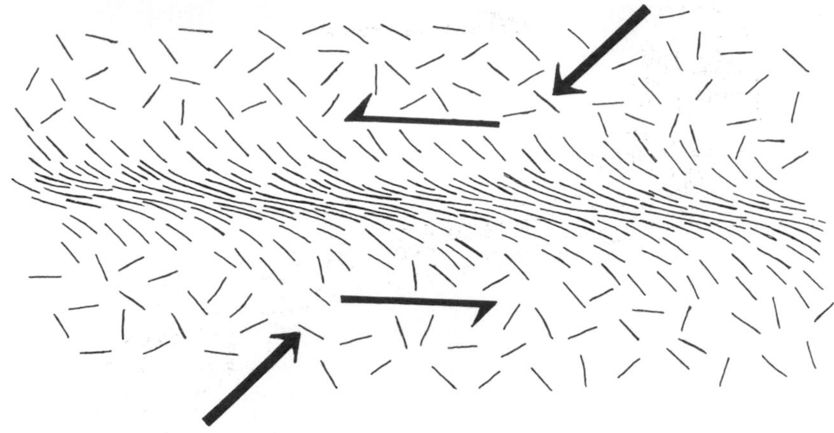

Fig. 7.14 Shear zone within an initially homogeneous rock. Arrows indicate the shortening direction and shear sense. Mylonitic foliation is initiated *c.* 45° to the shear-zone wall, but becomes nearly parallel to it in the most mylonitized parts.

c-axis. The method of measurement is outlined in section 6.3.2, and measuring and plotting 300–400 *c*-axes is usually one day's work. There are no 'blind spots', and measurements are easily made in well-recrystallized rocks. In some rocks, the following problems arise:

1. If grains are smaller than the thickness of the thin section, interference from adjacent grains does not permit accurate location of the *c*-axis with the U-stage. A sensitive-tint plate and three orthogonal sections may provide a qualitative measure of LPO, otherwise X-ray methods are required.
2. The orientation of quartz is usually measured grain by grain, but sometimes it is preferable to measure at a fixed spacing interval. This may be one approach to the problem of subgrains with a high degree of misorientation when it is difficult to decide where one grain starts and ends.
3. Quartz ribbons in mylonite are often replaced by a very fine-grained mass of recrystallized grains, usually of similar orientation to the host. These may be too small to measure individually, and in any case one must decide whether to measure the orientation of ribbons or the orientation of the individual fine grains. They constitute two populations.
4. The *c*-axis orientation of bent grains, particularly ribbons, may vary by some tens of degrees. Measure what seems to be the mean orientation for each bent ribbon, or measure orientations at a fixed spacing interval.

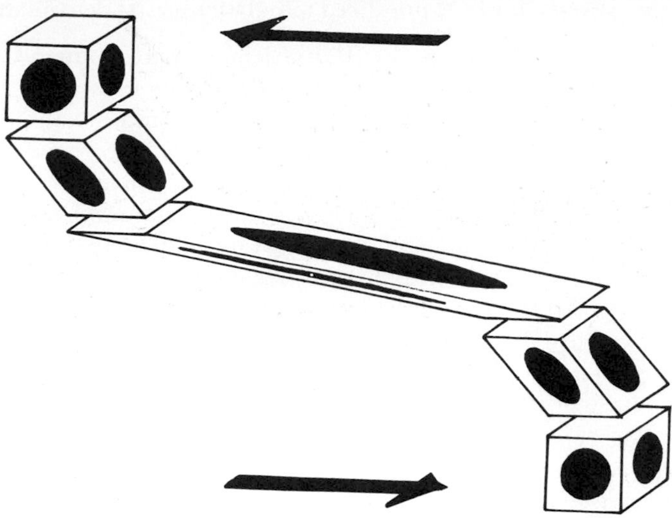

Fig. 7.15 Non-coaxial plane strain, or simple shear strain, in a shear zone such as that of Fig. 7.14. The cubes with circular shapes represent the unstrained state; the other forms are derived from such a cube. Equidimensional forms (such as cubes or spheres) develop into essentially linear shapes which are ribbon-like in section. Arrows indicate the sense of shear.

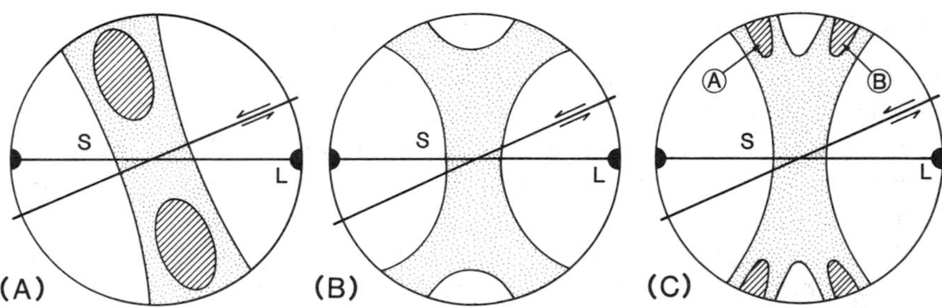

Fig. 7.16 Idealized contoured equal-area projection patterns for quartz *c*-axes. *S* = foliation, *L* = lineation. (A): Girdle, often with a double maximum, perpendicular to a shear zone, oblique to the foliation. (B): Type-1 crossed girdle more-or-less perpendicular to the foliation. (C): Type-1 crossed girdle with two maxima (A and B), produced in simulations by Etchecopar and Vasseur (1987); discussed in text.

Interpretation of quartz preferred-orientations

Metamorphic quartz commonly goes into solution, crystallizes, recrystallizes, and/or deforms plastically. Unlike the sheet-silicates, quartz does not typically display any marked crystallographically controlled shape anisotropy, and this means that quartz preferred-orientations originate by P- and/or G-type processes, not M-type.

Quartz SPOs and LPOs produced mainly by plastic deformation

Quartz ribbons in mylonitic rocks clearly demonstrate simultaneous SPO and LPO development as a result of plastic deformation, and much recent research has been focused on them. Bending of grains and the development of undulatory extinction with subgrain boundaries parallel to the *c*-axis indicate the operation of slip along the direction of the *a*-axes. In fact $\{0001\}\langle a \rangle$ is known to be the dominant slip mechanism under greenschist-facies conditions, and one therefore intuitively expects *c*-axes to be at a high angle to the ribbons of greenschist-facies mylonites (Figs 7.1B and 7.3B); this is typically the case and easily detected using a sensitive-tint plate. The dominant basal slip may be accompanied by prism slip along the *a*-axes.

An understanding of the geometry of a shear zone is important to understanding how mylonites form. Strain, for example, is likely to be a non-coaxial plane strain (no extension or shortening in the intermediate direction), and foliation due to shortening is initiated at a moderate angle to the shear-zone wall but rotates towards parallelism with the wall with increasing strain (Figs 7.14 and 7.15). Owing to the rotational component, one intuitively expects the quartz LPO to be asymmetrical to the shear zone. However, in some cases (Fig. 7.16A), the asymmetry is slight, and a girdle of *c*-axes is developed at a high angle to the shear zone (e.g., Law *et al.*, 1990). In other cases, a type-1 crossed girdle develops (Fig. 7.16B), asymmetrical to the shear plane but more-or-less symmetrical with respect to the foliation (e.g., Law *et al.*, 1986). In some mylonitic zones, transitions from type-1 crossed girdles to the single girdle are present.

The simulated fabric developments of both Lister and Hobbs (1980) and Etchecopar and Vasseur (1987) produce type-1 cross-girdle patterns similar to those found naturally, but neither simulations were able to produce the single girdle of *c*-axes by slip alone. Etchecopar and Vasseur (1987) therefore bring recrystallization into their model. They point out that the two concentrations of *c*-axes that make up their crossed-girdle pattern (A and B in Fig. 7.16C) are fundamentally different. Population A has an orientation suited to perpetual strain and easy slip on the basal plane. In contrast, population B has an unfavourable orientation because further strain involves slip within the grains in the opposite sense to that of the overall shear zone; large amounts of grain-boundary sliding are also required, and this is difficult with elongate grains. Consequently, the B population of grains becomes 'locked' in position. Recrystallization produces equidimensional subgrains which are more easily 'unlocked' by rotation from orientation *B* to *A*. Another mechanism for eliminating B grains is by their annihilation during recrystallization; paradoxically, this occurs because grains that develop the highest finite strains (population A) rid themselves of dislocations more easily than the 'locked' grains which strain heterogeneously and consequently have the highest levels of stored

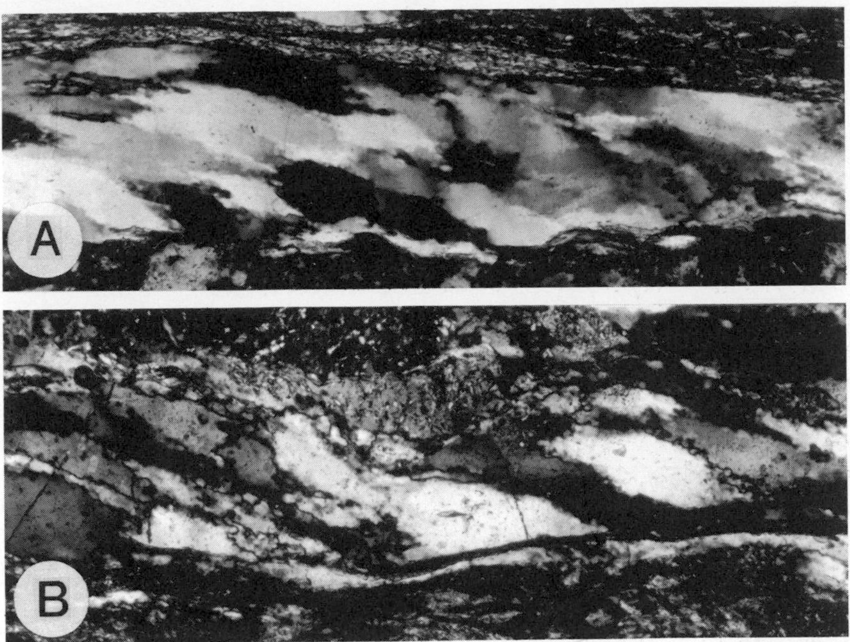

Fig. 7.17 Recrystallized quartz ribbons in which new grains form secondary ribbons (at a small angle to the original ribbon, and at a higher angle to σ_1). Mylonite, Alpine Fault, NZ. (Lengths of view measure 0.85 mm (A) and (B).)

strain energy. The interaction of dynamic recrystallization and plastic deformation can be explored further in the graphic computer simulations of Jessell (1988a and b), available on disc for Macintosh computer.

Many natural quartz ribbons provide evidence for dynamic recrystallization during shear. Not only are some ribbons replaced by a fine-grained mosaic of new grains (the fine grain size indicates a high differential stress – section 5.2), but the new grains may themselves be strained into secondary embryo ribbons (Fig. 7.17). The secondary ribbons are at a higher angle to the shear-zone wall than the older, more rotated, ribbons.

The single girdle pattern of mylonitic rocks is routinely used as a shear-sense indicator. The girdle always has the same sense of obliquity to the foliation, and the angle between the two increases with progressive strain (Figs 7.18A–B). The sense of shear is not always so clear in the case of the type-1 crossed girdle; it tends to be symmetrical to the foliation, and diffuse patterns may lack any obvious asymmetry. However, in the ideal case, the asymmetrical type-1 pattern (Fig. 7.18C) can also be used to determine shear sense.

Mylonitic rocks are rather special in that the bulk of the finite strain

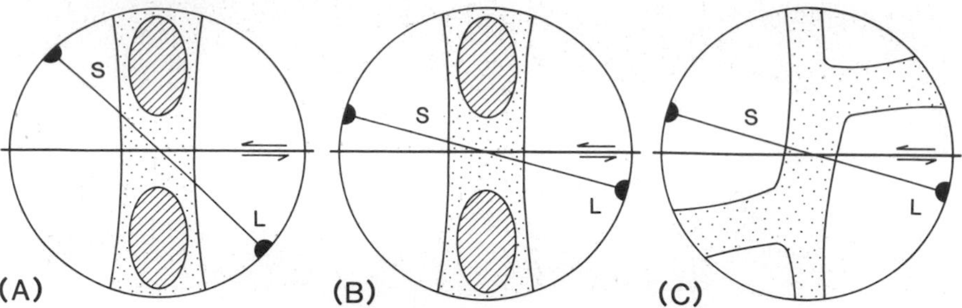

Fig. 7.18 Idealized contoured equal-area projection patterns for quartz *c*-axes. S = foliation, L = lineation. (A) and (B): Girdles perpendicular to a shear zone, but the angle between the girdle and foliation approaches 90° with increasing strain (A to B). (C): Asymmetrical type-1 crossed girdle in relation to foliation and a shear zone.

develops during one strain episode in a rather special site of deformation, a shear zone, and the action of plastic deformation is usually obvious. Most quartz-bearing metamorphic rocks do not display such obvious P-type fabrics, and typical metamorphic histories are more diverse and complex than those of a shear zone. Strain may be other than plane strain, recrystallization generally eliminates the grain shape record of strain, and at higher temperatures, slip systems other than $\{0001\}\langle a \rangle$ may become dominant.

Simulated P-type LPOs for various combinations of slip system and strain geometry are given by Lister and Hobbs (1980). The principal patterns include a small circle of *c*-axes at a small angle about the shortening direction for uniaxial compression, a type-1 crossed-girdle pattern for plane strain, and a small circle at a large angle about the extension direction (also known as a cleft girdle) for uniaxial extension (Fig. 7.19). Mainprice *et al.* (1986) found evidence for prism slip along the *c*-axes at $T > 650°C$ in the presence of abundant water: fabric patterns with *c*-axes parallel to the extension direction should result. If we adopt a 'relaxed' Etchecopar and Vasseur (1987) type approach, allowing heterogeneous strain and contributions from strain mechanisms other than plastic deformation, we can easily predict that slip planes rotate until they are at a high angle to the shortening direction, and slip directions rotate towards the extension direction. Girdles of poles to slip planes or slip directions spread towards the intermediate strain axis if the strain is not a simple uniaxial shortening or extension.

Quartz LPOs respond rapidly to strain, and Lister and Hobbs (1980) report that the main elements of a fabric pattern become evident with a shortening of just 30%. This means that patterns that record the early stages of a prolonged strain history may be completely overprinted by the late strain events.

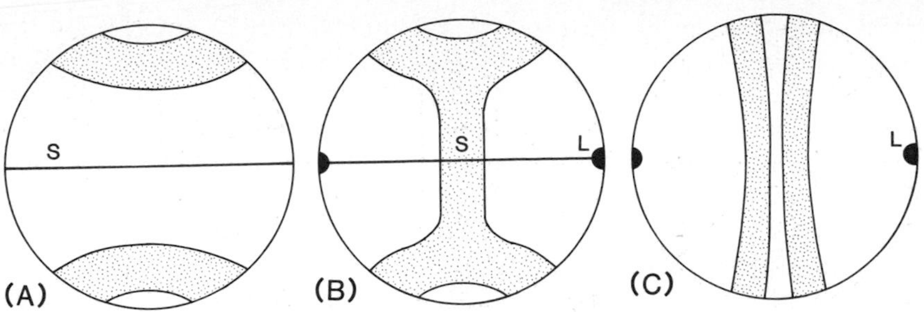

Fig. 7.19 Idealized contoured equal-area projection patterns for quartz *c*-axes, as produced in coaxial deformation simulations by Lister and Hobbs (1980). *S* = foliation, *L* = lineation. (A): Small-circle pattern in uniaxial compression. (B): Type-1 crossed girdle in plane strain. (C): Cleft girdle in uniaxial extension.

Quartz SPOs and LPOs produced mainly by G-type processes

In slates, partial solution of detrital quartz grains in seams rich in sheet-silicates (section 5.3.4) leaves a residue of thin or very thin quartz plates with a strong SPO: plates are oriented parallel to cleavage. No LPO is developed, and the *c*-axis orientations are essentially those of the original sedimentary clasts. Dissolved silica may be entirely lost to such a system (Bell and Cuff, 1989), precipitated in local Q-domains (section 5.3.4), or precipitated either side of the residual quartz grains as a fibrous pressure-shadow or as a coherent overgrowth. Spectacular examples of precipitation in cherts occur in the New South Wales slate belt (De Roo, 1989) where fibrous rocks developed by pervasive crack–seal deformation. De Roo did not measure the quartz LPO, but we can guess that ribbons of the 'stretched crystal fibre type' (section 5.3.5) have *c*-axis orientations that mimic the orientations of grains in the original chert, or if ribbons competed for space, an LPO with *c*-axes parallel to fibre length, or possibly with *c*-axes in a small circle 40° about the fibre length if $[10\bar{1}2]$ is the fastest growth direction (as described by Cox and Etheridge, 1983).

Other quartz ribbons ascribed to growth include those described by Stel (1991). In deformed granitoids, they lack any LPO and were formed by growth in spaces created by dilational conjugate shear fractures. The ribbons are parallel to the intermediate strain axis, and were formed under conditions transitional between the magmatic and metamorphic. It is not clear whether the quartz grew from residual magma or very high-temperature solutions.

Recrystallization, whether dynamic or static, usually acts to remove an SPO. As illustrated in Fig. 2.10, recrystallization resets the grain-shape record of strain. This means that a quartzite with an apparently unstrained equidimensional mosaic of quartz may actually have undergone a severe strain. It also means that a prominent quartz SPO may record only the final increment of a much larger finite strain. In rare instances, recrystallization generates an SPO: thus Urai *et al.* (1986) describe ribbons that formed by coalescence of grains during recrystallization.

Recrystallization also acts to change or modify P-type LPOs. Hobbs (1968) showed how small new grains within a large strained crystal tend to have *c*-axes 20–40° from the host *c*-axis, the general effect being to weaken pre-existing fabric patterns. Etchecopar and Vasseur (1987) suggest that recrystallization changes LPO in an indirect way by facilitating grain-boundary rotations. On the other hand, Knipe and Law (1987) and Jessell (1988a) suggest grains that store high levels of strain energy (grains not in an orientation suitable for easy slip) are annihilated during recrystallization. This has the important effect of simplifying *c*-axes patterns so that easy slip orientations dominate.

A relatively uncommon quartz LPO pattern has *c*-axes close to the extension direction. Although it may sometimes result from slip along the *c*-axis direction (Mainprice *et al.*, 1986), in metacherts and metatuffs from Bluff, New Zealand, the evidence is that the fabric has a growth origin (Shelley, 1980). The Bluff quartz grew from glass or chert during mild strain associated with intrusion of a gabbroic complex; no correlation exists between grain shape and *c*-axis orientation, and euhedral quartz ejecta are not plastically deformed. Gapais and Barbarin (1986) also suggest a G-type origin (either crystallization or recrystallization) for a similar *c*-axes orientation in deformed granite.

It has already been noted that a positive correlation exists between quartz LPO strength and modal quartz content in mica schists (section 7.1); the presence of sheet-silicates decreases the efficacy of the P-type process in quartz. In addition, solution transfer often acts on a large scale, and most quartz in the greenschist-facies Haast Schists of New Zealand, for example, has been reprecipitated during metamorphism. The G-type fabrics should characterize the formation of quartz-rich segregation layers, but may be obliterated or modified during subsequent dynamic recrystallization. The very low-grade G-type fabrics described by De Roo (1989) are already slightly modified by recrystallization, and the process has gone further in the Haast Schists. Nevertheless, the presence of feldspar growth fabrics in Haast Schist segregation layers, plus the lack of obvious plastic deformation of quartz around feldspar grains, led Shelley (1989a) to suggest that the quartz SPO was of G-type, the LPO of P-type.

Some final remarks

The P- and G-type mechanisms may act conjointly or in sequence to produce the quartz fabrics of natural rocks. There seems little doubt that

G-type fabrics dominate in very low-grade slates and associated rocks, and that P-type mechanisms dominate in mylonites and quartzites. However, in mylonites and quartzites the intervention of recrystallization is necessary to explain natural fabrics (Etchecopar and Vasseur, 1987; Knipe and Law, 1987; Jessell, 1988a), and in rocks subject to solution transfer, G-type fabrics are often overprinted by later plastic deformation.

The visible effects of plastic deformation should not be used as evidence for P-type orienting mechanisms as it is commonplace for crystals already oriented by some other means to be slightly strained by plastic deformation. For example, the familiar plastic deformation (bending) of feldspar laths in a dyke does not give evidence that trachytic texture is a P-type fabric (obviously it is not). Neither should strain mechanisms be confused with mineral orienting mechanisms: they need not be the same. For example, high-temperature strain of a quartzite may proceed mainly by plastic deformation, but the accompanying or succeeding dynamic or static recrystallization may substantially revise the LPO. On the other hand, in schists dominated by solution transfer, the bulk of the strain may be achieved by water loss, pressure solution, crystallization and cracking of the rock; on top of this, plastic deformation may impose a conspicuous LPO and SPO that are relatively unimportant in the total history of the rock.

7.4 CALCITE AND DOLOMITE

Calcite is one of the most plastic of rock-forming minerals, not only under 'metamorphic' conditions, but at room temperature as well. It deforms by twin-gliding and slip at all temperatures. Dolomite deforms by slip at low-moderate temperatures and twin-gliding at moderate-high temperatures.

Slip and twin-glide systems

Principal slip and twin-glide systems are summarized in Table 7.1 and Fig. 7.20. The information comes from experimental work summarized in Turner and Weiss (1963) and Nicolas and Poirier (1976). Calcite and dolomite are trigonal so the rhombohedra *e*, *r* and *f*, each comprise three planes which can be labelled e_1, e_2, e_3, etc. Note that the slip and twin-glide systems for calcite and dolomite are quite different.

Twin-glide systems have a unique sense of shear along the glide direction (unlike slip systems). This means that calcite twins only when the direction of shortening is at a high angle to *c* (Fig. 7.21A); for dolomite it must be at a low angle (Fig. 7.21B). Note that twinning effects an instantaneous rotation of the *c*-axis through 52.5° or 54°; this contrasts with the gradual rotation that occurs during slip.

Table 7.1 Principal twin-glide and slip systems, and interfacial angles, for calcite and dolomite

	Slip systems	Twin-gliding	Interfacial angles
Calcite	$r_1[f_2:f_3]$[a] (total of three equivalent systems) $f_1[r_2:f_3]$ or $[r_3:f_1]$ (total of six equivalent systems)	$e_1[e_1:r_2]$ (total of three equivalent systems)	c to $r = 44.5°$ c to $e = 26°$ c to $f = 64°$
Dolomite	$c\langle a\rangle$ (total of three equivalent systems)	$f_1[f_1:a_2]$ (total of three equivalent systems)	c to $r = 44°$ c to $f = 62.5°$

[a] Slip or twin-glide directions given in this form refer to the line of intersection of the two planes cited

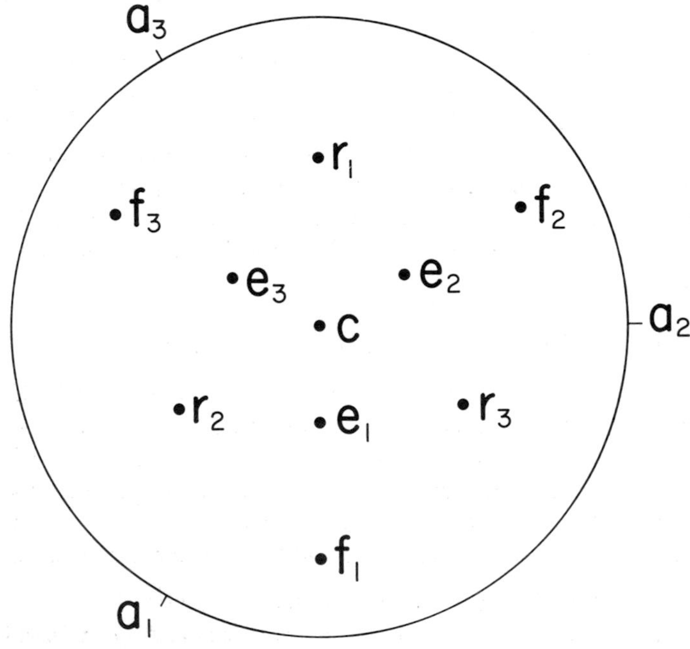

Fig. 7.20 Equal-area projection (lower hemisphere) of the poles of important planes and axes in calcite. A plot for dolomite would be almost identical (with some angles slightly changed).

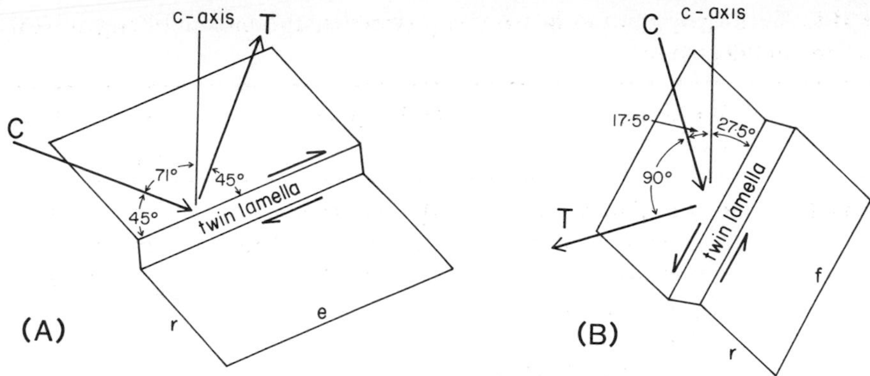

Fig. 7.21 Crystal sections to show the angular relationships between twin lamellae, *c*-axes and compression (*C*) and tension (*T*) directions, (A) in calcite, and (B), in dolomite.

In calcite, twinning always occurs in preference to slip, provided that crystals are suitably oriented. Complete twinning of calcite is an extremely effective way of producing an LPO (*c*-axes cannot remain close to the extension direction). Once grains have been reoriented by twinning, deformation continues by slip on *r* and *f*. Both slip systems are reported active over a wide range of temperature, but *f* usually seems to be common or dominant at $T > 500°C$. Twin lamellae rotate as passive markers away from their ideal crystallographic positions as a result of later slip on *r* or *f*. An accurate assessment of the rotation (possibly several degrees) can be used to determine the nature of the slip system responsible.

In dolomite, basal slip dominates at $T < 400°C$, twinning on *f* at $T > 500°C$. Twinning does not occur at $T < 300°C$ and is usually not well developed until $T > 400°C$.

Measuring calcite and dolomite *c*-axes and twin lamellae

The *c*-axes should be measured as outlined in section 6.3.2. Cleavage and/or twin-lamellae orientations can also be measured, thus allowing a complete determination of lattice orientation. Up to three twin-lamellae orientations can be recorded for each grain (follow the procedure given in section 6.3.2); these are used for determination of 'compression' and 'tension' directions (shown below). In addition, note the following:

1. Bent grains should be dealt with as noted for quartz.
2. Subdivide twinned grains into 'twin lamellae' and 'untwinned host'. Measure the *c*-axis orientation of the host and the planar orientation of the lamellae. If one twin set and the host each constitute 50% of the

crystal, the decision as to which *c*-axis orientation should be measured is arbitrary.

3. Some twins are so thin that they appear more like a cleavage when rotated into a sharp vertical position. Cleavage and twins are easily distinguished because the angles *c* to *e* and *c* to *r* are quite different (26° and 44.5°).
4. The twin lamellae of heavily twinned grains may overlap in thin-section view, obscuring the optics. This is a particular problem for grains with the *c*-axis close to vertical. It may be possible to plot sufficient twin-lamellae orientation data to work out the *c*-axis orientation of the host, using stereographic projection, but in many cases it is impossible to measure *c*. If this is a general problem, then it clearly invalidates the significance of the pole diagram. Solutions may be: to measure *c*-axes in another orientation section, or combine data from two sections of different orientation.
5. Extinction positions for calcite are often not as sharp as for quartz, and it may be desirable to follow the alternative procedure for locating a uniaxial *c*-axis given in section 6.3.2.

Determination of 'compression' and 'tension' directions using twin lamellae

Twin-gliding operates with a unique sense of shear, and so Turner (1953) showed how one could determine ideal 'compression' and 'tension' directions for any set of twin lamellae. Both directions lie in the same plane as *c* and the pole to *e*, as shown in Fig. 7.21A. When plotted on an equal-area net, the contoured maxima, or mean orientations, represent the possible positions of σ_1 and σ_3. Unfortunately, for calcite, there are two principal problems in applying this method in any general way:

1. Any substantial deformation completely twins all calcite which is then in an orientation unfavourable for further twinning. Owing to this, visible twinning in complexly deformed material such as marble usually represents a late minor strain unrelated to the principal deformation; in weakly deformed limestone it usually does relate to the principal strain episode.
2. The ideal 'tension' direction is so close to the *c*-axis orientation that a *c*-axes preferred-orientation generates spurious 'tension' directions.

To determine 'compression' and 'tension' directions plot the U-stage measurements for *c* and *e*, construct the great circle that joins them; 45° along the great circle from *e* in the direction away from *c* is the 'compression' direction, and the 'tension' direction is 90° from that along the same great circle. Transfer poles to separate nets for accumulated data plots. A similar procedure is used for dolomite, suitably modified for the actual orientations of *c*, *f* and the 'compression' and 'tension' directions (Fig. 7.21B).

The procedure can be rather tedious and time consuming, and a computerized version of the Turner method is available (Shelley, 1989c). I recommend that the students plot the data first by hand, and once the procedures are fully understood, move on to computer aids. An alternative graphical approach to determining the position of σ_1 from *e*-pole and *c*-axis data is described by Dietrich and Song (1984). My recent experience of work on 38 samples of marble is that the Turner and Dietrich and Song methods produce equally good results for σ_1, especially using computer software (STEREO) to determine the mean Turner 'compression' direction. The advantage of the Turner method is that it may also allow a determination of σ_3. Owing to the influence of *c*-axes orientations, the Turner 'tension' axes are almost always in spurious positions, but if a girdle of Turner 'compression' axes exists, then the girdle axis equals σ_3.

Interpretation of calcite SPOs and LPOs

The ease with which calcite deforms plastically ensures that almost all natural preferred orientations are P-type. The only common G-type fabrics are the SPO of stylolitic limestones and marbles where grains have been partially dissolved, and the SPO and LPO of veins. The ability of calcite to deform plastically and the lack of crystallographic controlled shape anisotropy in marble militates against the development of M-type fabrics.

Again, because of the ease with which fabrics develop (and redevelop), they may relate only to the later episodes of complex strain histories. Quartz does not react to strain as quickly as calcite; thus in the New Zealand Arthur Marble and associated rocks (Shelley, 1991), quartz fabrics relate to D_2, calcite SPOs relate mainly to D_2, LPOs were substantially modified by D_3, and twinning is related only to D_3. Of course, if the deformation history is not complex, the calcite fabric will be directly related to it.

The SPOs of calcite are common (Fig. 2.12A) and produced by plastic deformation. However, recrystallization and grain-coarsening reset the grain shapes to equidimensional forms. The example in Fig. 2.12A results from plastic deformation of an already coarsened and recrystallized marble.

In uniaxial compression, calcite develops an LPO with *c*-axes close to the shortening direction and at a high angle to foliation (Fig. 7.22A). Turner *et al.* (1956) interpreted this LPO in terms of twinning, which produces a broad maximum parallel to σ_1, and slip on *r*, which tends to

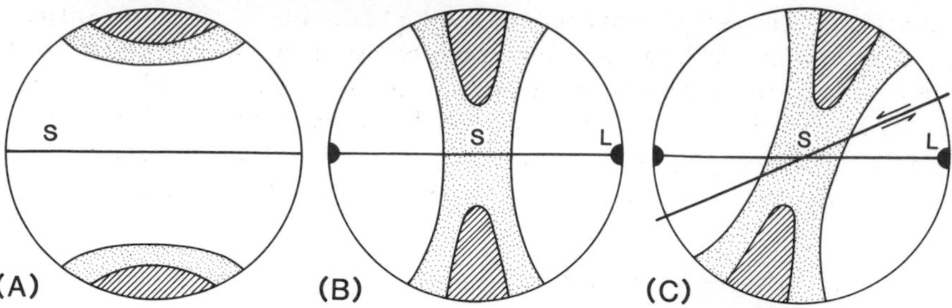

Fig. 7.22 Idealized contoured equal-area projection patterns for calcite *c*-axes. S = foliation, L = lineation. (A): Maximum perpendicular to foliation (uniaxial compression). (B): Girdle with maximum (plane strain); the extension direction is parallel to the girdle axis and L. (C): Relationship of girdle to foliation and shear zone (simple shear); σ_1 is 45° to the shear zone, and the finite extension direction is marked by L which rotates towards the shear plane with increasing strain.

move the *c*-axes into a small circle about σ_1. A strain of <20% is required to rotate *c*-axes into this general position. In plane strain, a girdle develops between σ_1 and σ_2 (Fig. 7.22B), and in L-tectonites a simple girdle about the lineation can be expected. Recrystallization tends to weaken the SPO and LPO formed by plastic deformation.

The development and evolution of calcite fabrics in shear zones is discussed by Schmid *et al.* (1987). Natural shear zone fabrics can be matched with those produced experimentally involving twinning as the dominant orienting mechanism. In simple shear, *c*-axes lie close to σ_1, at an angle to foliation, and extending as a girdle through σ_2 (Fig. 7.22C). Note that the sense of obliquity of this *c*-axes girdle is the opposite of quartz *c*-axes girdles produced in shear zones (Fig. 7.18).

Dolomite is less plastic than calcite, and at T < 400°C slips only on {0001}. The LPOs usually consist of a *c*-axes maximum perpendicular to foliation, and with a girdle about the extension direction. This fabric is consistent with {0001} slip, but requires relaxation of von Mises' criterion. Above 500°C, twinning would be expected to dominate fabrics and rotate *c*-axes away from the shortening direction, the opposite effect to {0001} slip. These observations do not correlate well with experimentally produced fabrics (Wenk, 1985b), and natural dolomite preferred-orientations require further investigation.

7.5 PLAGIOCLASE

Feldspar is the most abundant crustal mineral and one might expect a wealth of literature on the subject of its preferred orientation. There isn't, because 1. feldspar is considerably less plastic than calcite or quartz; 2. metamorphic feldspar is often more-or-less equidimensional so that SPOs are not common or conspicuous; 3. feldspar is biaxial which makes U-stage work laborious; and 4. most feldspar is triclinic, which makes correlation between optics and crystallography difficult or imprecise. Most work in the last 10–15 years has dealt specifically with plagioclase.

Measuring plagioclase LPOs

Optical methods for measuring plagioclase LPOs are time consuming, but there is no practical alternative. X-ray methods are unsatisfactory, and the only possible option is the expensive (and inaccessible to many of us) neutron diffraction analysis (Wenk *et al.*, 1986).

Optical methods serve no useful purpose unless the features measured can be related to crystal axes. There are two ways of achieving this, both of which have their limitations. Firstly, some near correlations exist between optical and crystallographic directions, but only for some compositions. Low-temperature albite has *X* 20° from *a*, *Y* 16–18° from *c*, and *Z* 14° from *b*, which means pole diagrams for *X*, *Y* and *Z* are an approximation of diagrams for *a*, *c* and *b*. This near relationship holds roughly for sodic plagioclase up to andesine. For very calcic plagioclase, *Y* comes close to *a*. Otherwise there are no near correlations. Secondly, one can measure the orientation of (001) cleavage traces and (010) cleavage or twin lamellae. Individually, they provide limited crystallographic data. Together they give the position of *a* (their line of intersection), and combined with *X*, *Y* and *Z*, they provide a complete determination of lattice orientation. Kruhl (1987a) and Benn and Mainprice (1989) have described computer programs to speed the processing of such combined data. The limitations are that not all grains are oriented so that (001) and (010) can be viewed, and in my experience with untwinned fine-grained greenschist-facies albite, many grains provide no trace of the position of either (001) or (010). One is then forced to use the near correlations between optics and crystallography.

Whichever approach is adopted, the optic directions *X*, *Y* and *Z* must be measured. The fact that 2V is always high means no difficulty in locating *X*, *Y* and *Z* for grains of any orientation. Follow the procedure in section 6.3.2. For many grains, only two directions will be observed directly, and the third must be determined using stereographic projection. This is best done while U-stage work is in progress, as it provides a check that the other two are 90° apart. If the first approach is adopted, separate plots for *X*, *Y* and *Z* are produced. If the second approach is taken, refer to Kruhl (1987a) and Benn and Mainprice (1989) for further details.

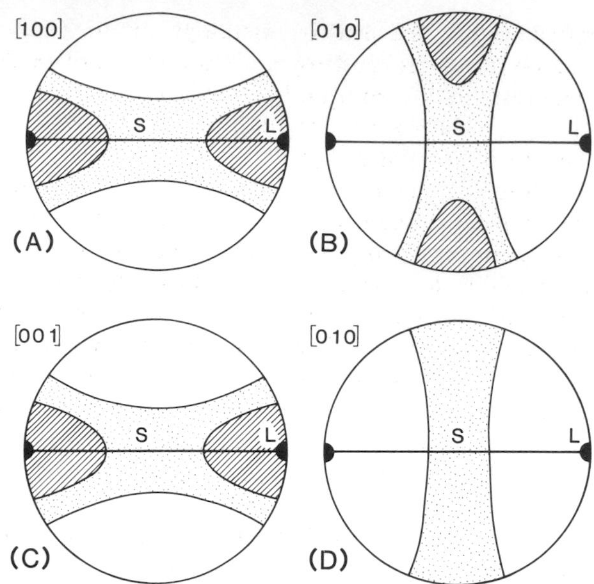

Fig. 7.23 Idealized contoured equal-area projection patterns illustrating plagioclase preferred-orientations. S = foliation, L = lineation. (A) and (B): Type-1 (meta-igneous) patterns for [100] and [010] poles. (C) and (D): Type-2 (metamorphic) patterns for [001] and [010] poles.

Interpretation of plagioclase preferred-orientations

P-, M-, and G-type fabric origins have been proposed for plagioclase LPOs and SPOs. In terms of crystallographic orientation, two principal types of LPO exist:

Type 1: *a*-axes parallel to the extension direction (lineation), and possibly in a girdle parallel to foliation; poles to (010), or *b*, are at a high angle to the foliation and possibly in a girdle around a lineation (Figs 7.23A and B).

Type 2: *c*-axes parallel to the extension direction (lineation), and possibly in a girdle parallel to foliation; poles to (010), or *b*, may form a girdle about the lineation (Figs 7.23C and D).

Both types may be combined with an SPO where tabular grain shapes help to define the foliation, and grain lengths the lineation.

(a) LPO type 1 and the 'tiling' effect

Typical of meta-igneous rocks, this LPO is often strong and typically M-type. Plagioclase phenocrysts are tabular on (010) and elongate parallel to the *a*-axis (less commonly *c*), so igneous flow fabrics are of this type, and they may simply be retained by meta-igneous rocks. During metamorphism, matrix minerals tend to be deformed more readily than the feldspar phenocrysts (e.g., in granite gneisses and mylonites), either because of the greater ductility of matrix minerals like quartz and mica, or because of their finer grain size: this may produce a feldspar LPO where none existed before, or enhance an existing one. In meta-igneous rocks where igneous shapes are retained, the LPO and SPO is obvious, even in the field, but in some orthogneisses, igneous shapes are lost or modified, and the fabric becomes less obvious. A spectacularly strong LPO of this type is described by Shelley (1979): the long-axes of plagioclase laths in deformed crystal tuffs were mechanically rotated towards the extension direction. The resulting *a*-axes pattern is as strongly defined as typical mica preferred-orientations. Subsequent recrystallization during the thermal peak of metamorphism destroyed almost all primary grain shapes and established equidimensional mosaics; these, however, did retain the LPO.

The SPOs of euhedral or subhedral relic igneous feldspar may indicate the sense of shear by what is known as 'tiling' (Fig. 8.4), as described, for example, by Blumenfeld and Bouchez (1988) for deformed granitic rocks. Tiling can be discerned only when the position of the shear-zone wall (to which the tiling is oblique) is known. The same tiling effect develops in response to velocity gradients in magma (Blanchard *et al.*, 1979), so feldspars in dykes are oblique to dyke walls, for example; it may therefore be difficult to decide whether tiling is magmatic or metamorphic in a deformed igneous rock. The angle between tiles and shear plane decreases with increasing strain, and for any particular strain the angle also decreases with decreasing aspect ratio (Fernandez *et al.*, 1983); at high shear strains the sense of obliquity may actually reverse, especially for grains of low aspect ratio.

(b) LPO type 2

This typifies newly crystallized plagioclase in schists; LPOs are weak compared with type 1, but SPOs may be moderately well developed. Two contrasting interpretations may both be valid: 1. the fabric is P-type due to slip mainly on (010)[001], which rotates *c*-axes towards the extension direction; 2. the fabric is G-type, formed through competitive anisotropic growth because metamorphic feldspar grows most quickly parallel to the *c*-axis.

Several slip systems have been documented for naturally deformed plagioclase, but the principal one is (010)[001] (Ji and Mainprice, 1988). Evidence for plastic deformation is widespread in amphibolite-facies and higher-grade metamorphic rocks where bent twins, deformation bands and recrystallization are common; provided that von Mises' criterion can be relaxed by other contributory strain mechanisms, slip on (010) rotates the (010) plane towards the foliation (or about the lineation in L-tectonites), and the [001] slip direction rotates the *c*-axis towards the extension direction. Natural plagioclase fabrics are explained in this way by Kruhl (1987b) and Ji and Mainprice. Kruhl also suggests dominant (001)[100] slip in the higher-temperature range of his rocks, and Ji and Mainprice suggest subsidiary (010)[100] slip; both these systems have the potential to produce fabrics with *a*-axes parallel to lineation. A maximum of poles to (010), asymmetrical to foliation, but perpendicular to the shear-zone wall, has been described by Ji *et al.* (1988) who propose its use as a shear-sense indicator in high-grade rocks where plagioclase has been plastically deformed. The sense of asymmetry is actually the same as in the 'tiling' effect, described above.

The evidence for G-type fabrics comes in greenschist-facies schists from New Zealand, including quartzofeldspathic schists of the chlorite zone (Shelley, 1977, 1989a). Quartz and plagioclase (albite) is segregated into layers, and in some samples, euhedral growth forms of plagioclase survive subsequent mild deformation (Fig. 7.24). Grains have aspect ratios of up to 8/1 in quartz-rich layers, up to 12/1 in sheet-silicate layers, but there is no correlation between aspect ratio and lattice orientation: this implies that if slip were the strain and orienting mechanism, multiple slip systems (not just (010)[001]) must have been active. Plagioclase grains are generally unstrained and untwinned except for simple growth albite twins. If multiple slip systems were responsible for grain shapes and orientation, the albite twins should have been passively rotated into irrational orientations (Fig. 7.25); in every case they remain exactly on (010). If slip had been the strain mechanism, albites in quartz-rich layers should be more elongate than those embedded in sheet-silicates: the opposite is the case. The plagioclase has not been recrystallized, but often forms individual crystals surrounded by quartz; they could only have formed by crystallization, presumably during growth of the segregation layering. For all these reasons, plus the fact that plagioclase is generally observed to be non-plastic at low grades of metamorphism, I proposed a growth mechanism for development of the LPO and SPO. Franke and Ghobarkar (1980, 1982) show that plagioclase grows most quickly parallel to *c* at metamorphic temperatures, and a schist undergoing strain by fracturing and solution transfer may develop a competitive growth fabric with plagioclase *c*-axes parallel to the extension direction.

Perhaps the most intriguing fact is that plagioclase LPO in the New Zealand schists remains the same through the transition into the amphibolite facies, where deformation pericline twinning and other strain features indicate that plastic deformation comes into play. The important

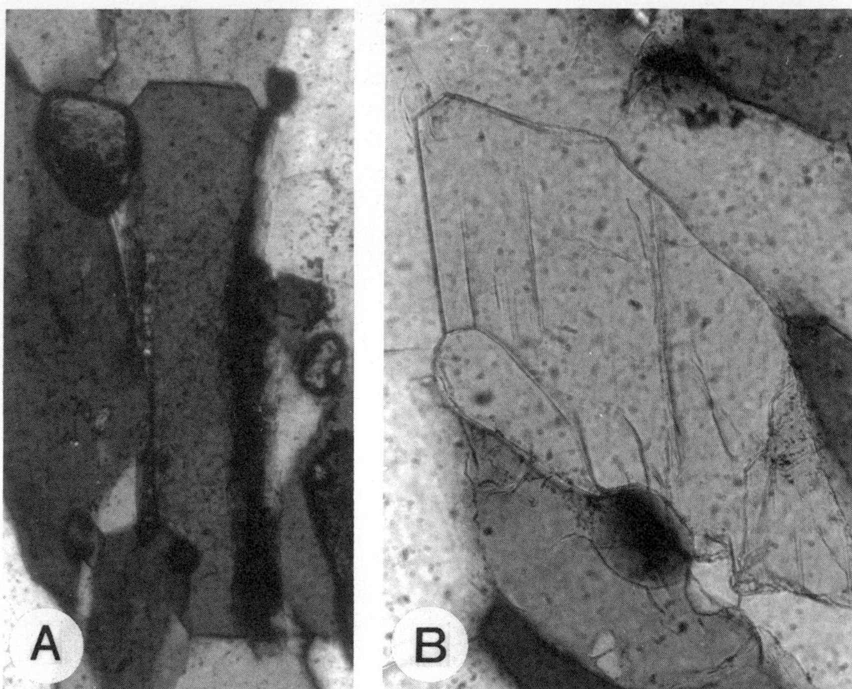

Fig. 7.24 The euhedral aspect of albite crystallized with a preferred orientation in a lower greenschist-facies schist from New Zealand. Lengths of view measure 0.44 mm (A) and 0.3 mm (B). Reproduced with permission from Shelley (1989a), copyright Pergamon Press PLC.

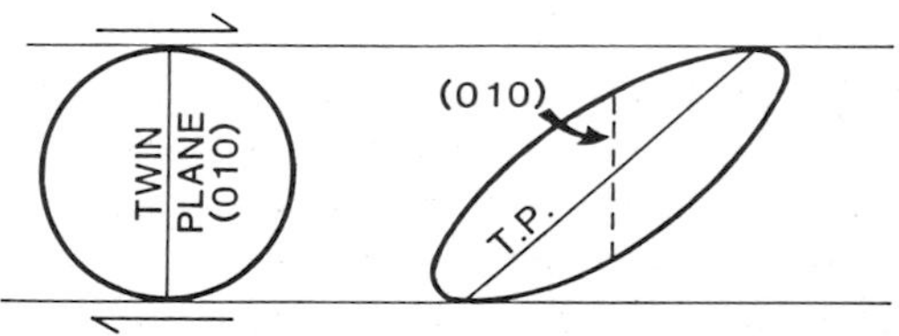

Fig. 7.25 Diagram to show how slip on a plane at a high angle to (010) will rotate the albite twin plane away from (010), an effect not observed in the Haast Schist, NZ. Reproduced with permission from Shelley (1989a), copyright Pergamon Press PLC.

question is: 'Does one necessarily have to invoke plastic deformation as the orienting mechanism, simply because plastic deformation was active?' The answer to this must be 'no'; there are abundant examples where plastic deformation is active but not important as an orienting mechanism (e.g., bent feldspar microphenocrysts as part of trachytic texture). The evidence for G-type plagioclase fabrics at low grades is very strong indeed, but there is no evidence of a direct link between plastic deformation as cause and LPO as effect, even at high grades. More work, perhaps on the relationship between grain shapes and LPOs, is necessary to make that direct linkage.

Plagioclase deformation twinning

Twinning in the carbonates is an important strain and orienting mechanism. This is not the case for plagioclase. Although deformation twinning is common in moderate- to high-grade metamorphic rocks, it represents a much smaller strain than carbonate twinning, and the reorientation of crystal axes is very small. However, plagioclase twinning can be used for determination of 'compression' and 'tension' directions in a way similar to the Turner method for calcite (Lawrence, 1970).

7.6 OLIVINE

The upper mantle consists mainly of olivine, and understanding its mode of deformation is important to the development of models for the lithosphere and plate tectonics. Experimental and natural evidence for plastic deformation of olivine is well established, and given the general lack of crystallographic-controlled shape anisotropy in deformed or recrystallized peridotite, olivine preferred-orientations are likely to be P-type, possibly G-type as a result of recrystallization, but not M-type.

Measuring olivine LPOs

Olivine is orthorhombic which means there is a direct correlation between optics and crystallography ($X = b$, $Y = c$, $Z = a$). The 2V is usually high which means there is no difficulty in measuring the orientation of X, Y and Z for grains of any orientation. Follow the procedure in section 6.3.2. If only two optical directions are observed directly, the orientation of the third is determined using stereographic projection; this is best done while U-stage work is in progress, as it provides a check that the other two are 90° apart. Compile data in separate plots for X, Y and Z (= b, c and a).

If olivine is present as a number of generations and/or textural types, then the LPO for each population should be measured separately. In particular, some peridotites exhibit relic grains of early coarse material surrounded by later fine-grained material. The relic grains are variously known as porphyroclasts or palaeoblasts, the new grains are known as neoblasts. In other cases, coarser material may represent selective growth of some grains during recrystallization.

Interpretation of olivine LPOs

(a) P-type fabrics

Numerous slip systems are known from experimental deformation and observations of dislocations in natural olivine (Mercier, 1985), but the evidence of natural fabrics is that the system (010)[100] is most important. Mercier (1985) notes that the conditions for the activity of particular slip systems, as observed in experimental work (Carter and Avé Lallemant, 1970), cannot be extrapolated to natural systems, and cites as an example the slip direction [001] which becomes naturally active at higher temperatures rather than the expected lower temperatures (<1000°C).

Activity of the [100] slip direction is often marked optically by deformation bands or kink boundaries parallel to (100); the kinks indicate movement of edge dislocations. In contrast, [001] slip is characterized by a greater degree of screw dislocation activity, so that (001) boundaries represent a twisting rather than simple kinking of the lattice. Combinations of subgrains bound by (100) and (001) form a columnar structure parallel to [010].

Dominance of a particular slip system is indicated by the fabric patterns of peridotites with strong SPOs produced by plastic deformation. Thus [100] close to the extension direction (lineation) and [010] at a high angle to foliation and/or forming a girdle about the lineation (Figs 7.26A and B) indicates (010)[100] slip. If [001] is at a high angle to foliation (Fig. 7.26C), (001)[100] slip is indicated, and so on. The [001] slip most commonly acts in co-ordination with [100] slip and causes both these directions to spread out as a girdle parallel to the foliation (Figs 7.26D–F).

Although plastic deformation undoubtedly occurs, slip systems of olivine are sufficiently restricted in orientation that von Mises' criterion cannot be complied with; heterogeneous strain and the action of other strain mechanisms need to be invoked. In the Etchecopar and Vasseur (1987) model, grain-boundary sliding is facilitated by recrystallization (as already discussed for quartz), and given this relaxed situation, the model successfully simulates the features of the most common natural-fabric patterns by (010)[100] slip, possibly combined with secondary (010)[001] slip. In simple shear, their simulations produce asymmetrical fabrics with (010) at a high angle to the shear-zone wall (Fig. 7.27). [100] lies in the plane of the shear zone at a small angle to the grain-shape lineation. This is similar to shear-zone fabrics in natural rocks.

(b) G-type fabrics

Recrystallization may contribute to, or adjust, what is essentially a P-type fabric pattern. In the Etchecopar and Vasseur (1987) model it produces

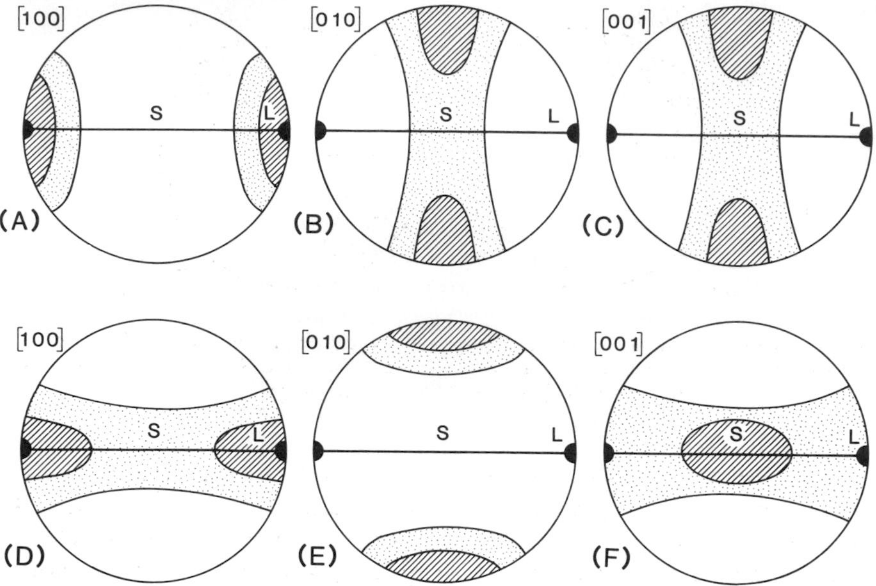

Fig. 7.26 Idealized contoured equal-area projection patterns illustrating olivine preferred-orientations. *S* = foliation, *L* = lineation. (A) and (B): Patterns for [100] and [010] poles resulting from dominant (010)[100] slip. (A) and (C): Patterns for [100] and [001] poles if (001)[100] slip is dominant. (D), (E) and (F): Patterns for [100], [010] and [001] poles resulting from a combination of [100] and [001] slip.

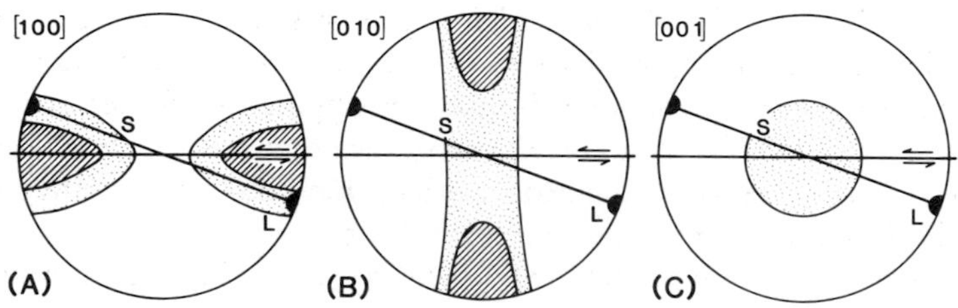

Fig. 7.27 Idealized contoured equal-area projection patterns illustrating olivine preferred-orientations simulated for simple shear by Etchecopar and Vasseur (1987). *S* = foliation, *L* = lineation. (A), (B) and (C): Patterns for [100], [010] and [001] poles respectively.

equidimensional grain shapes which allow easier grain-boundary sliding. In the experiments of Avé Lallemant and Carter (1970), neoblasts within palaeoblasts have a greater spread of lattice orientation than the host palaeoblasts, and there appears to be an undefined host control on neoblast orientation. During static recrystallization, euhedral neoblasts dominated by (010) replace large deformed palaeoblasts and retain the elements of the P-type palaeoblastic LPO (Mercier, 1985).

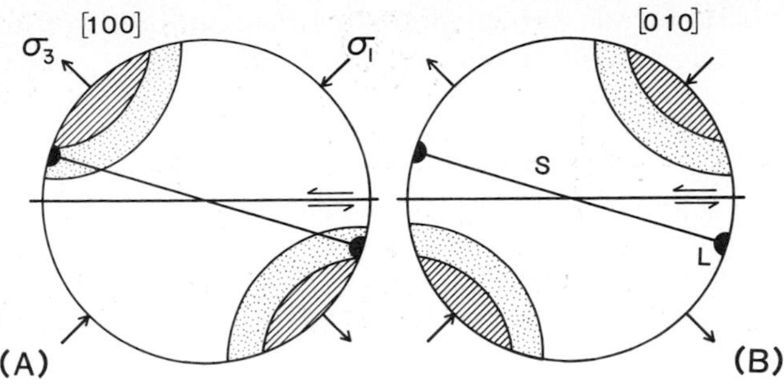

Fig. 7.28 Idealized contoured equal-area projection patterns illustrating stress-related olivine preferred-orientations developed during dynamic recrystallization. *S* = foliation, *L* = lineation. (A) and (B): Patterns for [100] and [010] poles respectively.

More importantly, it is observed that independent stress-related G-type LPOs develop during dynamic recrystallization. This was observed by Avé Lallemant and Carter (1970) where neoblasts growing along palaeoblast boundaries develop an LPO with [010] parallel to σ_1. In non-coaxial shear, Kunze and Avé Lallemant (1981) produced dynamically recrystallized olivine with a stress-related LPO of [010] parallel to σ_1 and [100] parallel to σ_3 (Fig. 7.28). The stress-related fabrics are similar to P-type fabrics but oblique to them in non-coaxial strain. Karato (1988 and 1989) shows that the stress-related LPOs develop only during migration (not rotation) recrystallization; he notes too a subsidiary maximum of [100] parallel to σ_1. The G-type LPO represents 'hard' orientations that cannot deform plastically in the stress field, and the fact that such orientations are energetically favoured is quite the opposite of the situation with quartz (Jessell, 1988a and b) where 'soft' orientations are favoured and consume 'hard' grains. This implies that olivine 'hard' grains remain relatively dislocation free compared to 'soft' grains.

Given that dynamic recrystallization and plastic deformation are usually closely associated processes, it may be difficult to distinguish P- and G-type olivine fabrics in some deformed peridotites, especially those affected by simple coaxial strain.

Olivine fabrics and seismic anisotropy

The mantle and lower crust are anisotropic to seismic waves because of mineral LPOs (Mainprice and Nicolas, 1989). In the upper mantle the cause is olivine LPO, and because the mantle cannot be sampled directly, it is important to examine the possibility of studying mantle structure by correlating seismic anisotropy with olivine fabrics. According to Karato (1989):

1. Lithospheric mantle has an anisotropy that can be correlated with the P-type fabrics found in ophiolitic peridotites and ascribed to mantle flow at oceanic spreading centres (Nicolas and Violette, 1982). [100] is parallel to the inferred palaeo-spreading direction.
2. Asthenospheric mantle older than 80 Ma has an anisotropy that can be correlated with stress-related G-type fabrics due to horizontal flow.

Karato (1989) emphasizes the limitations of our knowledge of strain mechanisms at deep levels in the mantle and concedes that a P-type fabric of as yet unknown origin may be found to explain asthenospheric anisotropy.

7.7 AMPHIBOLES

Amphiboles are the least plastic of all the common rock-forming minerals, but they exhibit a pronounced crystallographic-controlled shape anisotropy, especially at low to medium grades of metamorphism. Strong SPOs and LPOs with crystal lengths and *c*-axes parallel to the foliation and extension direction are well known (Fig. 7.29), and they often form the basis of the best examples of linear schists. At very high grades of metamorphism, slip may possibly become a significant strain and orienting mechanism (Rooney *et al.*, 1975), although natural P-type fabrics have not yet been recognized. The vast majority of amphibole fabrics are M- or G-types, but there is a dearth of literature on the subject.

Measuring amphibole LPOs

Most common amphiboles are biaxial and monoclinic with Z at a small angle (4–34°) to c, and lying in the (010) plane; $Y = b$. The orthorhombic amphiboles have $Z = c$ and $Y = b$. A moderate to high 2V is typical.

1. For orthorhombic amphiboles, measure the orientations of X, Y and Z (= a, b and c) for a complete determination of lattice orientation. A quicker procedure might be 3., below.
2. Complete determination of lattice orientation for the common monoclinic amphiboles is easily done, although it requires measurement of at least one cleavage orientation, two of X, Y and Z, and an auxilliary stereographic plot to find the positions of a, b and c for each grain. The third optical direction is always 90° from the other two. The XZ plane = (010) and contains the *c*-axis where the great circles to (010) and one or two cleavages intersect. The *a*-axis lies on the (010) great circle approximately 105° from c; the 105° must be measured from c through Z, not the other way. $Y = b$ which is also the pole to (010).
3. If determination of *b*- and c-axes orientations is sufficient, then this can

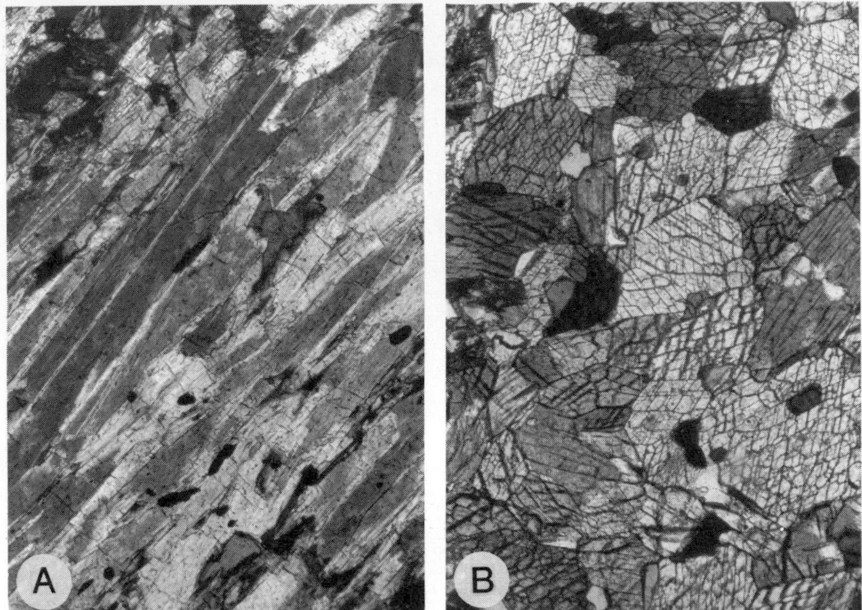

Fig. 7.29 Amphibole preferred-orientation in thin section. (A): Section parallel to lineation showing strongly oriented crystals with long-axes (=*c*-axes) that define the lineation. (B): Section perpendicular to lineation and looking down *c*-axes of strongly oriented amphibole. Amphibolite, Takaka Hill, NZ. (View lengths measure 0.85 mm (A) and (B).)

be done much more quickly, especially in sections perpendicular to foliation. Measure the orientation of two cleavage planes, plot their poles in stereographic projection, and draw the great circle containing both poles. The pole to the great circle is *c*. Bisect the obtuse angle between the two cleavage poles along the great circle to find the *b*-axis. Compile plots of poles to *c* and *b* on separate equal-area nets. The *b* and *c* axes normally lie at a small angle to foliation with *c* parallel to the lineation, and therefore this method works for most grains, especially in sections perpendicular to lineation. For grains that do not exhibit both cleavages when the U-stage is rotated, it is necessary to resort to methods 1. or 2., above.

4. The less-common amphiboles kataphorite, riebeckite, crossite, eckermannite and arfvedsonite, have $Z = b$ or larger angles of Z on to c. Some have a high dispersion, which means that determination of X, Y and Z is not feasible. The optical orientation and feasibility of measuring optical directions must be assessed to see whether method 2. can be suitably adapted; otherwise method 3. is used.

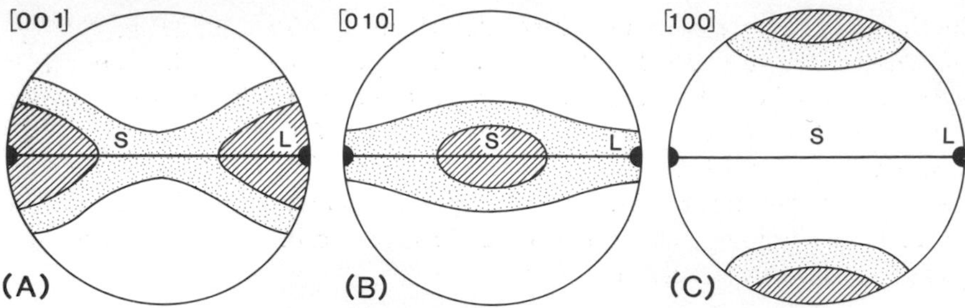

Fig. 7.30 Idealized contoured equal-area projection patterns illustrating amphibole preferred-orientations. *S* = foliation, *L* = lineation. (A), (B) and (C): Patterns for [001], [010] and [100] poles respectively.

Interpreting amphibole LPOs and SPOs

Very little is known about amphibole preferred-orientations, except that *c*-axes and prism lengths normally define the lineation, both *b*- and *c*-axes tend to lie within the foliation, and *a*-axes are concentrated at a high angle to foliation (Fig. 7.30).

(a) M-type fabric origin

Amphibole is universally known as a rigid, non-plastic mineral, and therefore an obvious explanation for its LPO and SPO is that amphibole prisms rotate as passive markers while the matrix is strained ductilely. This is the explanation of Ildefonse *et al.* (1990) for glaucophane preferred-orientations in eclogite-facies mica schists from the Italian Alps. The degree of preferred orientation increases with increase in glaucophane aspect ratio, as expected in models such as that of Fernandez (1987). The ductile matrix includes strongly lineated quartz.

(b) G-type fabric origin

Although the M-type model appears to be feasible in some cases, it is not always so, and the following questions should be posed:

1. Given that amphiboles grow during metamorphism, with what orientation did they grow?
2. If the amphiboles are very elongate and tightly packed together, is it feasible for them to have been rotated into this orientation without severe breakage of grains?
3. What is the nature of the matrix? Is there evidence that it behaved plastically?

In the glaucophane-bearing schists described by Ildefonse *et al.* (1990), the answer to question 3. satisfies the M-type model, but in low-grade lineated amphibole–feldspar rocks, it seems to be unlikely that the feldspar matrix could be plastic enough to satisfy M-type processes.

Question 2. is seldom posed, but many amphibole-rich metamorphic rocks do contain tightly packed, very elongate grains (Fig. 7.29), and the problem is a serious one in terms of an M-type mechanism. There are two possible solutions to these problems:

1. Fabrics are G-type, so that competition for growth space in a rock undergoing strain led to the dominance of *c*-axes parallel to the extension direction. The process might involve crack–seal mechanisms (Fig. 7.7) plus competition for space during the hydraulic fracturing; such amphiboles are similar to the 'stretched crystals' of Ramsay and Huber (1983) and the quartz ribbons of De Roo (1989). Enhancement of an existing but lesser preferred orientation (Fig. 7.8) may also be a contributory factor.
2. Fabrics are essentially M-type, but formed by rotation of crystals in an embryonic state, or at least much smaller than the final grain size. The question to be posed is then: 'What was the mineralogy and physical state of the matrix at the time of amphibole nucleation?'.

Without more comprehensive studies of amphibole fabrics, these possible solutions remain untested.

8
Igneous rocks

Mineral preferred-orientations do not enter into the primary terminology of igneous rocks. Nevertheless, they characterize the majority of igneous rocks. In dykes and lavas, the presence of a preferred orientation is almost taken for granted, and is usually described rather imprecisely as 'trachytic' or 'flow texture', etc., as an adjunct to the rock name. Plutonic rock names are commonly qualified by 'foliated' or 'laminated' if the preferred orientation is conspicuous. The subject has gained less attention than it deserves.

8.1 MECHANISMS

1. The main mechanism that produces preferred orientations in igneous rocks is mechanical alignment of inequidimensional grains. Subcategories include sedimentation, compaction and simple shear (or laminar flow).
2. Crystal growth in preferred directions also produces mineral preferred-orientations, as, for example, in crescumulate texture (section 4.2.2). The foliation of layered plutonic rocks has also been ascribed to primary crystal growth by McBirney and Noyes (1979), but this view is not generally supported.
3. Plastic deformation is normally regarded as a mechanism exclusively in the metamorphic realm, but a possible exception is plastic deformation associated with 'balloon tectonics'.

The various categories and subcategories are expanded on below. The M- and G-type mechanisms dominate, and because recrystallization and solution processes are relatively unimportant, igneous fabrics seldom display an LPO without SPO, or SPO without LPO. The two go together.

8.1.1 Sedimentation

Inequidimensional crystals take on a preferred orientation if they sink through a liquid and settle on the floor of a magma chamber. If the liquid is flowing across the floor, one might expect sedimentary-like layers to be characterized by cross-stratification, imbrication or tiling of crystals (indicating the flow direction – cf. Fig. 8.4), and an alignment of crystal long-axes in the direction of flow.

Whether or not layering in plutonic bodies is the result of sedimentary processes has been discussed in section 4.2.2. Recent literature throws increasing doubt on the idea. Possibly the densest minerals, such as olivine, do sometimes settle out in this way; and even if crystallization is restricted to a bottom boundary layer in a magma chamber, it may be that within that layer the densest minerals settle out, as proposed by Sørensen and Larsen (1987). However, it is now evident that the idea cannot be generally applied, and never without serious consideration of other possible mechanisms. For example, in the case of feldspar, which has a density similar to magma, it seems unlikely that it will sink easily, yet feldspar commonly exhibits very pronounced preferred orientations, and to explain them we need to look elsewhere.

8.1.2 Compaction

The compaction of sediments, and resultant loss of water, is a well-known phenomenon. Rather less well known is the phenomenon of compaction in igneous rocks (McKenzie, 1987). Nevertheless, igneous petrologists have been well aware of the need to explain extraction of natural melts from a crystal mush (deriving basaltic magma from partially melted mantle, for example), and in earlier literature this was often referred to as 'filter pressing'. Compaction of the residual mass of crystals is an inevitable complement to this process. The degree of compaction depends on the effective viscosity of the network of crystals (which has pores filled with fluid), and the thickness and permeability of the unconsolidated zone; the compaction is driven by the density difference between crystals in the mush and the fluid in the pores. McKenzie (1985) notes that melt extraction is feasible if it forms an interconnecting network: this is so for a mush of any volume % melt fraction provided the dihedral angles (section 4.2.2) are $<60°$ (this is not strictly the same as wet grain boundaries since the melt is restricted to grain edges). If the melt fraction is small, plastic deformation or textural maturation of the crystal network may also be required to effect further extraction of melt.

The process may be slow. Thus Higgins (1991) shows that a 1 m layer of anorthosite would take 200 000 years to compact, whereas a 1000 m layer would do so in 2000 years (on the basis of grain size 1 cm, density contrast $0.5 \times 10^3\,kg\,m^{-3}$, porosity 10%, viscosity of magma 10^2 Pa s, crystal network viscosity 10^{18} Pa s). The magnitude of these figures points to the importance of investigating compaction processes and preferred orientation development more thoroughly.

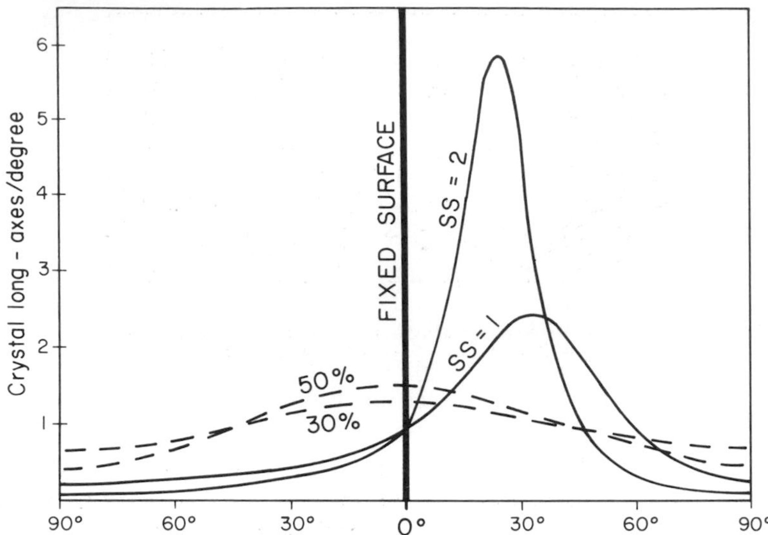

Fig. 8.1 Ideal angular relationships between crystal long-axes and either side of a fixed surface such as a magma chamber floor or dyke wall. Dashed curves show distribution for initially randomly oriented crystals subject to 30% and 50% compaction. Solid curves show distribution after simple shear with shear strains (SS) of 1.0 and 2.0 (after Higgins, 1991).

On the grand scale, Irvine (1980) and McKenzie (1987) discuss the compaction of layered complexes, alluding particularly to intrusions, such as the Great Dyke of Zimbabwe, that have an overall synformal disposition of layers. For that example, McKenzie estimates the compaction to have involved expulsion of a column of magma 0.5 km thick, hence producing the sagging (synformal) layers. Interestingly, Irvine (1980) notes the development of a vertical (perpendicular to layering) alignment of olivine crystals which he ascribes to the flow of magma upwards and out of the crystal mush.

On the scale of ordinary layering and foliation, Higgins (1991) shows the value of actually measuring crystal orientations. Even massive anorthositic layers, which at first sight appear to lack preferred orientation, exhibit a weak fabric pattern which can be interpreted in terms of a 50% compaction: feldspar long-axes, plotted against a reference surface (Fig. 8.1), exhibit a broad, low peak parallel to that surface. In contrast, well-foliated anorthosites have a very sharply defined peak that requires explanation in terms of simple shear (see below).

A more specific example of compaction is found in the Klokken Intrusion (Greenland), described by Parsons and Becker (1987). At the interface where granular syenite lies on laminated syenite, sedimentary-like load structures are developed (Fig. 8.2). The feldspars in laminated

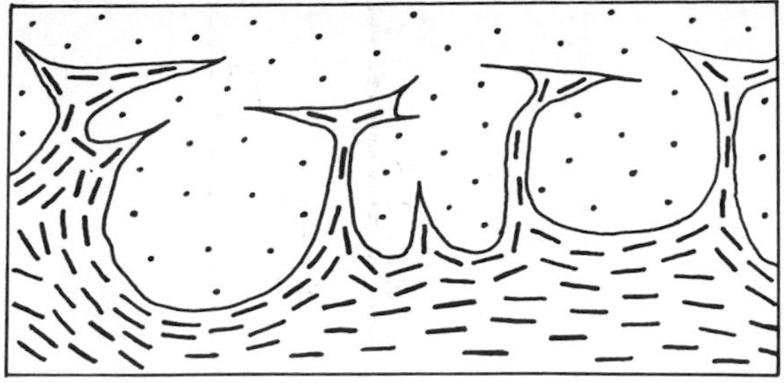

Fig. 8.2 Schematic view of load structures in granular syenite and flame structures in laminated syenite, as described by Parsons and Becker (1987). The load structures range up to more than 1 m across.

syenite are wrapped around the base of loads, and are in close parallel alignment in flame structures between loads. If these structures are interpreted like sedimentary loads, the implication is that the underlying syenite released fluids during load formation: the lamination, or at least the details of it near the interface, was formed during compaction.

8.1.3 Laminar flow

Inequidimensional crystals, carried in suspension, become oriented only if a velocity gradient exists, as for example at the base of a lava flow or near the walls of a dyke (Fig. 8.3). If flow patterns are irregular, flow is described as turbulent; preferred orientations will be correspondingly irregular. A regular pattern of velocity gradients is called laminar flow (Fig. 8.3), and it is geometrically equivalent to simple shear in metamorphic rocks. Boundary conditions between laminar and turbulent flow are discussed by Spera (1980). Laminar flow can be induced along the boundaries of a large intrusion, at the walls and floor of a magma chamber as a response to gravitational instability of crystal-rich magma, and along the walls of dykes or volcanic conduits, for example.

Two-dimensional models to describe how inequidimensional crystals rotate during simple shear have been developed by Fernandez *et al.* (1983) and Fernandez (1987); a three-dimensional model for laminar flow is given by Blanchard *et al.* (1979). In contrast to compaction (geometrically a pure shear), where crystals simply rotate towards a stable orientation with long-axes perpendicular to the shortening direction, simple shear (or laminar flow) causes crystals to tumble (Fig. 8.4A). The tumbling is most prevalent among the least inequidimensional grains, and presumes that crystals are free of constriction from their neighbours. Crystals of high aspect ratio, or crystals constrained by their neighbours, show a strong

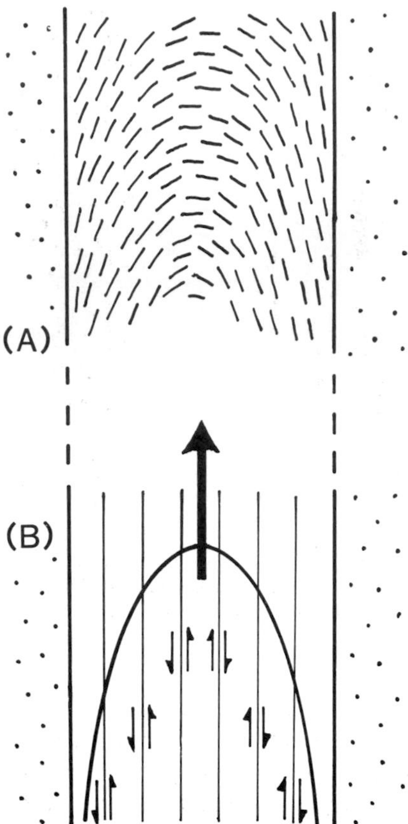

Fig. 8.3 Cross-sectional view of dyke, parallel to the flow direction, showing the typical curved pattern of feldspar or other platy minerals (A), and its relation to the velocity profile and laminar flow (B).

tendency to rotate towards the flow plane, but with the maximum concentration at some small angle to the flow plane. According to Fernandez (1987), crystals with an aspect ratio greater than 10/1 become aligned exactly parallel to the principal long-axis of the strain ellipsoid. The first weak preferred orientation of such crystals can be expected at 45° to the flow plane: as shear strain increases, the alignment becomes stronger and the angle to the flow plane smaller. For a shear strain of 2.0, the angle is close to 20° (Fig. 8.1). Higgins (1991) argues that tiling of crystals (which means they are no longer free to tumble) protects the ends of crystals facing the flow source, thus creating an effective aspect ratio greatly in excess of 10/1 (Fig. 8.4B); in such cases, the foliation defines the orientation of the strain ellipsoid.

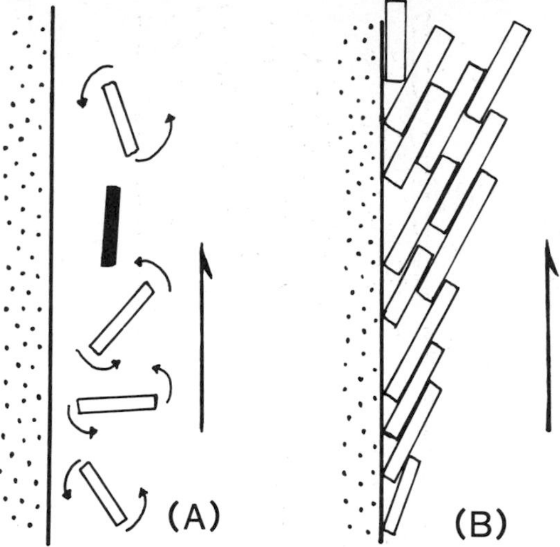

Fig. 8.4 (A): Tumbling of crystals due to laminar flow adjacent to a solid interface such as a dyke wall or chamber floor. The crystal (black) subparallel to the interface is in a metastable position. (B): Mutual interference of crystals causes tiling and prevents further tumbling.

Three-dimensional modelling by Blanchard *et al.* (1979) shows that platy crystals form a girdle about the flow direction as well as a tiling pattern subparallel to the flow plane (Fig. 8.5); the presence of such girdles has been confirmed in dykes of the Lyttelton volcano (Shelley, 1985a). The Blanchard *et al.* (1979) model presumes Newtonian flow and crystals that are free of constraint. If the crystals are prismatic rather than platy, then the long-axes rotate towards the flow direction (Bhattacharyya, 1966); this is certainly the case for amphiboles in the Lyttelton dykes.

(a) Measuring igneous preferred-orientations in foliated rocks: some examples

The majority of igneous preferred-orientations probably result from some combination of compaction and laminar flow. It is imprecise to describe the resulting texture and structure simply in terms of 'trachytic texture' or 'foliation (lamination)', and misleading to describe foliation simply as parallel to layering or the boundary of the igneous body.

Benn and Allard (1989) document SPOs and LPOs of feldspar, pyroxene and olivine in layered gabbro from the Oman ophiolite complex; fabrics are systematically oblique to layering, and provide a measure of the flow direction. The need for oriented samples is clear: foliation is oblique to layering in sections parallel to the flow direction; the same foliation is

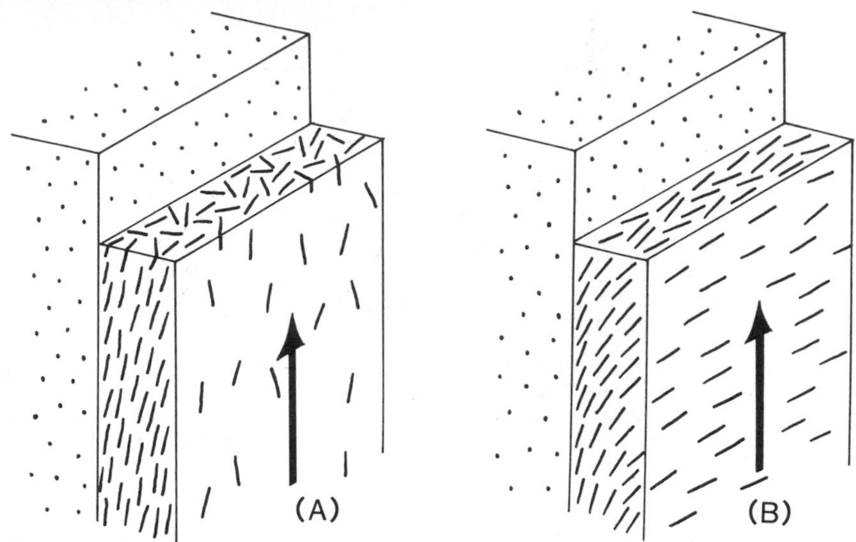

Fig. 8.5 Orientations of feldspar laths in dykes adjacent to the wall (stippled). (A): Tiling and well-developed girdle of poles to (010) around the flow direction. Lineation parallel to flow direction. (B): Poorly developed girdle combined with marked tiling pattern and intersection lineation perpendicular to the flow direction.

parallel to layering in sections perpendicular to the flow line (Benn and Allard, 1989, Fig. 12). The following LPOs were measured: olivine: [010] perpendicular to foliation; [001] parallel to lineation, but also perpendicular to the lineation within the foliation (note that [100] is not parallel to the lineation as in typical P-type peridotite fabric patterns); clinopyroxene: [010] perpendicular to foliation; [001] parallel to lineation; plagioclase: poles to (010) perpendicular to foliation, [100] parallel to lineation, [001] has a secondary maximum parallel to lineation too. Benn and Allard (1989) demonstrate that these LPOs reflect the alignment of crystal long-axes parallel to the lineation and short-axes perpendicular to foliation. Although one can measure SPOs in the Oman Complex to determine the obliquity between foliation and layering, Benn and Allard (1989) point out that textural maturation (Hunter, 1987: section 4.2.2) acts to reduce the strength of SPOs; if maturation has reached an advanced stage, the somewhat lengthier procedure of measuring LPOs becomes essential.

The Lyttelton radial dykes (Shelley, 1985a) provide another documented example of flow fabrics: a girdle pattern of poles to (010) defines the flow line, and the obliquity of the girdle relative to the dyke walls defines the sense of flow along that line (Fig. 8.6). The resulting textural patterns are quite diverse in three dimensions: sections perpendicular to foliation and

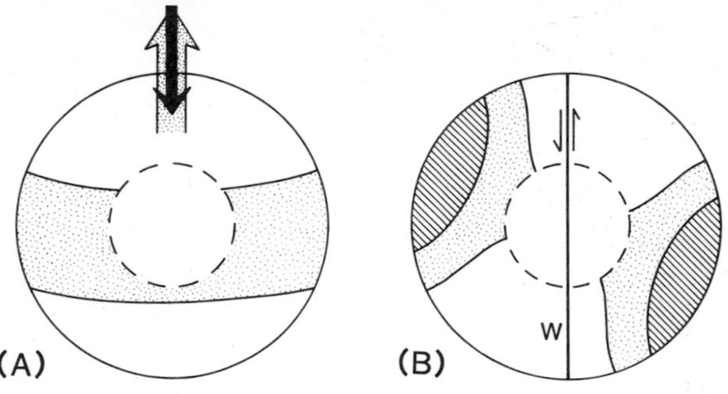

Fig. 8.6 Idealized contoured equal-area lower-hemisphere projection patterns of poles to (010) of feldspar. Dashed lines indicate the central 'blind spot'. (A): Section parallel to dyke wall. Girdle axis is the flow line, and its asymmetry gives the sense of shear (arrows). (B): Section parallel to flow direction. *W* = orientation of dyke wall. Asymmetry indicates the sense of laminar shear.

parallel to the flow line show a tiled trachytic texture; in contrast, sections perpendicular to the flow line may exhibit a weak trachytic texture, but often an apparently random pattern of feldspar plates (Fig. 8.5A). The third section, parallel to foliation, is actually the best to start with, because it provides a means of determining the orientation of the girdle of (010) poles (sections 6.2 and 6.3.2 – use the second procedure described under 'planar features').

Grain and particle alignments also develop in pyroclastic flow deposits as a result of laminar flow. Elston and Smith (1970) described the main features: long-axes of elongate shards, pumice fragments, crystals, and streaked-out vesicle and gas pockets are statistically aligned parallel to the flow direction; particles are imbricated (tiling pattern) so that they dip towards the source (similar to Fig. 8.5B); fork-shaped fragments (e.g., streaked-out shards) point away from source; symmetrical three-rayed shards have one ray pointing towards source; spindle-shaped particles have their blunt ends pointing towards the source; rigid particles may collide with soft fragments, bending them into arcuate shapes; small particles pile up on the source side of larger particles (a blocking effect). The last five features are summarized in Fig. 8.7. Ui *et al.* (1989) examined the flow behaviour of several large-scale pyroclastic flows and concluded that flow directions are from the source caldera or fissure in proximal to medial deposits, and in most plateau deposits. However, in medial to distal deposits, especially in small-scale deposits or where particles have very high aspect ratios, the flow directions may be topographically controlled instead (e.g., parallel to valley channels). Flow directions in lavas can be analysed using the criteria established by Elston and Smith (1970), as demonstrated by Smith and Rhodes (1972), for example.

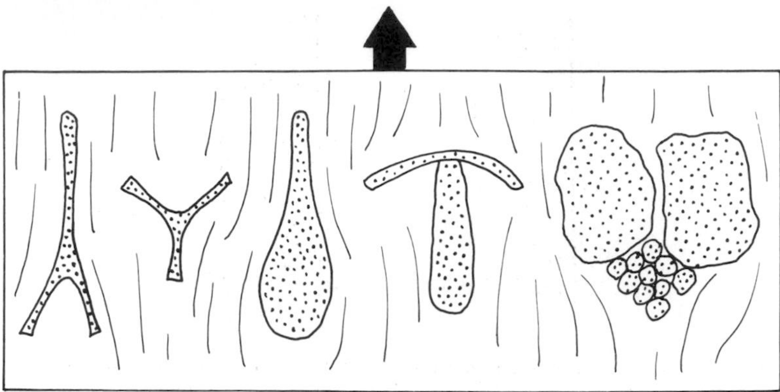

Fig. 8.7 Some of the patterns in pyroclastic flow deposits that may indicate the flow direction (after Elston and Smith, 1970). From left to right, oriented fork-shaped shards, oriented three-rayed symmetrical shards, oriented spindle-shaped fragments, deformation of a soft fragment by a rigid one (penetration effect), small particles 'blocked' by larger fragments. Arrow = flow direction.

Dykes provide an opportunity to examine fabrics that result from opposing senses of shear, at either wall (Fig. 8.3). The opposing obliquities of feldspar plates and flow plane can be explained by laminar flow and 'drag' against the two walls. Streaked-out vesicles exhibit the same sorts of obliquity, and in terms of laminar flow mechanisms, give an approximate idea of the strain since vesiculation. However, streaked-out vesicles and feldspar plates often form a continuous curve from one side of the dyke to the other (Fig. 8.3). This cannot be explained by laminar flow since the angle between plates or vesicles and the flow plane should never exceed 45°. Coward (1980) discusses the problem and suggests that shortening parallel to the bulk flow direction might result from changes in the width of the dyke fissure (causing fluctuations in pressure and flow velocity). This common pattern requires further study.

The degree of compaction versus the degree of laminar flow, responsible for any one fabric pattern, may be difficult to judge. Higgins (1991) showed that plagioclase fabric patterns in strongly foliated anorthosite could be ascribed entirely to laminar flow, or to some combination of laminar flow and compaction. Interlayered massive anorthosite has a weak fabric that indicates 50% compaction (Fig. 8.1), and it is tempting to suggest a similar amount of compaction in the foliated rocks. McKenzie (1987) points out that neither simple shear nor pure shear deformation of magma can aid compaction and extraction of fluids, so there is no reason to suggest strongly foliated rocks undergo a greater degree of compaction because of laminar flow.

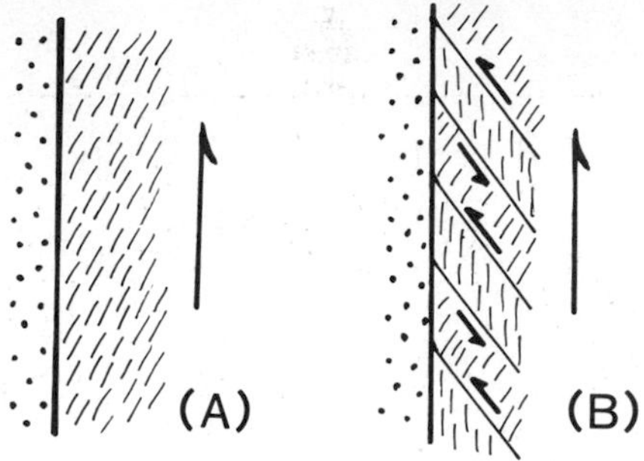

Fig. 8.8 (A): Tile-pattern due to laminar flow adjacent to a solid interface such as a dyke wall or chamber floor. (B): Continued laminar flow crenulates the foliation along secondary shears.

Deformation of thin layers around phenocrysts as a result of laminar flow has been described by Vernon (1987). Rotation causes asymmetrical microfolds to form on either side of the crystal (as in Fig. 5.47B), and one must be careful not to confuse such deformation with draping due to compaction around phenocrysts.

8.1.4 Crystal growth in preferred directions

Crystals that grow on a solid substrate nearly always have long axes at a high angle to the interface. This is the familiar texture of veins and vesicle linings. Crescumulate texture, where crystals grow on the sides of a magma chamber (section 4.2.2), is also of this type. Presuming an initial random orientation of crystal nuclei, those that can grow fastest at a high angle to the interface have a competitive advantage. Consequently a combined SPO and LPO develops as favoured orientations outgrow the others (cf. Fig. 7.5).

While one can generally assume a G-type origin for fabrics where crystals are at a high angle to layering or the boundaries of an intrusion, Irvine (1980) interpreted one instance of olivine crystals at a high angle to layering as the result of upward flow of magma extracted from the crystal mush.

Crystals often continue growing after foliation first develops. This is obviously the case where feldspars are completely intergrown, as in the foliated anorthosites described by Higgins (1991) or the trachytic-textured dykes from Lyttelton volcano (Shelley, 1985a). The sequence of events in the dykes included: 1. foliation development due to laminar flow (Fig. 8.8A); 2. crenulation of the foliation as laminar flow continued (Fig. 8.8B); and 3. continued growth of feldspar, which forms an interlocking, un-

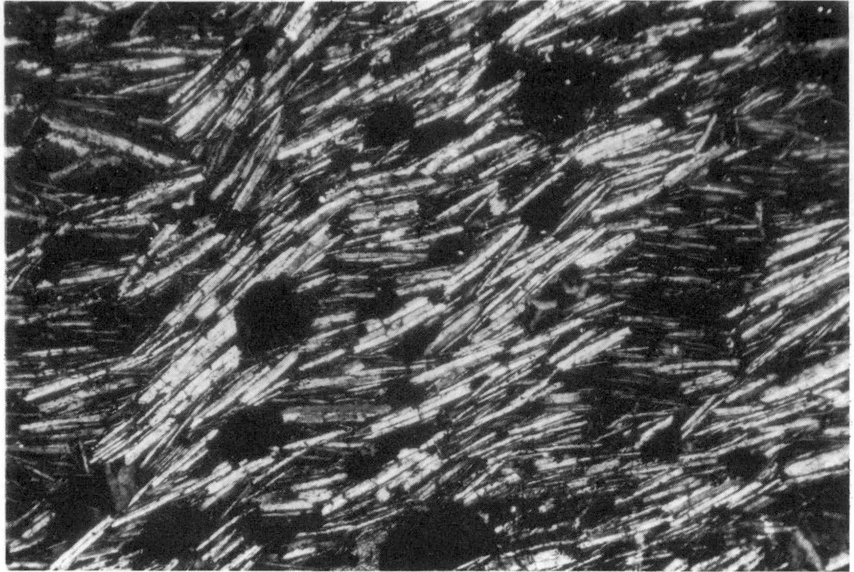

Fig. 8.9 Interlocking array of unstrained crystals along hinges of crenulations in a basaltic trachyandesite dyke sheared as shown in Fig. 8.8B. Lyttelton volcano, NZ. (Length of view measures 3.3 mm.)

strained, array of crystals (Fig. 8.9). The fact that growth continued, and possibly enhanced the SPO and LPO, should not obscure the essentially M-type fabric origin of the foliation.

The idea of McBirney and Noyes (1979) that foliation in layered gabbros could result from layer-parallel growth seems to be unlikely and unnecessary in view of the compaction and laminar flow mechanisms.

8.1.5 Plastic deformation: balloon tectonics

This is discussed in section 4.2.4 in association with granitic plutons. The idea is that an outer skin of a pluton solidifies and is then deformed as more magma is pumped into the pluton from below, a process likened to inflating a balloon. Provided that the outer skin is hot enough to be plastic, a secondary foliation parallel to the boundaries of the pluton is formed. It may be imposed on or may enhance an existing primary foliation. The preferred orientation mechanisms include plastic deformation of quartz, and the mechanical rotation of inequidimensional micas and feldspars.

A possible example of balloon tectonics is found in the Ardara pluton in Donegal, Eire (Holder, 1979), but the evidence has been reinterpreted by

Paterson *et al.* (1989 and 1991), who conclude that there is little evidence for inflation, as such.

Balloon tectonics and/or plastic deformation are strictly metamorphic phenomena. However, given the fact that diapirs become strained during ascent (Schmeling *et al.*, 1988), and that no sharp boundary exists between magma flow and high-temperature solid-state deformation, it would seem unduly pedantic not to include the possibility of solid-state foliation development as an integral part of a closely related sequence of essentially magmatic events.

REFERENCES

Allard, B. and Sotin, C. (1988) Determination of mineral phase percentages in granular rocks by image analysis on a microcomputer. *Computers and Geosciences*, **14**, 261–9.

Allison, I., Barnett, R.L. and Kerrich, R. (1979) Superplastic flow and changes in crystal chemistry of feldspars. *Tectonophysics*, **53**, T41–T46.

Anderson, A.T. Jun. (1984) Probable relations between plagioclase zoning and magma dynamics, Fuego Volcano, Guatemala. *Am. Min.*, **69**, 660–76.

Anderson, A.T. Jun., Swihart, G.H., and Artioli, G. *et al.* (1984) Segregation vesicles, gas filter-pressing, and igneous differentiation. *J. Geol.*, **92**, 55–72.

Anderson, J.E. (1969) Development of snowflake texture in a welded tuff, Davis Mountains, Texas. *Geol. Soc. Am. Bull.*, **80**, 2075–80.

Aranda-Gómez, J.J. and Ortega-Gutiérrez, F. (1987) Mantle xenoliths in Mexico. in *Mantle Xenoliths* (ed. P.H. Nixon), Wiley, Chichester, pp. 75–84.

Ashworth, J.R. (1976) Petrogenesis of migmatites in the Huntly–Portsoy area, northeast Scotland. *Min. Mag.*, **40**, 661–82.

Ashworth, J.R. (1985) Introduction. in *Migmatites* (ed. J.R. Ashworth), Blackie, Glasgow, pp. 1–35.

Ashworth, J.R. (1986) Myrmekite replacing albite in prograde metamorphism. *Am. Min.*, **71**, 895–9.

Ashworth, J.R. and McLellan, E.L. (1985) Textures. in *Migmatites* (ed. J.R. Ashworth), Blackie, Glasgow, pp. 180–203.

Aspen, P., Upton, B.G.J. and Dickin, A.P. (1990) Anorthoclase, sanidine and associated megacrysts in Scottish alkali basalts: high-pressure syenitic debris from upper mantle sources? *European J. Min.*, **2**, 503–17.

Atherton, M.P. and Edmonds, W.M. (1966) An electron microprobe study of some zoned garnets from metamorphic rocks. *Earth Planet. Sci. Lett.*, **1**, 185–93.

Avé Lallemant, H.G. and Carter, N.L. (1970) Syntectonic recrystallisation of olivine and modes of flow in the upper mantle. *Geol. Soc. Am. Bull.*, **81**, 2203–20.

Baker, B.H. and McBirney, A.R. (1985) Liquid fractionation. Part III: Geochemistry of zoned magmas and the compositional effects of liquid fractionation. *J. Volc. Geotherm. Res.*, **24**, 55–81.

Barber, D.J., Heard, H.C., and Wenk, H.R. (1981) Deformation of dolomite single crystals from 20–800°C. *Phys. Chem. Min.*, **7**, 271–86.

Barker, D.S. (1970) Compositions of granophyre, myrmekite, and graphic granite. *Geol. Soc. Am. Bull.*, **81**, 3339–50.

Barrière, M. (1976) Flowage differentiation: limitation of the 'Bagnold effect' to the narrow intrusions. *Contrib. Min. Pet.*, **55**, 139–45.

Becke, F. (1908) Über Myrmekit. *Min. Pet. Mitt.*, **27**, 377–90.

Behrmann, J.H. (1985) Crystal plasticity and superplasticity in quartzite: a natural example. *Tectonophysics*, **115**, 101–29.

Bell, K. (ed.) (1989) *Carbonatites: genesis and evolution*, Unwin Hyman, London, 618 pp.

Bell, T.H. (1985) Deformation partitioning and porphyroblast rotation in metamorphic rocks: a radical reinterpretation. *J. Metam. Geol.*, **3**, 109–18.

Bell, T.H. and Cuff, C. (1989) Dissolution, solution transfer, diffusion versus fluid flow and volume loss during deformation/metamorphism. *J. Metam. Geol.*, **7**, 425–47.

Bell, T.H. and Johnson, S.E. (1989) Porphyroblast inclusion trails: the key to orogenesis. *J. Metam. Geol.*, **7**, 279–310.

Bell, T.H. and Rubenach, M.J. (1980) Crenulation cleavage development–evidence for progressive bulk inhomogeneous shortening from 'millipede' microstructures in the Robertson River Metamorphics. *Tectonophysics*, **68**, T9–T15.

Bell, T.H. and Rubenach, M.J. (1983) Sequential porphyroblast growth and crenulation cleavage development during progressive deformation. *Tectonophysics*, **92**, 171–94.

Bell, T.H., Rubenach, M.J. and Fleming, P.D. (1986) Porphyroblast nucleation, growth and dissolution in regional metamorphic rocks as a function of deformation partitioning during foliation development. *J. Metam. Geol.*, **4**, 37–67.

Benn, K. and Allard, B. (1989) Preferred mineral orientations related to magmatic flow in ophiolite layered gabbros. *J. Pet.*, **30**, 925–46.

Benn, K. and Mainprice, D. (1989) An interactive program for determination of plagioclase crystal axes orientations from U-stage measurements: an aid for petrofabric studies. *Computers and Geosciences*, **15**, 1127–42.

Bergman, S.C. (1987) Lamproites and other potassium-rich igneous rocks: a review of their occurrence, mineralogy and geochemistry. in *Alkaline*

Igneous Rocks. (eds J.G. Fitton and B.G.J. Upton), Geol. Soc. London Spec. Publ. 30, pp. 103–90.

Berthé, D., Choukroune, P. and Jegouzo, P. (1979) Orthogneiss, mylonite and non-coaxial deformation of granites: the example of the South Armorican shear zone. *J. Struct. Geol.*, **1**, 31–42.

Bhattacharyya, D.S. (1966) Orientation of mineral lineation along the flow direction in rocks. *Tectonophysics*, **3**, 29–33.

Billings, M.P. and Sharpe, R.P. (1937) Petrofabric study of a fossiliferous schist, Mt. Clough, New Hampshire. *Am. J. Sci.*, **34**, 277–92.

Birch, W.D. (1980) Mineralogy of vesicles in an olivine leucitite at Cosgrove, Victoria, Australia. *Min. Mag.*, **43**, 597–603.

Bishop, D.G. (1972a) Transposition structures associated with cleavage formation in the Otago Schists. *NZ J. Geol. Geophys.*, **15**, 360–71.

Bishop, D.G. (1972b) Progressive metamorphism from prehnite–pumpellyite to greenschist facies in the Dansey Pass area, Otago, New Zealand. *Geol. Soc. Am. Bull.*, **83**, 3177–97.

Black, P.M. (1977) Regional high-pressure metamorphism in New Caledonia: phase equilibria in the Ouégoa District. *Tectonophysics*, **43**, 89–107.

Blackerby, B.A. (1968) Convolute zoning of plagioclase phenocrysts in Miocene volcanics from the western Santa Monica Mountains, California. *Am. Min.*, **53**, 954–62.

Blake, D.H., Elwell, R.W.D. and Gibson, I.L., *et al.* (1965) Some relationships resulting from the intimate association of acid and basic magma. *Quart. J. Geol. Soc. Lond.*, **121**, 31–49.

Blanchard, J.P., Boyer, P. and Gagny, C. (1979) Un nouveau critère de sens de mise en place dans une caisse filonienne: le 'pincement' des minéraux aux épontes. *Tectonophysics*, **53**, 1–25.

Blumenfeld, P. and Bouchez, J.L. (1988) Shear criteria in granite and migmatite deformed in the magmatic and solid states. *J. Struct. Geol.*, **10**, 361–72.

Boles, J.R. and Coombs, D.S. (1977) Zeolite facies alteration of sandstones in the Southland Syncline, New Zealand. *Am. J. Sci.*, **277**, 982–1012.

Boullier, A.M. and Gueguen, Y. (1975) SP-mylonites: origin of some mylonites by superplastic flow. *Contrib. Min. Pet.*, **50**, 93–104.

Bowen, N.L. (1913) The melting phenomena of the plagioclase feldspars. *Am. J. Sci. (4th series)*, **35**, 577–99.

Bowen, N.L. (1915) The crystallisation of haplobasaltic, haplodioritic, and related magma. *Am. J. Sci. (4th series)*, **40**, 161–85.

Bowen, N.L. (1922) The behavior of inclusions in igneous magmas. *J. Geol.*, **30**, 513–70.

Bowen, N.L. and Anderson, O. (1914) The system MgO-SiO_2. *Am. J. Sci. (4th series)*, **37**, 487–500.

Brouxel, M. (1991) Geochemical consequences of flow differentiation in a multiple injection dike (Trinity ophiolite, N. California). *Lithos*, **26**, 245–52.

Brown, W.L. and Parsons, I. (1989) Alkali-feldspars: ordering rates, phase transformations and behaviour diagrams for igneous rocks. *Min. Mag.*, **53**, 25–42.

Browne, P.R. (1978) Hydrothermal alteration in active geothermal fields. *Ann. Rev. Earth Planetary Sci.*, **6**, 229–50.

Burnham, C.W. (1979) The importance of volatile constituents. in *The Evolution of the Igneous Rocks (Fiftieth Anniversary Perspectives)* (ed. H.S. Yoder Jun.), Princeton Univ. Press, New Jersey, pp. 439–82.

Burton, K.W. (1986) Garnet–quartz intergrowths in graphitic pelites: the role of the fluid phase. *Min. Mag.*, **50**, 611–20.

Bussell, M.A. (1985) The centred complex of the Rio Huaura: a study of magma mixing and differentiation in high-level magma chambers. in *Magmatism at a Plate Edge–the Peruvian Andes* (eds W.S. Pitcher, M.P. Atherton, and E.J. Cobbing *et al.*), Blackie, London, pp. 128–55.

Carlson, W.D. (1989) The significance of intergranular diffusion to the mechanisms and kinetics of porphyroblast crystallisation. *Contrib. Min. Pet.*, **103**, 1–24.

Carlson, W.D. and Rosenfeld, J.L. (1981) Optical determination of topotactic aragonite–calcite growth kinetics: metamorphic implications. *J. Geol.*, **89**, 615–38.

Carmichael, D.M. (1969) On the mechanism of prograde metamorphic reactions in quartz-bearing pelitic rocks. *Contrib. Min. Pet.*, **20**, 244–67.

Carpenter, M.A. (1980) Composition and cation order variations in a sector-zoned blueschist pyroxene. *Am. Min.*, **65**, 313–20.

Carswell, D.A. (ed.) (1989) *Eclogite Facies Rocks*, Blackie, Glasgow. 396 pp.

Carter, N.L. and Avé Lallemant, H.G. (1970) High temperature flow of dunite and peridotite. *Geol. Soc. Am. Bull.*, **81**, 2181–202.

Cas, R.A.F. and Wright, J.V. (1987) *Volcanic Successions, Modern and Ancient*, Allen and Unwin, London. 528 pp.

Cashman, K.V. and Ferry, J.M. (1988) Crystal size distribution (CSD) in rocks and the kinetics and dynamics of crystallisation. III: Metamorphic crystallisation. *Contrib. Min. Pet.*, **99**, 401–15.

Cashman, K.V. and Marsh, B.D. (1988) Crystal size distribution (CSD) in rocks and the kinetics and dynamics of crystallisation. II: Makaopuhi lava lake. *Contrib. Min. Pet.*, **99**, 292–305.

Chappell, B.W. and Stephens, W.E. (1988) Origin of infracrustal (I-type) granite magmas. *Trans. Roy. Soc. Edinburgh. Earth Sci.*, **79**, 71–86.

Chappell, B.W. and White, A.J.R. (1974) Two contrasting granite types. *Pacific Geol.*, **8**, 173–4.

Chester, F.M. (1989) Dynamic recrystallisation in semi-brittle faults. *J. Struct. Geol.*, **11**, 847–58.

Chopin, C. (1984) Coesite and pure pyrope in high-grade blueschists of the Western Alps: a first record and some consequences. *Contrib. Min. Pet.*, **86**, 107–18.

Chopin, C. and Schreyer, W. (1983) Magnesiocarpholite and magnesiochloritoid: two index minerals of pelitic blueschists and their preliminary phase relations in the model system $MgO-Al_2O_3-SiO_2-H_2O$. *Am. J. Sci.*, **283A**, 72–96.

Clarke, G.L., Collins, W.J. and Vernon, R.H. (1990) Successive overprinting granulite facies metamorphic events in the Anmatjira Range, central Australia. *J. Metam. Geol.*, **8**, 65–88.

Clarke, L.B. and Le Bas, M.J. (1990) Magma mixing and metasomatic reaction in silicate–carbonate liquids at the Kruidfontein carbonatitic volcanic complex, Transvaal. *Min. Mag.*, **54**, 45–56.

Coombs, D.S. (1954) The nature and alteration of some Triassic sediments from Southland, New Zealand. *Trans. R. Soc. NZ*, **82**, 65–109.

Coombs, D.S. (1961) Some recent work on the lower grades of metamorphism. *Australian J. Sci.*, **24**, 203–15.

Cooper, A.F. (1979) Petrology of ocellar lamprophyres from western Otago, New Zealand. *J. Pet.*, **20**, 139–63.

Corrigan, G.M. (1982) Supercooling and the crystallisation of plagioclase, olivine, and clinopyroxene from basaltic magmas. *Min. Mag.*, **46**, 31–42.

Coward, M.P. (1980) The analysis of flow profiles in a basaltic dyke using strained vesicles. *J. Geol. Soc. Lond.*, **137**, 605–15.

Cox, K.G., Bell, J.D. and Pankhurst, R.J. (1979) *The Interpretation of Igneous Rocks*, Allen and Unwin, London. 450 pp.

Cox, S.F. and Etheridge, M.A. (1983) Crack–seal fibre growth mechanisms and their significance in the development of oriented layer silicate microstructures. *Tectonophysics*, **92**, 147–70.

Daly, J.S., Cliff, R.A. and Yardley, B.W.D. (eds) (1989) *Evolution of Metamorphic Belts*, Geol. Soc. London Spec. Publ. 43, 566 pp.

Dawson, J.B. (1987) The kimberlite clan: relationship with olivine and leucite lamproites, and inferences for upper-mantle metasomatism. in *Alkaline Igneous Rocks* (eds J.G. Fitton, and B.G.J. Upton), Geol. Soc. London Spec. Publ. 30, pp. 95–101.

Deer, W.A., Howie, R.A. and Zussman, J. (1982) *Rock-Forming Minerals. (Volume 1A*, 2nd ed, *Orthosilicates)*, Longman, London. 919 pp.

De Roo, J.A. (1989) Mass transfer and preferred orientation development during extensional microcracking in slate-belt folds, Elura Mine, Australia. *J. Metam. Geol.*, **7**, 311–22.

Dietrich, D. and Song, H. (1984) Calcite fabrics in a natural shear environment, the Helvetic nappes of western Switzerland. *J. Struct. Geol.*, **6**, 19–32.

Dixon, J.D. and Williams, G. (1983) Reaction softening in mylonites from the Arnaboll thrust, Sutherland. *Scott. J. Geol.*, **19**, 157–68.

Donaldson, C.H. and Henderson, C.M.B. (1988) A new interpretation of round embayments in quartz crystals. *Min. Mag.*, **52**, 27–33.

Dorais, M.J., Whitney, J.A. and Roden, M.F. (1990) Origin of mafic enclaves in the Dinkey Creek Pluton, Central Sierra Nevada Batholith, California. *J. Pet.*, **31**, 853–81.

Dougan, T.W. (1983) Textural relations in melanosomes of selected specimens of migmatitic pelitic schists: implications for leucosome-generating process. *Contrib. Min. Pet.*, **83**, 82–98.

Dowty, E. (1976) Crystal structure and crystal growth: II. Sector zoning in minerals. *Am. Min.*, **61**, 460–9.

Dowty, E. (1980a) Crystal growth and nucleation theory and the numerical simulation of igneous crystallisation. in *Physics of Magmatic Processes* (ed. R.B. Hargraves), Princeton University Press, New Jersey, pp. 419–85.

Dowty, E. (1980b) Synneusis reconsidered. *Contrib. Min. Pet.*, **74**, 75–84.

Drury, M.R. and Urai, J.L. (1990) Deformation-related recrystallisation processes. *Tectonophysics*, **172**, 235–53.

Duchesne, J.C., Wilmart, E. and Demaiffe, D. *et al.* (1989) Monzonorites from Rogaland (Southwest Norway): a series of rocks coeval but not comagmatic with massif-type anorthosites. *Precambrian Res.*, **45**, 111–28.

Dunnet, D. (1969) A technique of finite strain analysis using elliptical particles. *Tectonophysics*, **7**, 117–36.

Dymek, R.F. and Schiffries, C.M. (1987) Calcic myrmekite: possible evidence for the involvement of water during the evolution of andesine anorthosite from St-Urbain, Quebec. *Can. Min.*, **25**, 291–319.

Eby, G.N. (1990) The A-type granitoids: a review of their occurrence and chemical characteristics and speculations on their petrogenesis. *Lithos*, **26**, 115–34.

Eggleton, R.A. and Banfield, J.F. (1985) The alteration of granitic biotite to chlorite. *Am. Min.*, **70**, 902–10.

Elston, W.E. and Smith, E.I. (1970) Determination of flow direction of rhyolitic ash-flow tuffs from fluidal textures. *Geol. Soc. Am. Bull.*, **81** 3393–406.

England, P.C. and Richardson, S.W. (1977) The influence of erosion upon the mineral facies of rocks from different metamorphic environments. *J. Geol. Soc. Lond.*, **134**, 201–13.

Essene, E.J. (1989) The current status of thermobarometry in metamorphic rocks. in *Evolution of Metamorphic Belts* (eds J.S. Daly, R.A. Cliff and B.W.D. Yardley), Geol. Soc. Lond. Spec. Publ. 43, pp. 1–44.

Etchecopar, A. (1977) A plane kinematic model of progressive deformation in a polycrystalline aggregate. *Tectonophysics*, **39**, 121–39.

Etchecopar, A. and Vasseur, G. (1987) A 3-D kinematic model of fabric development in polycrystalline aggregates: comparison with experimental and natural examples. *J. Struct. Geol.*, **9**, 705–17.

Etheridge, M.A. and Hobbs, B.E. (1974) Chemical and deformational controls on recrystallisation of mica. *Contrib. Min. Pet.*, **43**, 111–24.

Etheridge, M.A. and Lee, M.F. (1975) Microstructure of slate from Lady Loretta, Queensland, Australia. *Geol. Soc. Am. Bull.*, **86**, 13–22.

Fenn, P.M. (1986) On the origin of graphic granite. *Am. Min.* **71**, 325–30.

Fernandez, A. (1987) Preferred orientation developed by rigid markers in two-dimensional simple shear strain: a theoretical and experimental study. *Tectonophysics*, **136**, 151–8.

Fernandez, A., Feybesse, J.L. and Mezure, J.F. (1983) Theoretical and experimental study of fabrics developed by different shaped markers in two-dimensional simple shear. *Bull. Soc. Géol. France*, **25**, 319–26.

Ferry, J.M. (1983) Regional metamorphism of the Vassalboro Formation, south-central Maine, USA: a case study of the role of fluid in metamorphic petrogenesis. *J. Geol. Soc. Lond.*, **140**, 551–76.

Fisher, R.V. and Schmincke, H.U. (1984) *Pyroclastic Rocks*, Springer, Berlin. 472 pp.

Flood, R.H. and Vernon, R.H. (1988) Microstructural evidence of orders of crystallisation in granitoid rocks. *Lithos*, **21**, 237–45.

Folk, R.L., Andrews, P.B. and Lewis, D.W. (1970) Detrital sedimentary rock classification and nomenclature for use in New Zealand. *NZ J. Geol. Geophys.*, **13**, 937–68.

Franke, W. and Ghobarkar, H. (1980) Die Morphologie von Albit beim Wachstum aus überkritischer Phase. *Z. Phys. Chem.*, **122**, 43–51.

Franke, W. and Ghobarkar, H. (1982) The morphology of hydrothermally grown plagioclase. *Crystal Res. Tech.*, **17**, 459–64.

French, W.J. and Cameron, E.P. (1981) Calculation of the temperature of crystallisation of silicates from basaltic melts. *Min. Mag.*, **44**, 19–26.

Frey, M. (1987) Very low-grade metamorphism of clastic sedimentary rocks. in *Low Temperature Metamorphism* (ed. M. Frey), Blackie, Glasgow, pp. 9–58.

Frey, M. and Kisch, H.J. (1987) Scope of subject. in *Low Temperature Metamorphism* (ed. M. Frey), Blackie, Glasgow, pp. 1–8.

Frondel, C. (1934) Selective incrustation of crystal forms. *Am. Min.*, **19**, 316–29.

Frondel, C. (1962) *Dana's 'The System of Mineralogy'* (7th edn, volume III), *Silica Minerals*, Wiley, New York. 334 pp.

Fyson, W.K. (1975) Fabrics and deformation of Archaean metasedimentary rocks, Ross Lake–Gordon Lake area, Slave Province, Northwest Territories. *Can. J. Earth Sci.*, **12**, 765–76.

Fyson, W.K. (1980) Fold fabrics and emplacement of an Archaean granitoid pluton, Cleft Lake, Northwest Territories. *Can. J. Earth Sci.*, **17**, 325–32.

Gapais, D. and Barbarin, B. (1986) Quartz fabric transition in a cooling syntectonic granite (Hermitage Massif, France). *Tectonophysics*, **125**, 357–70.

Gapais, D. and White, S.H. (1982) Ductile shear bands in a naturally deformed quartzite. *Textures and Microstructures*, **5**, 1–17.

George, W.O. (1924) The relation of physical properties of natural glasses to their chemical composition. *J. Geol.*, **32**, 353–72.

Gittins, J. (1989) The origin and evolution of carbonatite magmas. in *Carbonatites: Genesis and Evolution* (ed. K. Bell), Unwin Hyman, London, pp. 580–600.

Gorai, M. (1951) Petrological studies on plagioclase twins. *Am. Min.*, **36**, 884–901.

Grant, S.M. (1988) Diffusion models for corona formation in metagabbros from the Western Grenville Province, Canada. *Contrib. Min. Pet.*, **98**, 49–63.

Gratier, J.P. (1987) Pressure solution–deposition creep and associated tectonic differentiation in sedimentary rocks. in *Deformation of Sediments and Sedimentary Rocks*, (eds M.E. Jones and R.M.F. Preston), Geol. Soc. Lond. Spec. Publ. 29, pp. 25–38.

Gray, D.R. (1979) Microstructure of crenulation cleavages: an indicator of cleavage origin. *Am. J. Sci.*, **279**, 97–128.

Greenwood, H.J. (1967) Wollastonite: stability in H_2O–CO_2 mixtures and occurrences in a contact-metamorphic aureole near Almo, British Columbia, Canada. *Am. Min.*, **52**, 1669–80.

Greig, J.W. (1927) Immiscibility in silicate melts. *Am. J. Sci. (5th series)*, **13**, 1–44.

Griffin, W.L. (1972) Formation of eclogites and the coronas in anorthosites, Bergen Arcs, Norway. *Geol. Soc. Am. Mem.*, **135**, 37–63.

Halsor, S.P. (1989) Large glass inclusions in plagioclase phenocrysts and their bearing on the origin of mixed andesitic lavas at Toliman Volcano, Guatemala. *Bull. Volc.*, **51**, 271–80.

Hanmer, S. (1990) Natural rotated inclusions in non-ideal shear. *Tectonophysics*, **176**, 245–55.

Harley, S.L., Hensen, B.J. and Sheraton, J.W. (1990) Two-stage decompression in orthopyroxene–sillimanite granulites from Forefinger Point, Enderby Land, Antarctica: implications for the evolution of the Archaean Napier Complex. *J. Metam Geol.*, **8**, 591–613.

Hausback, B.P. (1987) An extensive, hot, vapor-charged rhyodacite flow, Baja Califormnia, Mexico. in *The Emplacement of Silicic Domes and Lava Flows*, (ed. J.H. Fink), Geol. Soc. Am. Spec. Pap., 212, pp. 111–18.

Heitzmann, P. (1987) Calcite mylonites in the Central Alpine 'root zone'. *Tectonophysics*, **135**, 207–15.

Helz, R.T. (1987) Diverse olivine types in lava of the 1959 eruption of Kilauea Volcano and their bearing on eruption dynamics. in *Volcanism in Hawaii*, (eds R.W. Decker, T.L. Wright and P.H. Stauffer), US Geol. Surv. Prof. Pap., 1350, pp. 691–722.

Hibbard, M.J. (1981) The magma mixing origin of mantled feldspars. *Contrib. Min. Pet.*, **76**, 158–70.

Hickmott, D.D. and Shimizu, N. (1990) Trace element zoning in garnet from the Kwoiek area, British Columbia: disequilibrium partitioning during garnet growth? *Contrib. Min. Pet.*, **104**, 619–30.

Higgins, M.D. (1991) The origin of laminated and massive anorthosite, Sept Iles layered intrusion, Québec, Canada. *Contrib. Min. Pet.*, **106**, 340–54.

Hobbs, B.E. (1968) Recrystallisation of single crystals of quartz. *Tectonophysics*, **6**, 353–401.

Hofmann, A.W. (1980) Diffusion in natural silicate melts: a critical review. in *Physics of Magmatic Processes* (ed. R.B. Hargraves), Princeton University Press, New Jersey, pp. 385–417.

Holdaway, M.J. (1971) Stability of andalusite and the aluminium silicate phase diagram. *Am. J. Sci.*, **271**, 97–131.

Holder, M.T. (1979) An emplacement mechanism for post-tectonic granites and its implications for their geochemical features. in *Origin of Granite Batholiths: Geochemical Evidence*, (eds M.P. Atherton and J. Tarney), Shiva, Orpington, pp. 116–28.

Hollister, L.S. (1966) Garnet zoning: an interpretation based on the Rayleigh fractionation model. *Science*, **154**, 1647–51.

Houghton, B.F. (1982) Low-grade metamorphism of the Takitimu Group, western Southland, New Zealand, *NZ J. Geol. Geophys.*, **25**, 1–19.

Hunter, R.H. (1987) Textural equilibrium in layered igneous rocks. in *Origins of Igneous Layering* (ed. I. Parsons), Reidel, Dordrecht, pp. 473–503.

Huppert, H.E., Sparks, R.S.J. and Wilson, J.R. *et al.* (1987) Laboratory experiments with aqueous solutions modelling magma chamber processes. II. Cooling and crystallisation along inclined planes. in *Origins of Igneous Layering* (ed. I. Parsons), Reidel, Dordrecht, pp. 539–68.

Hutchison, C.S. (1974) *Laboratory Handbook of Petrographic Techniques*, Wiley, New York, 527 pp.

Hutton, C.O. and Turner, F.J. (1936) Metamorphic zones in northwest Otago. *Trans. R. Soc. NZ*, **65**, 405–6.

Hutton, D.H.W. (1982) A tectonic model for the emplacement of the Main Donegal Granite, NW Ireland. *J. Geol. Soc. Lond.*, **139**, 615–31.

Ildefonse, B., Lardeaux, J.M. and Caron, J.M. (1990) The behavior of shape preferred orientations in metamorphic rocks: amphiboles and jadeites from the Monte Mucrone area (Sesia–Lanzo zone, Italian Western Alps). *J. Struct. Geol.*, **12**, 1005–11.

Irvine, T.N. (1970) Heat transfer during solidification of layered intrusions. I. Sheets and sills. *Can. J. Earth Sci.*, **7**, 1031–61.

Irvine, T.N. (1980) Magmatic infiltration metasomatism, double-diffusive fractional crystallization, and adcumulus growth in the Muskox intrusion and other layered intrusions. in *Physics of Magmatic Processes* (ed. R.B. Hargraves), Princeton University Press, New Jersey, pp. 325–83.

Irvine, T.N. (1987) Layering and related structures in the Duke Island and Skaergaard intrusions: similarities, differences, and origins. in *Origins of Igneous Layering* (ed. I. Parsons), Reidel, Dordrecht, pp. 185–245.

Jaeger, J.C. (1957) The temperature in the neighborhood of a cooling intrusive sheet. *Am. J. Sci.*, **255**, 306–18.

Jaeger, J.C. (1959) Temperatures outside a cooling intrusive sheet. *Am. J. Sci.*, **257**, 44–54.

Jaffe, H.W., Robinson, P. and Tracey, R.J. *et al.* (1975) Orientation of pigeonite exsolution lamellae in metamorphic augite: correlation with composition and calculated optimal phase boundaries. *Am. Min.*, **60**, 9–28.

Jessell, M.W. (1988a) Simulation of fabric development in recrystallising aggregates – I. Description of the model. *J. Struct. Geol.*, **10**, 771–8.

Jessell, M.W. (1988b) Simulation of fabric development in recrystallising aggregates – II. Example model runs. *J. Struct. Geol.*, **10**, 779–93.

Ji, S. and Mainprice, D. (1988) Natural deformation fabrics of plagioclase: implications for slip systems and seismic anisotropy. *Tectonophysics*, **147**, 145–63.

Ji, S., Mainprice, D. and Boudier, F. (1988) Sense of shear in high-temperature movement zones from the fabric asymmetry of plagioclase feldspars. *J. Struct. Geol.*, **10**, 73–81.

Jiang, J. and Lasaga, A.C. (1990) The effect of post-growth thermal events on growth-zoned garnet: implications for metamorphic P–T history calculations. *Contrib. Min. Pet.*, **105**, 454–9.

Johannes, W. (1985) The significance of experimental studies for the formation of migmatites. in *Migmatites* (ed. J.R. Ashworth), Blackie, Glasgow, pp. 36–85.

Johannes, W. (1988) What controls partial melting in migmatites? *J. Metam. Geol.*, **6**, 451–65.

Johannes, W. and Gupta, L.N. (1982) Origin and evolution of a migmatite. *Contrib. Min. Pet.*, **79**, 114–23.

Johnson, C.D. and Carlson, W.D. (1990) The origin of olivine–plagioclase coronas in metagabbro from the Adirondack Mountains, New York. *J. Metam. Geol.*, **8**, 697–717.

Johnson, M.R.W. (1961) Polymetamorphism in movement zones in the Caledonian thrust belt of northwest Scotland. *J. Geol.*, **69**, 417–32.

Johnson, S.E. (1990) Lack of porphyroblast rotation in the Otago Schists, New Zealand: implications for crenulation cleavage development, folding and deformation partitioning. *J. Metam. Geol.*, **8**, 13–30.

Johnston, A.D. and Beckett, J.R. (1986) Compositional variation of coexisting olivine, orthopyroxene and Fe/Mg-ferrite as a function of T and f_{O2}: a geothermometer and oxygen barometer. *Contrib. Min. Pet.*, **94**, 323–32.

Karato, S. (1988) The role of recrystallisation in the preferred orientation of olivine. *Phys. Earth Planet. Inter.*, **51**, 107–22.

Karato, S. (1989) Seismic anisotropy: mechanisms and tectonic implications. in *Rheology of Solids and of the Earth* (eds S. Karato and M. Toriumi), Oxford Univ. Press, pp. 393–422.

Karlsson, H.R. and Clayton, R.N. (1991) Analcime phenocrysts in igneous rocks: primary or secondary? *Am. Min.*, **76**, 189–99.

Keller, J. (1989) Extrusive carbonatites and their significance. in *Carbonatites: Genesis and Evolution* (ed. K. Bell), Unwin Hyman, London, pp. 70–88.

Kerrick, D.M. (1987) Fibrolite in contact aureoles of Donegal, Ireland. *Am. Min.*, **72**, 240–254.

Kirkpatrick, R.J. (1977) Nucleation and growth of plagioclase, Makaopuhi and Alae lava lakes, Kilauea Volcano, Hawaii. *Geol. Soc. Am. Bull.*, **88**, 78–84.

Knipe, R.J. and Law, R.D. (1987) The influence of crystallographic orientation and grain boundary migration on microstructural and textural evolution in an S–C mylonite. *Tectonophysics*, **135**, 155–69.

Kogarko, L.N. (1987) Alkaline rocks of the eastern part of the Baltic Shield (Kola Peninsula). in *Alkaline Igneous Rocks* (eds J.G. Fitton and B.G.J. Upton), Geol. Soc. London Spec. Publ. 30, pp. 531–44.

Komar, P.D. (1972) Flow differentiation in igneous dikes and sills: profiles of velocity and phenocryst concentration. *Geol. Soc. Am. Bull.*, **83**, 3443–8.

Komor, S.C. and Elthon, D. (1990) Formation of anorthosite–gabbro rhythmic phase layering: an example at North Arm Mountain, Bay of Islands Ophiolite. *J. Pet.*, **31**, 1–50.

Kouchi, A. and Sunagawa, I. (1985) A model for mixing basaltic and dacitic magmas as deduced from experimental data. *Contrib. Min. Pet.*, **89**, 17–23.

Kretz, R. (1966) Grain-size distribution for certain metamorphic minerals in relation to nucleation and growth. *J. Geol.*, **74**, 147–73.

Kruhl, J.H. (1987a) Computer-assisted determination and presentation of crystallographic orientations of plagioclase, on the basis of universal-stage measurements. *Neues Jb. Min. Abh.*, **157**, 185–206.

Kruhl, J.H. (1987b) Preferred lattice orientations of plagioclase from amphibolite and greenschist facies rocks near the Insubric Line (Western Alps). *Tectonophysics*, **135**, 233–42.

Kunze, F.R. and Avé Lallemant, H.G. (1981) Non-coaxial experimental deformation of olivine. *Tectonophysics*, **74**, T1–T13.

Laird, M.G. (1970) Vertical sheet structures – a new indicator of sedimentary fabric. *J. Sed. Pet.*, **40**, 428–34.

Langmuir, C.H. (1989) Geochemical consequences of *in situ* crystallisation. *Nature*, **340**, 199–205.

Law, R.D., Casey, M. and Knipe, R.J. (1986) Kinematic and tectonic significance of microstructures and crystallographic fabrics within quartz mylonites from the Assynt and Eriboll regions of the Moine thrust zone, NW Scotland. *Trans. Roy. Soc. Edinburgh, Earth Sci.*, **77**, 99–125.

Law, R.D., Schmid, S.M. and Wheeler, J. (1990) Simple shear deformation and quartz crystallographic fabrics: a possible natural example from the Torridon area of NW Scotland. *J. Struct. Geol.*, **12**, 29–45.

Lawrence, R.D. (1970) Stress analysis based on albite twinning of plagioclase feldspars. *Geol. Soc. Am. Bull.*, **81**, 2507–12.

Leake, B.E. (1964) The chemical distinction between ortho- and para-amphibolite. *J. Pet.*, **5**, 238–54.

Le Bas, M.J. (1977) (ed.) *Carbonatite–Nephelinite Volcanism*. Wiley, London, 347 pp.

Le Bas, M.J. (1987) Nephelinites and carbonatites. in *Alkaline Igneous Rocks* (eds J.G. Fitton and B.G.J. Upton), Geol. Soc. London Spec. Publ. 30, pp. 53–83.

Le Bas, M.J., Le Maitre, R.W. and Streckeisen, R. *et al.* (1986) A chemical classification of volcanic rocks based on the total alkali–silica diagram. *J. Pet.*, **27**, 745–50.

Le Maitre, R.W. (ed.) (1989) *A Classification of Igneous Rocks and Glossary of Terms*, Blackwell, Oxford. 193 pp.

Liou, J.G., Maruyama, S. and Cho, M. (1987) Very low-grade metamorphism of volcanic and volcaniclastic rock–mineral assemblages and mineral facies. in *Low Temperature Metamorphism* (ed. M. Frey), Blackie, Glasgow, pp. 59–113.

Lister, G.S. and Hobbs, B.E. (1980) The simulation of fabric development during plastic deformation and its application to quartzite: the influence of deformation history. *J. Struct. Geol.*, **2**, 355–70.

Lister, G.S. and Snoke, A.W. (1984) S–C mylonites. *J. Struct. Geol.*, **6**, 617–38.

Lister, G.S. and Williams, P.F. (1979) Fabric development in shear zones: theoretical controls and observed phenomena. *J. Struct. Geol.*, **1**, 283–97.

Lister, G.S., Paterson, M.S. and Hobbs, B.E. (1978) The simulation of fabric development in plastic deformation and its application to quartzite: the model. *Tectonophysics*, **45**, 107–58.

Lofgren, G. (1971) Experimentally produced devitrification textures in natural rhyolitic glass. *Geol. Soc. Am. Bull.*, **82**, 111–24.

Lofgren, G. (1980) Experimental studies on the dynamic crystallisation of silicate melts. in *Physics of Magmatic Processes* (ed. R.B. Hargraves), Princeton University Press, New Jersey, pp. 487–551.

Lonker, S.W. (1988) An occurrence of grandidierite, kornerupine, and tourmaline in southeastern Ontario, Canada. *Contrib. Min. Pet.*, **98**, 502–16.

Loomis, T.P. (1982) Numerical simulations of crystallisation processes of plagioclase in complex melts: the origin of major and oscillatory zoning in plagioclase. *Contrib. Min. Pet.*, **81**, 219–29.

Maaløe, S. and Printzlau, I. (1979) Natural partial melting of spinel lherzolite. *J. Pet.*, **20**, 727–41.

MacKenzie, W.S., Donaldson, C.H., and Guilford, C. (1982) *Atlas of Igneous Rocks and their Textures*, Longman, Harlow. 148 pp.

Mainprice, D. and Nicolas, A. (1989) Development of shape and lattice preferred orientations: application to the seismic anisotropy of the lower crust. *J. Struct. Geol.*, **11**, 175–89.

Mainprice, D., Bouchez, J.L. and Blumenfeld, P. *et al.* (1986) Dominant c slip in naturally deformed quartz: implications for dramatic plastic softening at high temperature. *Geology*, **14**, 819–22.

March, A. (1932) Mathematische Theorie der Regelung nach der Korngestalt bei affiner Deformation. *Z. Krist.*, **81**, 285–97.

Marsh, B.D. (1988) Crystal size distribution (CSD) in rocks and the kinetics and dynamics of crystallisation. I. Theory. *Contrib. Min. Pet.*, **99**, 277–91.

Marzouki, F., Kerrich, R. and Fyfe, W.S. (1979) Epidotisation of diorites at Al Hadah, Saudi Arabia: fluid influx into cooling plutons. *Contrib. Min. Pet.*, **68**, 281–4.

Mathews, W.H. (1951) A useful method for determining approximate composition of fine-grained igneous rocks. *Am. Min.*, **36**, 92–101.

Maxwell, J.C. (1962) Origin of slaty and fracture cleavage in the Delaware Water Gap area, New Jersey and Pennsylvania. in *Petrologic Studies: a*

Volume in Honor of A.F. Buddington (eds A.E.J. Engel, H.L. James and B.F. Leonard), Geological Society of America, pp. 281–311.

McBirney, A.R. (1984) *Igneous Petrology*, Freeman Cooper, San Francisco. 509 pp.

McBirney, A.R. (1989) The Skaergaard Layered Series: I. Structure and average compositions. *J. Pet.*, **30**, 363–97.

McBirney, A.R. and Noyes, R.M. (1979) Crystallisation and layering of the Skaergaard Intrusion. *J. Pet.*, **20**, 487–554.

McBirney, A.R., Baker, B.H. and Nilson, R.H. (1985) Liquid fractionation. Part I: Basic principles and experimental simulations. *J. Volc. Geotherm. Res.*, **24**, 1–24.

McGarvie, D.W., MacDonald, R. and Pinkerton, H. *et al.* (1990) Petrogenetic evolution of the Torfajökull Volcanic Complex, Iceland. II. The role of magma mixing. *J. Pet.*, **31**, 461–81.

McKenzie, D.P. (1985) The extraction of magma from the crust and mantle. *Earth Planet. Sci. Lett.*, **74**, 81–91.

McKenzie, D.P. (1987) The compaction of igneous and sedimentary rocks. *J. Geol. Soc. Lond.*, **144**, 299–307.

McLellan, E.L. (1983) Contrasting textures in metamorphic and anatectic migmatites: an example from the Scottish Caledonides. *J. Metam. Geol.*, **1**, 241–62.

McLellan, E.L. (1988) Migmatite structures in the Central Gneiss Complex, Boca de Quadra, Alaska. *J. Metam. Geol.*, **6**, 517–42.

Mercier, J.C.C. (1985) Olivine and pyroxenes. in *Preferred Orientation in Deformed Metals and Rocks: an Introduction to Modern Texture Analysis* (ed. H.R. Wenk), Academic Press, Orlando, pp. 407–30.

Mezger, K., Bohlen, S.R. and Hanson, G.N. (1990) Metamorphic history of the Archean Pikwitonei Granulite Domain and the Cross Lake Subprovince, Superior Province, Manitoba, Canada. *J. Pet.*, **31**, 483–517.

Miller, J.D. Jun. and Weiblen, P.W. (1990) Anorthositic rocks of the Duluth Complex: examples of rocks formed from plagioclase crystal mush. *J. Pet.*, **31**, 295–339.

Mitchell, R.S. (1985) *Dictionary of Rocks*, Van Nostrand Reinhold, New York. 228 pp.

Miyashiro, A. (1961) Evolution of metamorphic belts. *J. Pet.*, **2**, 277–311.

Moore, J.G. and Lockwood, J.P. (1973) Origin of comb layering and orbicular structure, Sierra Nevada Batholith, California. *Geol. Soc. Am. Bull.*, **84**, 1–20.

Morse, S.A. (1970) Alkali-feldspars with water at 5 kb pressure. *J. Pet.*, **11**, 221–51.

Mysen, B.O. (1988) *Structure and Properties of Silicate Melts* (Developments in Geochemistry, 4). Elsevier, Amsterdam. 354 pp.

Nakamura, Y. (1973) Origin of sector-zoning in igneous clinopyroxenes. *Am. Min.*, **58**, 986–90.

Nedelec, A., Nsifa, E.N. and Martin, H. (1990) Major and trace element geochemistry of the Archaean Ntem plutonic complex (south Cameroon): petrogenesis and crustal evolution. *Precambrian Res.*, **47**, 35–50.

Nekvasil, H. (1988) Calculated effect of anorthite component on the crystallisation paths of H_2O-undersaturated haplogranite melts. *Am. Min.*, **73**, 966–81.

Nekvasil, H. (1990) Reaction relations in the granite system: implications for trachytic and syenitic magmas. *Am. Min.*, **75**, 560–71.

Nelson, S.A. (1981) The possible role of thermal feedback in the eruption of siliceous magmas. *J. Volc. Geotherm. Res.*, **11**, 127–37.

Nicolas, A. and Poirier, J.P. (1976) *Crystalline Plasticity and Solid State Flow in Metamorphic Rocks*, Wiley, London. 444 pp.

Nicolas, A. and Violette, J.F. (1982) Mantle flow at oceanic spreading centers: models derived from ophiolites. *Tectonophysics*, **81**, 319–39.

Nixon, G.T. and Pearce, T.H. (1987) Laser-interferometry study of oscillatory zoning in plagioclase: the record of magma mixing and phenocryst recycling in calc-alkaline magma chambers, Iztaccihuatl volcano, Mexico. *Am. Min.*, **72**, 1144–62.

Norris, R.J. and Bishop, D.G. (1990) Deformed conglomerates and textural zones in the Otago Schists, South Island, New Zealand. *Tectonophysics*, **174**, 331–49.

O'Brien, H.E., Irving, A.J. and McCallum, I.S. (1988) Complex zoning and resorption of phenocrysts in mixed potassic mafic magmas of the Highwood Mountains, Montana. *Am. Min.*, **73**, 1007–24.

Oertel, G. (1985a) Reorientation due to grain shape. in *Preferred Orientation in Deformed Metals and Rocks: an Introduction to Modern Texture Analysis* (ed. H.R. Wenk), Academic Press, Orlando, pp. 259–65.

Oertel, G. (1985b) Phyllosilicate textures in slates. in *Preferred Orientation in Deformed Metals and Rocks: an Introduction to Modern Texture Analysis* (ed. H.R. Wenk), Academic Press, Orlando, pp. 431–40.

Olsen, S.N. (1985) Mass balance in migmatites. in *Migmatites* (ed. J.R. Ashworth), Blackie, Glasgow, pp. 145–79.

Olson, J.E. and Pollard, D.D. (1991) The initiation and growth of en echelon veins. *J. Struct. Geol.*, **13**, 595–608.

Ord, A. and Christie, J.M. (1984) Flow stresses from microstructures in mylonitic quartzites of the Moine Thrust zone, Assynt area, Scotland. *J. Struct. Geol.*, **6**, 639–54.

Orville, P.M. (1963) Alkali ion exchange between vapor and feldspar phases. *Am. J. Sci.*, **261**, 201–37.

Parsons, I. (ed.) (1987) *Origins of Igneous Layering*, Reidel, Dordrecht, 666 pp.

Parsons, I. and Becker, S.M. (1987) Layering, compaction and post-magmatic processes in the Klokken intrusion. in *Origins of Igneous Layering* (ed. I. Parsons), Reidel, Dordrecht, pp. 29–92.

Parsons, I. and Brown, W.L. (1984) Feldspars and the thermal history of igneous rocks. *NATO Advanced Study Institute Series C*, **137**, 317–71.

Passchier, C.W. and Simpson, C. (1986) Porphyroclast systems as kinematic indicators. *J. Struct. Geol.*, **8**, 831–43.

Paterson, S.R., Vernon, R.H. and Tobisch, O.T. (1989) A review of criteria for the identification of magmatic and tectonic foliations in granitoids. *J. Struct. Geol.*, **11**, 349–63.

Paterson, S.R., Brudos, T. and Fowler, K. (1991) Papoose Flat pluton: forceful expansion or postemplacement deformation? *Geology*, **19**, 324–7.

Pattison, D.R.M. and Harte, B. (1988) Evolution of structurally contrasting anatectic migmatites in the 3-kbar Ballachulish aureole, Scotland. *J. Metam. Geol.*, **6**, 475–94.

Pearce, T.H., Russell, J.K. and Wolfson, I. (1987) Laser-interference and Normarski interference imaging of zoning profiles in plagioclase phenocrysts from the May 18, 1980 eruption of Mount St. Helens, Washington. *Am. Min.*, **72**, 1131–43.

Peck, D.L., Wright, T.L. and Moore, J.G. (1966) Crystallisation of tholeiitic basalt in Alae lava lake, Hawaii. *Bull. Volc.*, **29**, 629–55.

Petersen, J.S. and Lofgren, G.E. (1986) Lamellar and patchy intergrowths in feldspars: experimental crystallisation of eutectic silicates. *Am. Min.*, **71**, 343–55.

Philippot, P. and Kienast, J.-R. (1989) Chemical–microstructural changes in eclogite facies shear zones (Monviso, Western Alps, north Italy) as indicators of strain history and the mechanism and scale of mass transfer. *Lithos*, **23**, 179–200.

Phillips, E.R. and Ransom, D.M. (1968) The proportionality of quartz in myrmekite. *Am. Min.*, **53**, 1411–13.

Phillips, E.R., Ransom, D.M. and Vernon, R.H. (1972) Myrmekite and muscovite developed by retrograde metamorphism at Broken Hill, New South Wales. *Min. Mag.*, **38**, 570–8.

Philpotts, A.R. (1972) Density, surface tension and viscosity of the immiscible phase in a basic alkaline magma. *Lithos*, **5**, 1–18.

Philpotts, A.R. (1982) Compositions of immiscible liquids in volcanic rocks. *Contrib. Min. Pet.*, **80**, 201–18.

Philpotts, A.R. and Lewis, C.L. (1987) Pipe vesicles – an alternate model for their origin. *Geology*, **15**, 971–74.

Pitcher, W.S. and Berger, A.R. (1972) *The Geology of Donegal*, Wiley, New York. 435 pp.

Pitcher, W.S., Atherton, M.P. and Cobbing, E.J. *et al.* (eds) (1985) *Magmatism at a Plate Edge – the Peruvian Andes*, Blackie, London. 328 pp.

Powell, D. (1966) On the preferred crystallographic orientation of garnet in some metamorphic rocks. *Min. Mag.*, **35**, 1094–109.

Putnis, A. and McConnell, J.D.C. (1980) *Principles of Mineral Behaviour*, Blackwell, Oxford. 257 pp.

Ramberg, H. (1981) *Gravity, Deformation and the Earth's Crust*, 2nd ed., Academic Press, London. 452 pp.

Ramsay, J.G. (1962) The geometry and mechanics of formation of 'similar' type folds. *J. Geol.*, **70**, 309–27.

Ramsay, J.G. (1980) The crack–seal mechanism of rock deformation. *Nature*, **284**, 135–9.

Ramsay, J.G. and Huber, M.I. (1983) *The Techniques of Modern Structural Geology. Strain Analysis*, (Vol. I), Academic Press, London. 307 pp.

Reeder, R.J. and Grams, J.C. (1987) Sector zoning in calcite cement crystals: implications for trace element distributions in carbonates. *Geochim. Cosmochim. Acta*, **51**, 187–94.

Ridley, J. (1985) The effect of reaction enthalpy on the progress of a metamorphic reaction. in *Metamorphic Reactions: Kinetics, Textures, and Deformation* (eds A.B. Thompson and D.C. Rubie), Springer, New York, pp. 80–97.

Rivers, T. and Fyson, W.K. (1977) Shape, size, and orientation of muscovite crystals in a schist of variable metamorphic grade. *Can. J. Earth Sci.*, **14**, 185–95.

Robin, P.Y.F. (1979) Theory of metamorphic segregation and related processes. *Geochim. Cosmochim. Acta*, **43**, 1587–600.

Robins, B., Haukvik, L. and Jansen, S. (1987) The organisation and internal structure of cyclic units in the Honningsvåg intrusive suite, North Norway: implications for intrusive mechanisms, double-diffusive convection and pore-magma infiltration. in *Origins of Igneous Layering* (ed. I. Parsons), Reidel, Dordrecht, pp. 287–312.

Robinson, P., Jaffe, H.W. and Ross, M. *et al.* (1971) Orientation of exsolution lamellae in clinopyroxenes and clinoamphiboles: considerations of optimal phase boundaries. *Am. Min.*, **56**, 909–39.

Rock, N.M.S. (1987) The nature and origin of lamprophyres: an overview. in *Alkaline Igneous Rocks* (eds J.G. Fitton and B.G.J. Upton), Geol. Soc. London Spec. Publ. 30, pp. 191–226.

Rooney, T.P., Riecker, R.E. and Gavasci, A.T. (1975) Hornblende deformation features. *Geology*, **3**, 364–66.

Rosenfeld, J.L. (1970) Rotated garnets in metamorphic rocks. *Geol. Soc.*

Am. Spec. Paper, **129**, 102 pp.

Ross, J.V., Avé Lallemant, H.G. and Carter, N.L. (1980) Stress dependence of recrystallised-grain and subgrain size in olivine. *Tectonophysics*, **70**, 39–61.

Rowe, K.J. and Rutter, E.H. (1990) Palaeostress estimation using calcite twinning: experimental calibration and application to nature. *J. Struct. Geol.*, **12**, 1–17.

Sampson, D.E. (1987) Textural heterogeneities and vent structures in the 600-year-old lavas of the Inyo volcanic chain, eastern California. in *The Emplacement of Silicic Domes and Lava Flows* (ed. J.H. Fink), Geol. Soc. Am. Spec. Pap., 212, pp. 89–101.

Sawka, W.N., Chappell, B.W. and Kistler, R.W. (1990) Granitoid compositional zoning by side-wall layer differentiation: evidence from the Palisade Crest Intrusive Suite, Central Sierra Nevada, California. *J. Pet.*, **31**, 519–53.

Sawyer, E.W. and Barnes, S.J. (1988) Temporal and compositional differences between subsolidus and anatectic migmatite leucosomes from the Quetico metasedimentary belt, Canada. *J. Metam. Geol.*, **6**, 437–50.

Sawyer, E.W. and Robin, P.Y.F. (1986) The subsolidus segregation of layer-parallel quartz–feldspar veins in greenschist to upper amphibolite facies metasediments. *J. Metam. Geol.*, **4**, 237–60.

Schaskolsky, M. and Schubnikov, A. (1933) Über die künstliche Herstellung gesetzmässiger Kristallverwachsungen des Kalialauns. *Z. Krist. Mineral.*, **85**, 1–16.

Schermerhorn, L.J.G. (1956) The granites of Transcosco (Portugal): a study of microclinitisation. *Am. J. Sci.*, **254**, 329–48.

Schmeling, H., Cruden, A.R. and Marquart, G. (1988) Finite deformation in and around a fluid sphere moving through a viscous medium: implications of diapiric ascent. *Tectonophysics*, **149**, 17–34.

Schmid, R. (1981) Descriptive nomenclature and classification of pyroclastic deposits and fragments: recommendations of the IUGS subcommission on the systematics of igneous rocks. *Geology*, **9**, 41–3.

Schmid, S.M., Panozzo, R. and Bauer, S. (1987) Simple shear experiments on calcite rocks: rheology and microfabric. *J. Struct. Geol.*, **9**, 747–78.

Schmidt, W. (1932) *Tektonik und Verformungslehre*. Borntraeger, Berlin, 208 pp.

Schreyer, W. (1973) Whiteschist: a high pressure rock and its geological significance. *J. Geol.*, **81**, 735–9.

Selverstone, J. and Chamberlain, C.P. (1990) Apparent isobaric cooling paths from granulites: two counterexamples from British Columbia and New Hampshire. *Geology*, **18**, 307–10.

Shelley, D. (1968) Ptygma-like veins in graywacke, mudstone, and low-grade schist from New Zealand. *J. Geol.*, **76**, 692–701.

Shelley, D. (1973) Myrmekites from the Haast Schists, N.Z. *Am. Min.*, **58**, 332–8.

Shelley, D. (1974) Mechanical production of metamorphic banding – a critical appraisal. *Geol. Mag.*, **111**, 287–92.

Shelley, D. (1975) Temperature and metamorphism during cleavage and fold formation of the Greenland Group, N of Greymouth. *J. R. Soc. NZ*, **5**, 65–75.

Shelley, D. (1977) Plagioclase preferred orientation in Haast Schist N.Z. *J. Geol.*, **85**, 635–44.

Shelley, D. (1979) Plagioclase preferred orientation, Foreshore Group metasediments, Bluff, New Zealand. *Tectonophysics*, **58**, 279–90.

Shelley, D. (1980) Quartz [0001]-axes preferred orientation, Bluff, New Zealand: origin elucidated by grain-size measurements. *Tectonophysics*, **62**, 321–37.

Shelley, D. (1985a) Determining paleo-flow directions from groundmass fabrics in the Lyttelton radial dykes, New Zealand. *J. Volc. Geotherm. Res.*, **25**, 69–79.

Shelley, D. (1985b) *Optical Mineralogy*, 2nd edn, Elsevier, New York, 321 pp.

Shelley, D. (1988) Radial dikes of Lyttelton Volcano – their structure, form, and petrography. *NZ J. Geol. Geophys.*, **31**, 65–75.

Shelley, D. (1989a) Plagioclase and quartz preferred orientations in a low-grade schist: the roles of primary growth and plastic deformation. *J. Struct. Geol.*, **11**, 1029–37.

Shelley, D. (1989b) *P*, *M* and *G* Tectonites: a classification based on origin of mineral preferred orientations. *J. Struct. Geol.*, **11**, 1039–44.

Shelley, D. (1989c) Calcstress: a program that calculates compression and tension directions from calcite U-stage data. *Computers and Geosciences*, **15**, 269–73.

Shelley, D. (1991) Structure, fabric, and metamorphism of Arthur Marble, Pikikiruna Range, Nelson, New Zealand. *NZ J. Geol. Geophys.*, **34**, 385–96.

Shimamoto, T. (1989) The origin of S–C mylonites and a new fault-zone model. *J. Struct. Geol.*, **11**, 51–64.

Sibson, R.H. (1975) Generation of pseudotachylyte by ancient seismic faulting. *Geophys. J. R. Astron. Soc.*, **43**, 775–94.

Sibson, R.H., White, S.H. and Atkinson, B.K. (1979) Fault rock distribution and structure within the Alpine Fault Zone: a preliminary account. in *The Origin of the Southern Alps* (eds R.I. Walcott and M.M. Cresswell), Bull. R. Soc. NZ 18, 55–65.

Sigurdsson, H. and Sparks, R.S.J. (1981) Petrology of rhyolitic and mixed magma ejecta from the 1875 eruption of Askja, Iceland. *J. Pet.*, **22**, 41–84.

Simpson, C. and Wintsch, R.P. (1989) Evidence for deformation-induced K-feldspar replacement by myrmekite. *J. Metam. Geol.*, **7**, 261–75.

Smith, C.S. (1964) Some elementary principles of polycrystalline microstructure. *Metall. Rev.*, **9**, 1–48.

Smith, D.C. (ed.) (1988) *Eclogites and Eclogite-Facies Rocks* (Developments in petrology, 12), Elsevier, Amsterdam. 524 pp.

Smith, E.I. and Rhodes, R.C. (1972) Flow direction determination of lava flows. *Geol. Soc. Am. Bull.*, **83**, 1869–74.

Smith, J.V. and Brown, W.L. (1988) *Feldspar Minerals*, 2nd edn (Vol. I), Springer-Verlag, Berlin. 828 pp.

Smith, R.K. and Lofgren, G.E. (1983) An analytical and experimental study of zoning in plagioclase. *Lithos*, **16**, 153–68.

Sorby, H.C. (1853) On the origin of slaty cleavage. *Edinburgh New Phil. J.*, **55**, 137–48.

Sørensen, H. and Larsen, L.M. (1987) Layering in the Ilimaussaq alkaline intrusion, South Greenland. in *Origins of Igneous Layering* (ed. I. Parsons), Reidel, Dordrecht, pp. 1–28.

Sorensen, S.S. (1988) Petrology of amphibolite-facies mafic and ultramafic rocks from the Catalina Schist, southern California: metasomatism and migmatization in a subduction zone metamorphic setting. *J. Metam. Geol.*, **6**, 405–35.

Spear, F.S. (1989) Relative thermobarometry and metamorphic P–T paths. in *Evolution of Metamorphic Belts* (eds J.S. Daly, R.A. Cliff and B.W.D. Yardley), Geol. Soc. London Spec. Publ. 43, pp. 63–81.

Spear, F.S., Kohn, M.J. and Florence, F.P. *et al.* (1990) A model for garnet and plagioclase growth in pelitic schists: implications for thermobarometry and P–T path determinations. *J. Metam. Geol.*, **8**, 683–96.

Spear, F.S. and Selverstone, J. (1983) Quantitative P–T paths from zoned minerals: theory and tectonic applications. *Contrib. Min. Pet.*, **83**, 348–57.

Spear, F.S., Selverstone, J. and Hickmott, D. *et al.* (1984) P–T paths from garnet zoning: a new technique for deciphering tectonic processes in crystalline terranes. *Geology*, **12**, 87–90.

Spera, F.J. (1980) Aspects of magma transport. in *Physics of Magmatic Processes* (ed. R.B. Hargraves), Princeton University Press, New Jersey, pp. 265–323.

Spooner, E.T.C. and Fyfe, W.S. (1973) Sub-sea-floor metamorphism, heat and mass transfer. *Contrib. Min. Pet.*, **42**, 287–304.

Spry, A. (1969) *Metamorphic Textures*, Pergamon, Oxford. 350 pp.

Stamatelopoulou-Seymour, K., Vlassopoulos, D. and Pearce, T.H. *et al.*

(1990) The record of magma chamber processes in plagioclase phenocrysts at Thera Volcano, Aegean Volcanic Arc, Greece. *Contrib. Min. Pet.*, **104**, 73–84.

Starkey, J. and Cutforth, C. (1978) A demonstration of the interdependence of the degree of quartz preferred orientation and the quartz content of deformed rocks. *Can. J. Earth Sci.*, **15**, 841–7.

Starkey, J. and Simigian, S. (1987) IMAGE: a FORTRAN V program for image analysis of particles. *Computers and Geosciences*, **13**, 37–59.

Stel, H. (1991) Linear dilation structures and syn-magmatic folding in granitoids. *J. Struct. Geol.*, **13**, 625–34.

Stel, H. and Breedveld, M. (1990) Crystallographic orientation patterns of myrmekitic quartz: a fabric memory in quartz ribbon-bearing gneisses. *J. Struct. Geol.*, **12**, 19–28.

Streckeisen, A. (1976) To each plutonic rock its proper name. *Earth Sci. Rev.*, **12**, 1–33.

Streckeisen, A. (1979) Classification and nomenclature of volcanic rocks, lamprophyres, carbonatites, and melilitic rocks: recommendations and suggestions of the IUGS subcommission on the systematics of igneous rocks. *Geology*, **7**, 331–5.

Sunagawa, I. (1987) Morphology of minerals. in *Morphology of Crystals (Part B)* (ed. I. Sunagawa), Terra, Tokyo, pp. 509–87.

Sutcliffe, R.H., Smith, A.R. and Doherty, W. *et al.* (1990) Mantle derivation of Archaean amphibole-bearing granitoid and associated mafic rocks: evidence from the southern Superior Province, Canada. *Contrib. Min. Pet.*, **105**, 255–74.

Swanson, S.E., Naney, M.T. and Westrich, H.R. *et al.* (1989) Crystallisation history of Obsidian Dome, Inyo Domes, California. *Bull. Volc.*, **51**, 161–76.

Symes, R.F., Bevan, J.C. and Jan, M.Q. (1987) The nature and origin of orbicular rocks from near Deshai, Swat Kohistan, Pakistan. *Min. Mag.*, **51**, 635–47.

Tait, S.R. and Kerr, R.C. (1987) Experimental modelling of interstitial melt convection in cumulus piles. in *Origins of Igneous Layering* (ed. I. Parsons), Reidel, Dordrecht, pp. 569–87.

Thompson, A.B. and Ridley, J.R. (1987) Pressure–temperature–time (P–T–t) histories of orogenic belts. *Phil. Trans. R. Soc. Lond.*, **A321**, 27–45.

Thompson, R.N. (1972) Melting behaviour of two Snake River lavas at pressures up to 35 kb. *Carnegie Inst. Wash. Geophys. Lab. Yearbook*, **71**, 406–10.

Toriumi, M. (1986) Mechanical segregation of garnet in synmetamorphic flow of pelitic schists. *J. Pet.*, **27**, 1395–408.

Treloar, P.J. and Kramers, J.D. (1989) Metamorphism and geochronology of granulites and migmatitic granulites from the Magnodi Mobile Belt, Zimbabwe. *Precambrian Res.*, **45**, 277–89.

Tsuchiyama, A. (1985) Dissolution kinetics of plagioclase in the melt of the system diopside–albite–anorthite, and origin of dusty plagioclase in andesite. *Contrib. Min. Pet.*, **89**, 1–16.

Tsuchiyama, A. (1986) Melting and dissolution kinetics: application to partial melting and dissolution of xenoliths. *J. Geophys. Res.*, **91B**, 9395–406.

Tullis, J. and Yund, R.A. (1987) Transition from cataclastic flow to dislocation creep of feldspar: mechanisms and microstructures. *Geology*, **15**, 606–9.

Turner, F.J. (1953) Nature and dynamic interpretation of deformation lamellae in calcite of three marbles. *Am. J. Sci.*, **251**, 276–98.

Turner, F.J. (1981) *Metamorphic Petrology*, 2nd edn, McGraw-Hill, New York. 524 pp.

Turner, F.J. and Weiss, L.E. (1963) *Structural Analysis of Metamorphic Tectonites*, McGraw-Hill, New York, 545 pp.

Turner, F.J., Griggs, D.T. and Clark, R.H. *et al.* (1956) Deformation of Yule marble. Part VII: Development of oriented fabrics at 300°C–500°C. *Geol. Soc. Am. Bull.*, **67**, 1259–94.

Turner, J.S. and Campbell, I.H. (1986) Convection and mixing in magma chambers. *Earth Sci. Rev.*, **23**, 255–352.

Tuttle, O.F. and Bowen, N.L. (1958) Origin of granite in the light of experimental studies in the system $NaAlSi_3O_8$–$KAlSi_3O_8$–SiO_2–H_2O. *Geol. Soc. Am. Mem.*, **74**.

Ui, T., Suzuki-Kamata, K. and Matsusue, R. *et al.* (1989) Flow behavior of large-scale pyroclastic flows – evidence obtained from petrofabric analysis. *Bull. Volc.*, **51**, 115–22.

Urai, J.L., Means, W.D. and Lister, G.S. (1986) Dynamic recrystallisation of minerals. in *Mineral and Rock deformation: Laboratory Studies. The Paterson Volume* (eds B.E. Hobbs and H.C. Heard), Am. Geophys. Union Geophys. Monograph 36, pp. 161–99.

Vallance, T.G. (1967) Mafic rock alteration and the isochemical development of some cordierite–anthophyllite rocks. *J. Pet.*, **8**, 84–96.

Vance, J.A. (1961) Polysynthetic twinning in plagioclase. *Am. Min.*, **46**, 1097–119.

Vance, J.A. (1969) On synneusis. *Contrib. Min. Pet.*, **24**, 7–29.

Vernon, R.H. (1975) Deformation and recrystallisation of a plagioclase grain. *Am. Min.*, **60**, 884–8.

Vernon, R.H. (1985) Possible role of superheated magma in the formation of orbicular granitoids. *Geology*, **13**, 843–5.

Vernon, R.H. (1986) K-feldspar megacrysts in granites – phenocrysts, not porphyroblasts. *Earth Sci. Rev.*, **23**, 1–63.

Vernon, R.H. (1987) A microstructural indicator of shear sense in volcanic rocks and its relationship to porphyroblast rotation in metamorphic rocks. *J. Geol.*, **95**, 127–33.

Vernon, R.H. (1989) Porphyroblast–matrix microstructural relationships: recent approaches and problems. in *Evolution of Metamorphic Belts* (eds J.S. Daly, R.A. Cliff and B.W.D. Yardley), Geol. Soc. London Spec. Publ. 43, pp. 83–102.

Vernon, R.H. and Collins, W.J. (1988) Igneous microstructures in migmatites. *Geology*, **16**, 1126–9.

Vernon, R.H., Etheridge, M.A. and Wall, V.J. (1988) Shape and microstructure of microgranitoid enclaves: indicators of magma mingling and flow. *Lithos*, **22**, 1–11.

von Mises, R. (1928) Mechanik der plastischen Formänderung von Kristallen. *Z. Angew. Math. Mech.*, **8**, 161–85.

Wager, L.R. (1959) Differing powers of crystal nucleation as a factor producing diversity in layered igneous intrusions. *Geol. Mag.*, **96**, 75–80.

Wager, L.R. and Brown, G.M. (1968) *Layered Igneous Rocks*, Oliver and Boyd, Edinburgh. 588 pp.

Walker, G.P.L. (1951) The amygdale minerals in the Tertiary lavas of Ireland. I. The distribution of chabazite habits and zeolites in the Garran Plateau area, County Antrim. *Min. Mag.*, **29**, 773–91.

Walker, G.P.L. (1960) Zeolite zones and dike distribution in relation to the structure of the basalts of eastern Iceland. *J. Geol.*, **68**, 515–28.

Walker, G.P.L. (1989) Spongy pahoehoe in Hawaii: a study of vesicle-distribution patterns in basalt and their significance. *Bull. Volc.*, **51**, 199–209.

Walniuk, D.M. and Morris, A.P. (1985) Quartz deformation mechanisms in metasediments from Prins Karls Forland, Svalbard. *Tectonophysics*, **115**, 87–100.

Walther, J.V. and Wood, B.J. (1984) Rate and mechanism in prograde metamorphism. *Contrib. Min. Pet.*, **88**, 246–59.

Walther, J.V. and Wood, B.J. (1986) Mineral–fluid reaction rates. in *Fluid–Rock Interactions during Metamorphism* (eds J.V. Walther and B.J. Wood), Springer, New York, pp. 194–211.

Wenk, H.R. (1985a) Measurement of pole figures. in *Preferred Orientation in Deformed Metals and Rocks: an Introduction to Modern Texture Analysis* (ed. H.R. Wenk), Academic Press, Orlando, pp. 11–47.

Wenk, H.R. (1985b) Carbonates. in *Preferred Orientation in Deformed Metals and Rocks: an Introduction to Modern Texture Analysis* (ed. H.R. Wenk), Academic Press, Orlando, pp. 361–84.

Wenk, H.R., Bunge, H.J. and Jansen, E. *et al.* (1986) Preferred orientation of plagioclase – neutron diffraction and U-stage data. *Tectonophysics*, **126**, 271–84.

Whalen, J.B. (1985) Geochemistry of an island-arc plutonic suite: the Uasilan–Yau Yau intrusive complex, New Britain, PNG. *J. Pet.*, **26**, 603–32.

Whalen, J.B., Currie, K.L. and Chappell, B.W. (1987) A-type granites: geochemical characteristics, discrimination and petrogenesis. *Contrib. Min. Pet.*, **95**, 407–19.

White, A.J.R. and Chappell, B.W. (1988) Some supracrustal (S-type) granites of the Lachlan Fold Belt. *Trans. R. Soc. Edinburgh Earth Sci.*, **79**, 169–81.

White, S. (1975) Tectonic deformation and recrystallisation of oligoclase. *Contrib. Min. Pet.*, **50**, 287–304.

Williams, P.F., Means, W.D. and Hobbs, B.E. (1977) Development of axial-plane slaty cleavage and schistosity in experimental and natural materials. *Tectonophysics*, **42**, 139–58.

Wilson, J.R. and Larsen, S.B. (1985) Two-dimensional study of a layered intrusion – the Hyllingen Series, Norway. *Geol. Mag.*, **122**, 97–124.

Wilson, M. (1989) *Igneous Petrogenesis*, Unwin Hyman, London. 466 pp.

Woolley, A.R. and Kempe, D.R.C. (1989) Carbonatites: nomenclature, average chemical compositions, and elemental distribution. in *Carbonatites: Genesis and Evolution* (ed. K. Bell), Unwin Hyman, London, pp. 1–14.

Yardley, B.W.D. (1977) An empirical study of diffusion in garnet. *Am. Min.*, **62**, 793–800.

Yardley, B.W.D. (1978) Genesis of the Skagit Gneiss migmatites, Washington, and the distinction between possible mechanisms of migmatites. *Geol. Soc. Am. Bull.*, **89**, 941–51.

Yardley, B.W.D. (1982) The early metamorphic history of the Haast Schists and related rocks of New Zealand. *Contrib. Min. Pet.*, **81**, 317–27.

Yardley, B.W.D. (1989) *An Introduction to Metamorphic Petrology*, Longman, Harlow. 248 pp.

Zwart, H.J. (1960) Relations between folding and metamorphism in the Central Pyrenees, and their chronological succession. *Geol. Mijnboum*, **22**, 163–80.

Zwart, H.J. (1962) On the determination of polymetamorphic mineral associations and its application to the Bosost area (Central Pyrenees). *Geol. Rundsch.*, **52**, 38–65.

Part Four

Glossary and Index

Glossary and index

For ease of use, the subject index, author index and glossary are consolidated in one alphabetical listing. Glossary entries are provided for rock names, textures, microstructures and other technical terms related to the main subject-matter of the book. Their purpose is to provide a quick, handy reference source and/or a simple definition without having to search through the main text. A small number of terms given glossary entries are not used in the text of the book, but are added here for completeness. Glossary entries are not provided for all geological terms (minerals and sedimentary rock names, for example, are not defined here).

A 7, 21
[Mineral grouping in the IUGS classification system for igneous rocks. A = alkali feldspar, including albite up to An5]
Acid 44
[Describes igneous rocks with wt% $SiO_2 > 63\%$]
Acid rocks
hydrothermal metamorphism 66–7
Adcumulate texture 39, 216, 224
[Describes an accumulation of crystals in an igneous rock, often of one phase only, with grain boundaries due to mutual interference during growth and removal of intercumulus liquid. Used for products of crystal settling and/or in situ *crystallization]*
Adularia sector zoning 135
Agate 205
[Layered variety of chalcedony]
Agglomerate 9
[Pyroclastic deposit made of bombs and/or blocks]
Agmatite, *see* Migmatite, agmatitic
Agpaite 36
[Peralkaline nepheline syenite with Zr–Ti rich minerals such as eudialyte]
Air-fall deposits 9–13
Alaskite 25
[Leucocratic alkali feldspar granite]
Albitite 31
[Syenite made almost entirely of albite]
Albitization 64, 68, 70
[Process whereby albite replaces more calcic plagioclase]
Alkali basalt, *see* Basalt
Alkali feldspar
crystallization 137–43
megacrysts 197, 229
overgrowths 239
poikilitic 229, 242
rounding 240
zoning 230
see also Feldspar
Alkali feldspar plagioclase nepheline symplectite 148
Alkaline igneous rocks (and provinces) 16, 19, 30, 35, 54, 148
[Igneous rocks relatively rich in Na_2O and/or K_2O, typically with modal feldspathoid and/or normative nepheline, and usually formed in intraplate settings]
Allard, B. 5, 375, 376
Allison, I. 281
Alnoite 17, 19
[Lamprophyre lacking feldspar or foids. Biotite, usually with olivine and pyroxene, is set in matrix of melilite (Table 1.1)]
Alteration 18, 30, 33, 42, 44, 63, 65, 138,

Microporphyritic texture 64
[Porphyritic only when viewed with the microscope]
Microstructures
igneous 170, 197–223, 233–44
metamorphic 245–313
Migmatite 24, 30, 60, 63, 92, 107–9, 111, 227, 276, 291–6
[Rock that appears to be partly igneous, partly metamorphic]
agmatitic 109
[Breccia-like migmatite]
diktyonitic 109
[Migmatite with mesosome veined by leucosome]
stromatic 110
[Layered migmatite]
see also Schollen
Miller, J.D. Jr 31
Mineral preferred orientations, *see* Preferred orientations
Minette 17, 191
[Lamprophyre (Table 1.1) with phenocrysts of biotite and possibly other mafic minerals set in a matrix of mafic minerals and alkali feldspar with subordinate plagioclase. It lacks foids]
Missourite 37
[Foidolite (Fig. 1.13) with leucite the dominant foid and $M > 70\%$]
Mitchell, R.S. 4
Miyashiro, A. 311
Modal mineralogy 5, 7
[Mineral composition of a rock, usually expressed in volume % for each mineral]
Monchiquite 17
[Lamprophyre with phenocrysts of any combination of amphibole, titanian augite, biotite, olivine, and a matrix-lacking feldspar but containing glass and foids, especially analcime (Table 1.1)]
Monzodiorite 30
[Plutonic rock made essentially of plagioclase ($An < 50\%$) with subordinate alkali feldspar plus pyroxene, hornblende or biotite (Fig. 1.13)]
foid 35
[Plutonic rock made essentially of feldspathoid, plagioclase ($An < 50\%$) and alkali feldspar, as defined in Fig. 1.13, plus mafic minerals such as pyroxene, amphibole (often alkaline varieties) and biotite]
Monzogabbro 30
[Plutonic rock made essentially of plagioclase ($An > 50\%$) with subordinate alkali feldspar plus mafic minerals such as pyroxene, hornblende or biotite (Fig. 1.13)]
foid 35
[Plutonic rock made essentially of feldspathoid, plagioclase ($An < 50\%$) and alkali feldspar, as defined in Fig. 1.13, plus mafic minerals such as pyroxene, amphibole (often alkaline varieties) and biotite]
Monzogranite 25
[One of the possible subdivisions of granite (Fig. 1.13)]
Monzonite 30
[Plutonic rock made essentially of sodic plagioclase and alkali feldspar (as defined in Fig. 1.13), with minor pyroxene or amphibole]
orthopyroxene 21
[Alternative name for mangerite]
quartz 27, 229
[Plutonic rock transitional between monzonite and granite with quartz equal to 5–20% of the felsic minerals]
Monzonitic texture 31
[In igneous rocks, poikilitic K-feldspar encloses euhedral plagioclase]
Monzonorite 21
[Plutonic rock (possibly charnockitic) similar to monzodiorite or monzogabbro but containing significant orthopyroxene]
Monzosyenite, foid 35
[Plutonic rock with alkali feldspar, sodic plagioclase and foids as defined in Fig. 1.13, plus mafic minerals]
Moore, J.G. 234, 238
see also Peck, D.L.
Morris, A.P. 336
Morse, S.A. 142
Mortar texture 76
[Term often used for fine-grained material in mylonites and mylonitic rocks, especially that surrounding porphyroclasts. It is appropriate for products of cataclasis, but not for products of dynamic recrystallization]

Igneous and Metamorphic Rocks under the Microscope